NITROGEN OXIDES
And Their Effects On Health

S.D. LEE, Editor

ANN ARBOR SCIENCE
P.O. BOX 1425 • ANN ARBOR, MICHIGAN 48106

PREFACE

Nitrogen oxides are ubiquitous in the environment, and there is considerable concern that increased utilization of coal will increase their number. The purpose of this volume is to review available health effects data and to assist standard-setting officials and organizations in their examination of available literature on NO_2 toxicity, so that safe levels can be defined and proposed standards evaluated.

This is particularly timely because the U.S. Environmental Protection Agency plans to propose NO_2 standards in the spring of 1980. It is gratifying to note that some of the information presented in this book has already been considered in NO_2 criteria documents.

The World Health Organization (WHO) task committee recently recommended that the maximum exposure level of NO_2 for the protection of public health be 190 to 320 $\mu g/m^3$ (0.10 to 0.17 ppm) for one hour, not to be exceeded more than once per month. This standard incorporates a minimum safety factor of 3 to 5. The committee cautioned that a greater margin of safety might be needed in areas where much of the population at risk is comprised of more sensitive subjects.

This volume emanates from the symposium entitled "Biological Studies of Environmental Pollutants, Part II. Health Effects of Nitrogen Oxides." The symposium was conducted in conjunction with the Chemical Congress, a joint conference between the American Chemical Society and the Chemical Society of Japan, held in Honolulu, April 1970.

A companion volume entitled *Generation of Aerosols and Facilities for Exposure Experiments*, compiled from presentations in Part I, was edited by Dr. Klaus Willeke of the Department of Environmental Health, University of Cincinnati, and published by Ann Arbor Science Publishers.

The introductory chapters provide an overview and establish a proper forum for comprehensive discussion. Subsequent chapters define nitrogen oxides as pollutants through presentations on NO_x measurement and monitoring techniques and the interactions of NO_x in the atmosphere. Animal toxicological studies and human studies, including clinical and epidemiological investigations, are presented.

This compilation will serve a useful purpose as a data base in regulatory activities, as well as an excellent resource book for scientists and students.

I am deeply indebted to the United States Environmental Protection Agency and Electric Power Research Institute for their financial support.

The kind support so willingly provided by the following individuals is gratefully acknowledged: Dr. Stephen Gage, Dr. Jerry Stara, Mr. Charles Ris, Dr. Lester Grant, Mr. Mike Berry, Dr. Gordon Hueter, Dr. Donald Gardner, Dr. Robert Lee, Dr. Tom Wagner, Mr. Gerald Rausa, Dr. Gary Glass and Mr. Fred Meadows of EPA, and Dr. James McCarroll of the Electric Power Research Institute.

In addition, I thank the American Chemical Society, Division of Environmental Chemistry, especially Mrs. Linda Deans, and the Chemical Society of Japan for providing this forum in an ideal environment.

A special appreciation is extended to Dr. Klaus Willeke for his tireless effort in performing administrative functions. I also want to thank the Japanese co-chairman Dr. Taichi Nakajima and his colleagues for their kind cooperation.

Si Duk Lee
Cincinnati, Ohio

Dr. S. D. Lee is Acting Deputy Director, Environmental Criteria and Assessment Office, U.S. Environmental Protection Agency, Cincinnati.

He received his PhD from the University of Maryland, specializing in animal nutrition and biochemistry. He continued his training in biochemistry at Duke University Medical Center under a postdoctoral fellowship from the National Institute of Health. He was awarded an Advance Research Fellowship from the American Heart Association for continuation of his work at Duke. His research work at EPA has been primarily devoted to the early identification of adverse effects of environmental pollutants using animal models to obtain necessary information for assessing possible health effects on human populations. More recently, he has been engaged in health assessment and criteria document preparation of various water-associated pollutants.

Dr. Lee is the author of numerous papers presented at national and international conferences and symposia, and has published over fifty articles in various professional journals. He is the editor of *Biochemical Effects of Environmental Pollutants*, and senior editor of *Assessing Toxic Effects of Environmental Pollutants*, published by Ann Arbor Science in 1977 and 1979, respectively.

TO MY FAMILY

CONTENTS

CHAPTER 1

RESEARCH FOR PROTECTION OF OUR ENVIRONMENT

Stephen J. Gage

Assistant Administrator
Office of Research and Development
United States Environmental Protection Agency
Washington, DC 20460

INTRODUCTION

The United States Environmental Protection Agency (EPA) was established on December 2, 1970, by a presidential order, and this action reflected a major commitment by this nation to control pollution and abate its effects on human health and the environment. The decision to create a federal Environmental Protection Agency to function as a single organization was a far-reaching response to the environmental challenge. For the first time, numerous environmental control programs previously scattered throughout various federal agencies were consolidated within a single independent agency.

The EPA was authorized to carry out a coordinated, integrated attack on the environmental problems associated with air and water quality, solid waste, pesticides, radiation and noise. The EPA serves primarily a regulatory function with responsibilities for the establishment and enforcement of environmental quality standards as specified in statutes enacted by Congress. Industry now recognizes that environmental protection is a necessary and integral part of business conduct. As a result, educational programs on environmental pollution and conservation are now incorporated in numerous academic curricula; environmental law and environmental engineering are examples of expanding new specialties. During the last decade, environmental protection has become a part of our society and daily lives, although careless release of pollutants into the environment still occurs from numerous sources. In addition, the accumulated damage resulting from years of environmental abuse and neglect will not be easily corrected.

By the late 1960s the lakes and waterways of our once pristine country had become choked with sewage, industrial waste and other forms of pollution. Air pollution became unbearable in many urban and industrial areas, and consequently the rate of pulmonary diseases increased markedly. Residues of pesticides, such as DDT, had been found in tissue samples taken from both wildlife and humans. Various forms of trash, such as empty cans and carcasses of automobiles, littered the landscape. Concerns also had been expressed about noise pollution, and potential problems associated with the use of radioactive materials and disposition of their waste.

In response to these problems, the EPA committed almost $11 billion in grants to states and communities for construction of wastewater treatment plants during its first five years of operation. During the same period, the EPA initiated more than 6000 enforcement actions against violators of existing air, water and pesticide laws. As a result of rigorous enforcement of the 1970 Clean Air Act amendments, current standards for two major automobile emissions, carbon monoxide and hydrocarbons, now require a reduction of nearly 85% from pre-1968 emission levels.

Regulatory actions which led to the decreased use of persistent pesticides have resulted in a significantly lower level of these pesticides being detected in human and wildlife tissues. At least 25 major American cities will be involved in some form of resource recovery from municipal trash by 1980. Federal standards and guidelines are being formulated to protect citizens from exposure to radiation and unnecessary noise.

The first steps toward achieving a cleaner environment for the protection of human health have been gratifying. We are learning to use modern technology for the service and maintenance of civilization. A greater respect is being developed for the fragile nature of the biosphere—the earth's surface layer of land, water and air on which our life depends. Industry is moving away from the concept that it cannot cope with the restrictions imposed by environmental controls. We have a running start in the clean-up of the past half century's neglects and mistakes. However, much work remains to be done. The air quality of metropolitan areas is still unsatisfactory. Many waterways still must be cleaned and protected, and we must always watch for new hazardous materials entering the environment.

The will of the American people regarding the necessity for ecological controls was formally expressed when the National Environmental Policy Act was signed by the President about a year prior to the creation of the EPA. It established a national priority to "maintain conditions under which man and nature can exist in productive harmony." Since then, new environmental laws have been passed at every level of American government demonstrating a pervasive concern for protection of human health and life-sustaining ecosystems.

The regulatory mission of EPA includes establishment, enforcement and monitoring of pollutant standards. The key to rational environmental control lies in the determination of tolerable levels of change in our environment, which has limited natural cleansing and buffering capabilities. Congress has passed several laws requiring EPA to define the levels which certain environmental pollutants must not exceed. State and local governments may develop additional controls or programs for various reasons but the EPA's direct responsibilities are restricted to protecting human health and welfare.

The EPA encourages voluntary compliance by private industries and communities, and encourages state and local governments to perform whatever enforcement activities are needed to meet EPA pollutant standards. If these agencies fail to produce effective plans for pollution abatement, the EPA must step in using the enforcement authorities contained in most of the major environmental laws passed by Congress.

The EPA also conducts several kinds of monitoring programs and activities. Some are broadly based monitoring programs for pollution profiles, to determine whether pollution levels and emissions are increasing or declining. Others oversee the effectiveness of the various abatement programs developed by the EPA and state and local governments.

RESEARCH AND DEVELOPMENT ACTIVITIES
AND THEIR PRIORITIES

Effective environmental regulation as directed by federal legislation requires accurate technical data on the various substances which present possible hazards to health and the environment. The research and development area of EPA supports the agency's primary functions by providing necessary information for developing and enforcing appropriate regulations, standards and control programs. EPA's research program is authorized under various key congressional acts which allocate over 20% of the agency's operating budget for scientific studies. These funds support the research activities of EPA's Washington, DC headquarters, 15 major laboratories and a number of smaller field stations. Research programs are also conducted as extramural projects through EPA grants and contracts with academic, private and industrial research organizations and through cooperative agreements with other federal, state and local agencies.

The Research and Development Program focuses on four principal areas:

1. The Office of Monitoring and Technical Support is responsible for: (1) developing reference standards for environmental measurements, monitoring equipment, techniques and systems; (2) developing agencywide quality assurance programs; (3) disseminating scientific and technical knowledge; and (4) providing technical support, including monitoring and analytical support, to the agency.

2. The Office of Health and Ecological Effects studies the implications of varying types and levels of pollution on human health and the environment. The major impetus behind these efforts to control toxic substances is the need to protect human health. The EPA's role is protective and preventative, not curative, an idea which is reflected in a recent statement by the EPA administrator:

When the cost of health care in America has risen to $140 billion a year, with most of this going for after-the-fact attempts at treatment and cure, it is obvious that we need to reorder our national and individual priorities. How much more health effective and cost effective it would be if more emphasis were placed on the prevention—on keeping harmful materials out of our air, water, and soil—and out of our people." [1]

The EPA is not mandated to treat diseases associated with pollution after they have become obvious, but to prevent such harmful pollution in the first place.

3. The Office of Energy, Mineral, and Industry is responsible for assessing, developing and demonstrating pollution control technology to mining, energy and other industrial activities. This office plans and administers a comprehensive federal energy and environmental control, research, development, and demonstration program.

4. The Office of Air, Land, and Water Use is responsible for planning a comprehensive research program on water supply and on municipal waste, including wastewater, solid and hazardous waste management. This office also is responsible for research on the transportation and disposition of pollutants in the environment, areawide environmental management and pollution from nonpoint sources such as urban runoff and agricultural and forestry activities.

In its program planning process EPA attempts to evaluate those environmental issues that are expected to become problems over the next two decades. When these problems can be clearly delineated, or where presidential or congressional guidance is explicit, discussions are detailed and precise. When our vision of the future is less clear, we rely on our experience and expertise to give shape and substance to potential problem areas. The agency simply cannot solve all the problems of the environment. Therefore, it must establish priorities, and the first priority is to protect people.

The topics which follow were selected to represent problem areas of major concern for the next few years. The list is not exhaustive; however, these topics reflect our best estimate of the trends and factors which should dictate the direction of future environmental research. Each of these eight topics—toxic substances, air pollution, energy and the environment, solid waste, global pollution, nonionizing radiation, nonpoint sources and watersheds, and measurements and monitoring—were selected because of their relationship to severe threats to human health and the natural environment.

Unfortunately, even as we address these few but very important topics, we are overlooking many other substantive problems. The importance of the other problems, however, should not be underestimated; we may instead be lacking the relevant information necessary to evaluate them. Therefore, our program planning is a dynamic process where priorities and problems change each year as new perceptions emerge. The foundations for such a process are our forecasts of environmental change and our best scientific judgment.

Detailed in the following discussion are relative priorities for dealing with each of the problem areas, a set of health and environmental protection goals to guide research efforts on these potential problems, and brief descriptions of each problem [2].

Toxic Substances

The goals of the toxic substance research effort are to reduce the threat of toxic substances to human health and the environment by improving screening tests for the toxicity of substances, to detect the presence of potential toxic substances through studies of historical health and disease patterns (epidemiology), to determine how each major toxic pollutant travels through the environment, how it changes, and how it affects humans at various dose levels, and, finally, to develop improved means of controlling these pollutants [2].

With advances in industrial technology and economy during the last few decades, thousands of new chemicals were introduced to our planet. These substances were developed for a variety of reasons, and people of all industrialized nations are exposed to them in the air they breathe, in the water they drink and in the food they eat.

Recognition of the pervasive threat of environmental toxicants to human health by our presidents and the Congress resulted in the Toxic Substances Control Act, the Resource Conservation and Recovery Act and the amendments to the Clean Air and Clean Water Acts. The EPA's mandate to protect public health demands that the highest priority research area be the prediction and control of toxic substances.

In order to control toxic substances, we must first predict their environmental behavior and effects. The estimation of exposure levels to target systems is dependent upon the availability of accurate data on chemical production, use, and disposal. Moreover, the behavior of a chemical in the environment is determined by physical and chemical characteristics, chemodynamics, biotransformation and bioaccumulation, and the identity of decomposition products. Ultimately, this information must be integrated into the best possible predictive model which allows for the evaluation of nonlinear interactions involving multiple causative factors.

Research tools such as epidemiology, toxicology and chemical analysis must be used to accurately predict the harmful biological effects of any chemical through any exposure route. The early development of rapid testing procedures is necessary to provide this prediction capability. Understanding of the complex biochemical mechanisms involved will be necessary and will require the evaluation of many different testing procedures to assure a greater likelihood of success.

A cautious extrapolation of animal experimentation data is necessary in evaluating the effects of chemical exposures. A critical evaluation must be made of such factors as dose-response relationships, species employed, chemical analogies with the test substance, and experimental design. In addition, the type of response (organ change, biochemical change, carcinogenesis, mutagenesis, teratogenesis) is a vital factor in defining the severity of potential consequences from exposure.

Air Pollution

The goals of the national air pollution research efforts are to improve our understanding of the formation and transport of atmospheric aerosol pollutants, the potentially serious risks that these aerosols present to the public health and the environment, and the possible alternatives for reducing human exposures to these pervasive pollutants [2].

A basic understanding of the effects of simple gaseous air pollutants such as carbon monoxide, ozone and sulfur oxides has already been developed, and our attention is now shifting to the complex of aerosols which pollute the air over large areas of the country.

Aerosols are formed from numerous sources. The increased utilization of coal, for example, is expected to contribute significantly to SO_x and NO_x emissions during the next 15 years. During the transformation of these gaseous pollutants to compounds of higher toxicity (e.g., sulfates and nitrates), they may go through a formative aerosol phase. In this phase, they combine with water in the atmosphere to form acids which then react with metallic air pollutants, to form the ultimate toxic pollutants responsible for adverse health effects. Adding to these aerosol problems are the complex secondary pollutants generated in the urban atmosphere by reactions of hydrocarbons and nitrogen oxides, largely produced by internal combusion engines.

In addition to their chemical properties, the biological hazard of suspended aerosols also depends upon their size. Larger aerosol droplets (10 μ in diameter) are readily deposited in the upper respiratory passages of the nose and throat, where they may subsequently be swallowed. Smaller droplets, however, may penetrate to the deep lung compartments and be retained for indefinite periods. These inhaled aerosols are left free to react with the respiratory epithelium, leading to exacerbation of chronic lung disease (e.g., emphysema) and may also play a role in the development of lung cancer.

Due to the increased presence of atmospheric aerosols and their potential threat to human health, understanding and controlling such aerosols must be our second highest priority research area. Our understanding of aerosol formation, reactivity and transport must be improved, along with elucidating the relationship among the exposure, type, and size of aerosols and the defense mechanisms of and ultimate damage to the human body. Although the experiments required to understand such subtle effects are more sophisticated than traditional inhalation experiments with gaseous pollutants, it is important that we accelerate our health effects studies on atmospheric aerosols.

In addition to their impacts on human health, aerosols combined with other air pollutants can lead to poor visibility and acidic precipitation. With an improved understanding of aerosol formation in the environment, maps which depict pollutant sources and environmental conditions can be used to predict exposure routes and patterns. These predictions can then be used to support more accurately targeted efforts to control pollutant emissions, and identify populations at greatest risk of health impairment.

Energy and the Environment

The goals of the national energy-related environmental research and development efforts are to understand the health and environmental impacts which could result from increased domestic energy production and use and to take the steps required to protect public health and the natural environment. These goals encompass our domestic energy resources (coal, oil, gas, nuclear, solar, geothermal, etc.) and both existing and new methods for energy processing and conversion [2].

The National Energy Plan requires conservation of oil and natural gas through greater reliance on coal and, to a lesser extent, on nuclear power and oil shale. Our main priority in this regard is to anticipate and control any health and environmental risks from toxic substances and atmospheric aerosols resulting from coal combustion and conversion processes. For example, fundamental research is needed to fully characterize the chemical nature and potential carcinogenicity of the complex mixtures of polycyclic hydrocarbons, aza arenes, and trace metals produced from new coal conversion technologies. In this manner, we can take advantage of a rare opportunity to use foresight rather than hindsight in managing our national development and protecting public health. An anticipated shift to utilization of massive amounts of coal demands that energy-related environmental problems be assigned a high priority for research. A close relationship exists between energy-related pollution and the previously discussed research priorities for toxic substances identification and air pollution.

Solid Waste

The goals of the solid waste research effort are to determine the risks associated with the handling and disposal of toxic solid waste and to develop means of adequately protecting public health by isolating the toxic materials from the environment, detoxifying the wastes before disposal, or recovering the energy and mineral content of the waste for recycling [2].

The quantitative increase of solid wastes over the next 15 years is expected to be much greater than that of airborne or waterborne pollutants. Pollutants which would otherwise have been discharged into the atmosphere and water are now being disposed in solid or semisolid form due to industrial and economic growth, and to altered control technology. A considerable portion of this waste, especially industrial sludges, contains known toxic substances.

Because of rapid growth in their quantity and growing awareness of the toxic nature of certain components of these wastes, this area is the fourth highest priority research area. Our responsibilities are to prevent toxic solid waste from being discharged into the environment by developing detoxification techniques, recovering waste as a resource, substituting products that result in less waste, and informing consumers about better waste management.

Global Pollution

The goals of the global pollution research efforts are to understand the dynamic global processes within the biosphere and to develop the capability to predict the movement and changes of pollutants in the biosphere and the impacts of these pollutants on human health, ecosystems, and the climate. With adequate predictive capability, control of pollutants which pose threats to a healthful earth can be undertaken [2].

The ability of the earth's biosphere to survive human abuse is contingent upon at least three issues: (1) the threat of depleting our protective ozone layer; (2) the increased release of carbon dioxide into the atmosphere; and (3) the increased exploitation and pollution of the oceans.

Due to the long-term importance of such subtle effects to all people, understanding and predicting the effects of human activity on the biosphere is the fifth highest priority research area. A better understanding of natural global cycles is necessary if we are to determine how these cycles are affected by human activities. For example, since the nitrogen cycle is emerging as a possible prototype for understanding the limits to the buffering capability of nature, we could focus increasingly on improving our understanding of the global nitrogen cycle.

Nonionizing Radiation

The goals of the nonionizing radiation research effort are to understand the nature and severity of the health effects caused by the pervasive sources of nonionizing radiation and to help to determine what exposure standard would protect public health [2].

Rapidly increasing levels of human exposure to nonionizing radiation from radio, television, electrical transmission and microwave sources, coupled with new data concerning potential adverse health effects, strongly support the expansion of research in this area. The question of whether more stringent control of electromagnetic radiation sources is necessary to protect public health ranks sixth in research priority. Our focus must be on understanding the impacts of the more subtle, nonthermal, effects of nonionizing radiation both alone and in combination with other stresses. Prime candidates for expanded research are the effects of radiation in the communications (microwave) and commercial (UHF-TV) radio frequency bands.

Nonpoint Sources of Pollution

The goals of the research efforts on nonpoint sources and watersheds are to understand how all nonpoint sources of pollution contribute to environmental degradation, especially in watersheds, and to determine how effective various monitoring and control techniques are in identifying and abating, respectively, the deleterious effects of nonpoint source pollution [2].

Major point sources of pollution such as industries and municipalities are beginning to bring under control their more health-threatening water pollutants. In terms of sheer volume, a more intractable pollution problem is the control of water pollutants arising from farming and forestry practices, poorly planned land uses, atmospheric pollution fallout, and other such non-point sources, which can have severe immediate and long-term effects on watersheds and eventually, upon groundwater.

Nonpoint sources will remain the primary sources of water pollution over the next two decades. Understanding and being able to predict the impacts of these sources on watersheds and ultimately human health is the seventh highest priority research area. The watershed integrates the impacts of many pollution sources, both point and nonpoint, and thus emerges as the key hydrological unit upon which to base our studies of water pollution.

Measurement and Monitoring

The goals of the measurement and monitoring research effort are to anticipate potential environmental problems, to support regulatory actions by developing an in-depth understanding of the nature and processes that impact health and ecology, to provide innovative means of monitoring compliance with regulations and to evaluate the effectiveness of health and environmental protection efforts through the monitoring of long-term trends [2].

The detection of the multitude and variety of different pollutants at various locations throughout the country is currently prohibitively expensive. More accurate and cost-effective procedures and equipment must be developed before a comprehensive monitoring system can be instituted. However, a disciplined, long-term monitoring effort to accurately describe our physical, chemical and biological environment must be undertaken to improve our understanding of the dynamics of that environment.

In this eighth highest priority research area, the discovery of casual environmental relationships, especially concerning toxic substances, can vastly improve our ability to anticipate pollution problems which may threaten human health, and focus our resources where they will be most effective in preventing those problems.

INTERAGENCY COORDINATION

For expeditious and economical resolution of these complex environmental problems, extensive interagency collaborations are currently underway, improving both the efficiency of federally funded environmental research and its relevance to the protection of human health and environmental quality.

Approximately three-fourths of the environmental research funded by the federal government is performed by agencies other than the EPA. The EPA, however, requires much of the information developed via such efforts to fulfill its regulatory mission. Duplication of effort and information gaps are avoided through a multitude of formal or ad hoc ties between the EPA and other agencies or interagency commissions (Table I).

At present, the Office of Research and Development maintains formal coordinating linkages with 39 other federal agencies and departments under

Table I. Interagency Research and Development Coordination

Department or Agency	Air Pollution	Water Pollution	Energy	Pesticides & Toxics	Radiation	Health Effects	Other[a]
Agriculture							
Federal Research Science and Education Administration	○ ▲	○ ▲	▲	○ ▲			
Food Safety Quality Service	○ ▲	○ ▲	▲	○ ▲			
Commerce							
National Bureau of Standards	○ ▲	○ ▲	○ ▲		○ ▲	○ ▲	
National Oceanic and Atmospheric Administration	○ ▲	○ ▲	○ ▲		○	○ ▲	○ ▲
Defense							
Corps of Engineers	○ ▲	○ ○ ▲	○ ▲		○ ▲		○ ▲
Energy	○	▲			○		
Health, Education, and Welfare							
Public Health Service	○	○ ▲		○	○	○	
Food and Drug Administration				○	○	○ ▲	
Bureau of Radiological Health		▲				○	
National Center for Toxicological Research	▲			○ ▲		○ ▲	
National Center for Health Statistics				○ ▲		○ ▲	
Disease Control Center						○ ▲	
National Institute of Occupational Safety and Health	○ ▲		○ ▲	○ ▲		○ ▲	
National Cancer Institute	○			○ ▲		○ ▲	
National Heart, Lung, and Blood Institute						○ ▲	
National Institute of Environmental Health Sciences	○	○	○ ▲	○ ▲		○ ▲	

Housing and Urban Development

Interior
Fish and Wildlife Service
Bureau of Land Management
National Marine Fisheries Service
United States Geological Survey
Heritage Conservation and Recreation Service

Labor

State

Transportation
Federal Aviation Administration
United States Coast Guard

Other Agencies
Council on Environmental Quality
National Aeronautics and Space Administration
Consumer Product Safety Commission
Tennessee Valley Authority
National Science Foundation
National Academy of Sciences/National Academy of Engineering
Nuclear Regulatory Commission
Great Lakes Basin Commission

[a]Including noise, solid waste, and policy research.
○ Coordination through committees.
▲ Coordination through joint research.

the auspices of the Interagency Regulatory Liaison Group and is cofunding more than 200 projects each year with those agencies. These projects range from health effects studies and technology development to monitoring support and data assessment.

INTERNATIONAL COORDINATION

> The goals of our international activities are to recognize the worldwide and long-range character of environmental problems and, where consistent with the foreign policy of the United States, lend support to initiatives, resolutions, and programs designed to improve international cooperation in anticipating and preventing a decline in the quality of our global environment [2].

The Office of Research and Development currently conducts, and will continue to pursue, a high level of cooperative and coordinated research with other nations and interntional organizations (Table II). The benefits of these activities are threefold. First, some of the problems we face, such as depletion of the ozone layer or protection of the Great Lakes, are inherently international, and a unilateral approach to their solution would be inadequate. Second, just as our technology is superior in some areas, other nations have areas of experience and expertise which allow for an efficient, mutually beneficial sharing of information. Third, the reduction or elimination of pollution anywhere on the globe is, in a larger sense, of benefit to all who live on this planet. As we improve and share our pollution control technologies, our entire environment becomes a more healthful place to live [1,3,4].

For example, EPA specialists participate in joint projects under bilateral agreements with Canada, the Federal Republic of Germany, Japan, Mexico, the USSR and excess foreign currency countries (e.g., Egypt, Poland, Yugoslavia, Pakistan and India). The EPA also participates in working groups associated with multilateral organizations such as the Organization for Economic Cooperation and Development, NATO's Committee on the Challenges of Modern Society, United Nations Environmental Program, World Health Organization, Pan American Health Organization, Commission of European Communities and the Economic Commission for Europe.

Agency participation in ongoing international activities reflects areas of current United States concern for potential environmental problems and emphasizes toxic substances, atmospheric pollution, water pollution, hazardous wastes, energy and monitoring.

Toxic Substances

Issues relating to toxic substances, including multimedia exposure to environmental chemicals and related health effects, significant perturbations to ecosystems, and disposal of toxic substances, are of primary concern to environmentalists worldwide. Therefore, the agency is working on several international initiatives in this area.

Implementation of the Toxic Substances Control Act requires cooperation in establishing international agreements on regulatory procedures (e.g., consistent testing protocols, quality control procedures and standard methods).

The agency is concentrating its efforts within major international organizations such as the Chemicals Group of the Organization for Economic Cooperation and Development (OECD). The EPA is participating actively in the chemical testing program of the Chemicals Group to coordinate testing methods and systems used to predict the effects of substances on humans and the environment before such substances enter the marketplace. The EPA's focus is on developing test methods for evaluating the long-term effects of chemicals on human health.

The EPA is consulting with the European Commission regarding the administrative details of toxic substances control, including the coordination of the Toxic Substances Control Act premanufacturing procedures and the evaluation of toxicity testing results. Also, in cooperation with the World Health Organization, the EPA will help develop an international plan of action to assess health risk from exposure to chemicals.

Japan has long been concerned with the issue of toxic substances. Under a United States-Japan bilateral agreement, the EPA has exchanged information in such areas as mercury removal from contaminated wastewater and sludges, removal of DDT and PCB in accumulated sediments, and the fate and effects of toxic substances in sediments. Areas for future cooperation include sharing test information, risk assessments, chemical import/export control, and the establishment of an international convention to control toxic substances.

The Japanese are also knowledgable in the area of removal and disposal of toxic sediments. They have initiated full scale remedial programs in several harbors and bays in Japan. Their dredging technology, which attempts to keep aquatic environmental damage to a minimum, could have direct applicability to the PCB and Kepone situations in the United States.

The EPA is also working with the Federal Republic of Germany to establish an environmental specimen bank. The purpose of such a specimen bank is the long-term storage of biological specimens and data for future analysis to determine the historical record of pollutant burdens.

Atmospheric Pollution

Atmospheric pollution is an important area for international consideration because airborne pollutants can travel across international boundaries, continents and oceans. Research on transport, transformation, and biological and health effects of air pollutants is, and will continue to be, an important aspect of the EPA's international program.

For example, the United States and Canada have an increasing number of cross-boundary air pollution problems that require analysis and resolution through bilateral contact or reference to the International Joint Commission. At present, there are at least eight major stationary source pollution problems between the two countries, none of which is amenable to easy solution. The major need is for better information on the potential impacts of sources

Table II. Worldwide Environmental Activities

Organization/Activity	Research, Development, and Demonstration Programs							
	Air Pollution[a]	Water Pollution[b]	Radiation	Pesticides	Noise	Waste Mgmt[c]	Toxic Subst	Energy
International Organizations								
Commission of European Communities (CEC)	•	•				•	•	•
Committee on Challenges to Modern Society (CCMS)	•	•			•	•	•	•
International Organization for Legal Metrology (OIML)	•	•	•	•	•	•	•	•
International Standards Organization (ISO)	•	•	•	•	•	•	•	•
Organization for Economic Cooperation and Development (OECD)	•	•	•	•	•	•	•	•
United Nations Economic Commission for Europe (ECE)	•	•				•	•	•
Food and Agriculture Organization (FAO)		•						
Intergovernmental Maritime Consultative Organization (IMCO)						•		
International Atomic Energy Agency (IAEA)	•	•	•	•	•	•	•	•
International Civil Aviation Organization (ICAO)	•	•			•			
World Health Organization (WHO)	•				•			•
World Meteorological Organization (WMO)	•	•	•	•		•	•	•
United Nations Educational, Scientific, and Cultural Organization (UNESCO)		•						
United Nations Environmental Program (UNEP)	•	•	•	•	•	•	•	•

Bilateral Cooperation

- Brazil
- Canada
- Federal Republic of Germany
- France
- France and United Kingdom[d]
- Iran
- Israel
- Japan
- Mexico
- Saudi Arabia
- Soviet Union
- United Kingdom

Scientific Activities Overseas Program

- Egypt
- India
- Pakistan
- Poland
- Yugoslavia

[a] Includes troposphere and stratosphere.

[b] Includes marine, estuarine, and freshwater environments.

[c] Includes hazardous, solid, and radioactive wastes.

[d] Tripartite agreement.

and their effects on health and welfare in both countries. The EPA has been assisting the Department of State and the Joint Commission to make these assessments. In addition, the EPA has offered to hold a joint United States-Canadian workshop in 1978 to begin a coordinated research effort to investigate long-range pollutant transport. The EPA and Environment Canada are working closely on problems of mobile source air pollution, principally to foster the more than $18 billion annual trade in our integrated automobile manufacturing and marketing structure.

Under the United States-Japan bilateral agreement, researchers of Japan and the EPA are exchanging information on meteorology, photochemical air pollution and, in particular, air pollution conditions that lead to production of photochemical oxidants (smog). The Japanese are shifting their oxidant control strategy from nitrogen oxide control alone to nitrogen oxide control combined with hydrocarbon control. Information from each country's air pollution studies including diffusion modeling, analysis and field measurements should, in the future, improve our understanding of air pollution movement, our ability to forecast air pollution, concentrations, and our stationary source air pollution control technologies.

Another excellent example of international cooperation in the area of health effects research is the ongoing isotopic lead experiment sponsored by the Common Market, Italian Federal Hydrocarbon Authority, and International Lead and Zinc Research Organization. For this study, gasoline stations in Torino sell gasoline with a unique lead isotope ratio, giving the lead a unique "fingerprint" which will allow accurate tracking through the ecosystem. The amount of lead in human blood attributable to automotive sources will be determined by measuring blood lead levels during the use of this special gasoline and after it is discontinued.

Water Quality

A landmark in international cooperation regarding improved water quality is the 1972 United States Canadian Great Lakes Water Quality Agreement. Now under joint review and revision, this agreement is a unique mechanism for coordinating national efforts toward the cleanup and restoration of the Great Lakes. These lakes constitute more than 80% of the fresh surface water area of the United States and 97% of its fresh surface water storage. The agreement sets joint water quality objectives, formulates remedial programs, and commits the two governments to providing sufficient funding to achieve the objectives. Both sides have worked closely through the International Joint Commission to support a water quality monitoring program and to set forth the research program necessary to guide and support surveillance activities.

Wastewater Treatment

To improve knowledge of wastewater treatment and disposal methods, the EPA is participating in international research involving source characterization of pollution, advanced wastewater treatment technology, process modification and analyses of sludges and their environmental behavior. One of the most important efforts is the study of advanced wastewater treatment being conducted under the auspices of NATO's Committee on the Challenges of Modern Society (CCMS). The United States, United Kingdom, Canada, Italy, France and Germany are studying such topics as the standardization of formats for international information exchange, the use of oxygen-enriched air to treat contaminated effluent, land spreading of sludge, nutrient removal, reverse osmosis, electrodialysis and ion exchange.

Details on these advanced methods of wastewater treatment used abroad are helping the EPA to determine which treatment methods may be feasible for use in the United States. For example, Japan has provided technological information applicable to the United States problems, such as their best available technology (BAT) research on tannery wastes. A large scale pilot plant operation in Japan is testing four different pure oxygen systems on the same tannery waste under the same conditions utilizing different operational techniques. The efficiency and effectiveness of each will be evaluated by the Japanese and a full scale system will be installed. These results will be of immense value to EPA effluent guideline and enforcement offices as well as of direct value to the tanning industry.

Hazardous Waste

The EPA is also working with various countries to assess risks and benefits associated with various methods of hazardous waste disposal. Of key interest is a NATO CCMS pilot study on the disposal of hazardous wastes. Phase I of the study provided the EPA with valuable insight into mine and landfill disposal practices and alternatives, and produced recommended procedures for hazardous wastes landfill disposal and surface treatment. Areas of potential future cooperation under a United States-Japan bilateral treaty include pyrolysis of solid waste, environmental effects of vinyl and polyvinyl chlorides, improved collection systems, management and technology, hazardous waste treatment and disposal technologies, recovery of past consumer waste, and management of information systems on industrial wastes.

Energy

With the increasing focus on coal as a future energy source, much can be gained from the experience of other countries in utilizing this fuel. An excellent example of this potential is a study currently underway in Yugoslavia, which is providing a full evaluation of the Kosovo Coal Gasification Plant. The pollutant data and evaluation obtained during this study will be used by the EPA to determine the environmental impact of gasification plants proposed for the United States and the criteria required for control technology development. Assessment of the environmental consequences of coal conversion technology is essential.

The EPA is also conducting four projects under a United States-Soviet Union Environmental Agreement which concern gaseous emissions, particulate abatement technology, process improvement and modification, ferrous metallurgy, and a new United States initiative; protection of the environment from coal preparation plant operations. Both countries are examining problems related to the abatement of sulfur dioxide emissions through various control techniques using lime/limestone, magnesia and ammonia scrubbing; dust collection technology; characterization of aerosols; demetallization pretreatment for hydrodesulfurization of petroleum residuals; preliminary coal cleaning; and complex methods for fuel utilization in power generating systems for the elimination of harmful emissions.

Monitoring

In addition to its activities on the Stratospheric Ozone Monitoring Program, the EPA has the major responsibility for fulfilling the international monitoring program for the Great Lakes, under the 1972 United States Canadian Water Quality Agreement. International coordination of monitoring activities is achieved through the surveillance subcommittee of the Water Quality Board. This subcommittee ensures that appropriate measurements are taken for use in the management models for nutrients. The resulting data are shared by the two countries.

The EPA is also active in the United Nations Environmental Program's Global Environmental Monitoring System. Primarily concerned with air, this system will link existing national monitoring activities. United States cooperation in the global water quality monitoring network is expected to increase as a result of the EPA's role as World Health Organization Collaborating Center for Environmental Pollution Control. The data from joint surveillance and monitoring of the Great Lakes will be incorporated into the Global Environmental Monitoring System.

A comprehensive environmental monitoring program is a prerequisite for complete United States participation in the establishment of a global monitoring system and our utilization of data from it. International coordination, as well as development of our own national monitoring capability, will increase our ability to control environmental pollution before it reaches crisis proportions [4].

CONCLUSIONS

There is accumulating evidence that the quality of our environment is improving markedly in certain respects. As pointed out in the Ninth Annual Report of the Council on Environmental Quality, the benefits to man and society of the improvement achieved to date have far exceeded the cost of implementation of environmental regulations. The conduct of basic and applied research to solve the complex pollution problems which threaten our health and welfare has been a fundamental part of EPA's mission. The support of broadly based research programs in the areas of toxic substances, environmental monitoring, ecological effects, control technology and resource and waste management has led to a better understanding of the problems which we face today and the approach to be taken in protecting the future quality of life. Through in-house research, extramural grants and international agreements, EPA has made a firm commitment to seek answers for our most challenging environmental problems.

Priorities have been set at EPA to direct a coordinated and focused research effort in those areas where the need for information is greatest. The primary challenge to be faced is in predicting and controlling toxic chemicals. In conjunction with traditional methods of toxicologic testing and epidemiologic investigation, new approaches must be developed which can quickly and inexpensively provide reliable toxicity data for decision-making purposes.

The health-related aspects of air pollution and increased demands for energy are closely tied to our concerns over toxic chemicals. Therefore, both of these areas are being accorded a high research priority. Additional areas which have been selected by EPA for active research include solid waste, global pollution, nonionizing radiation, nonpoint sources of pollution and environmental monitoring. All of these EPA-sponsored programs share a common goal in that continued improvement of environmental conditions, and anticipation of future environmental concerns, will ultimately be reflected in the quality of our life and that of subsequent generations.

REFERENCES

1. Costle, D. "EPA's International Committment," *EPA J.* 4(6) (June 1978).
2. "Research Outlook, 1978," U.S. EPA 600/a 78-001 (June 1978).
3. Popkin, A. B. "International Cooperation Begins at Home," *EPA J.* 4(6) (June 1978).
4. Brown, N. "Global Environmental Monitoring," *EPA J.* 4(6) (June 1978).

CHAPTER 2

HEALTH EFFECTS OF NITROGEN OXIDES
KEYNOTE ADDRESS

D. S. Barth

Visiting Professor, Biology Department
University of Nevada at Las Vegas
Las Vegas, Nevada 89154

INTRODUCTION

The purpose of this book is to review selected recent data on the health effects of nitrogen oxides (NO_x). Hopefully, it will provide additional information to assist environmental and occupational health protection organizations in their evaluation of all nitrogen dioxide (NO_2) toxicity data for the purpose of better defining acceptable levels of human exposure. The book first addresses the measurement and monitoring of nitrogen oxides, including atmospheric interactions and quality assurance procedures. Subsequent chapters present first animal toxicological studies and then human studies, including both clinical and epidemiological investigations.

Hopefully, this book will stimulate interest into the appropriateness and adequacy of measurement technologies and margins of safety of any existing or proposed standards. Also of concern are the influence on toxic manifestations of pollutant interactions with other environmental factors and recommendations for future research.

This chapter attempts to set forth a framework of factors to keep in mind when using this book, so that its purpose will be accomplished efficiently and expeditiously. Specifically, the chapter will be organized around the following subjects: (1) definition of human population to be protected; (2) significance of various health effect endpoints; (3) exposure assessment; (4) risk assessment; and (5) importance of measurement and monitoring.

DEFINITION OF HUMAN POPULATION TO BE PROTECTED

An evaluation of existing health effects data to define acceptable levels of human exposure to any environmental or occupational pollutant must consider the constituency and size of the population to be protected. For example, society probably cannot afford to completely protect the single most sensitivie or most susceptible exposed human individual from any toxic environmental or occupational pollutant. If one accepts this concept, one must immediately specify the most sensitive population group and the minimum size that should be completely protected (with an adequate margin of safety) by any environmental or occupational standard. In some instances, the word "completely" may be inappropriate. It may be necessary to accept some finite risk of an adverse effect occurring in one or a few members of the human population to be protected—even when the established standard is never exceeded. In this case, we must specify what risk we are willing to accept.

These considerations all relate to problems encountered by standard-setters when faced with establishing and defending whatever standards are deemed appropriate based on health effects data then available. In most instances, there are inadequate data available to support a consensual and unique standard. If scientists could conduct their research so that more information can be made available on the spectrums of response in different human populations to different environmental pollutants, the tasks of standard-setters could become easier.

Figure 1 presents a hypothetical relationship between numbers of exposed humans adversely affected and exposure to an environmental pollutant.

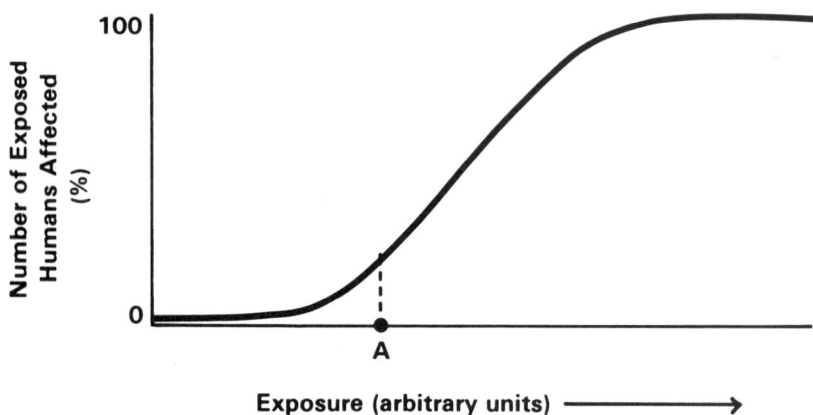

Figure 1. Hypothetical relationship between numbers of exposed humans affected and exposure to an environmental pollutant.

Scientific data must be generated so that such curves may be constructed for different sensitive populations at risk. The region of this curve of greatest interest and concern on this curve is that for exposures ranging from O to A since it is this region that defines responses for the most sensitive members of the study population and shows whether there is a threshold of exposure that must be exceeded before any adverse effects are observed. In Figure 1 the hypothetical curve is drawn as though there is no threshold.

Many factors combine to produce a wide difference between the response to an environmental pollutant of the most sensitive member of a human population at risk and that of the most resistant member. Some of these factors are:

- age, sex and race
- state of health
- nutritional status
- occupation
- existence of genetic defects
- extremes of temperature or humidity
- altitude of residence
- lifestyle and psychosocial stresses
- interactions with pharmaceuticals, drugs, smoking and alcohol use
- socioeconomic status

Obviously, all of the factors listed above are not mutually independent. The list merely illustrates that there are many factors to be considered in quantitating the range of responses between the most sensitive and the most resistant individuals in an exposed population. Ranges of response could probably be on the order of factors of 10 or higher for responses to individual environmental or occupational pollutants. The range could become even larger as a result of combination effects of multiple environmental pollutants, and in the general population as opposed to an occupational population because of the greater heterogeneity of the former. Two of many examples of enhanced sensitivity to environmental pollutants are: (1) children up to age 3, who are more sensitive than other age groups to environmental lead exposure; and (2) individuals with cardiopulmonary insufficiency, who are more sensitive to carbon monoxide.

The final point to be made relates to experimental design of health effects research. Although it may not be possible to control for, or measure the effects of, all the factors listed above in any single toxicological or human studies experiment, it should be possible to design research programs that address the effects of the factors one, or a few, at a time for specified pollutants or combinations of pollutants. In toxicological experiments, perhaps the most difficult task is the selection of an animal model or models to mimic human responses. This might best be achieved through the development of long-term research programs employing toxicological and human study techniques conjointly. A fragmented approach using only a single experimental technique to study "targets of opportunity" is unlikely to produce all the required answers for even a single environmental pollutant. Additional basic research to better understand mechanisms of action is also needed to that will most economically and quickly provide the required research data and information.

SIGNIFICANCE OF VARIOUS HEALTH EFFECT ENDPOINTS

Considering that environmental or occupational standards are usually set by nonscientists and must in general be acceptable to at least a majority of the citizenry, scientists must explain the significance of their findings on the effects of environmental or occupational pollutants on human health in terms understandable to an informed layman. If the significance to human health of an observed endpoint is uncertain or a matter of disagreement, that doubt should be clearly addressed when the findings are published. On the other hand, if the findings are commonly accepted by most scientists as being significant to human health, that information should also be published.

In general, responses of experimental animals or humans to environmental pollutants can cover the following spectrum:

- increased body burden
- physiological or biochemical changes of uncertain significance
- physiological or biochemical sentinels of disease
- clinical disease
- death

The most difficult responses to explain with respect to human health are the first two cited above. The fact that such explanations are difficult should not discourage scientists from discussing the various possibilities. Deduction, inference or pure speculation may be helpful. Such attempted explanations of significance may encourage other scientists to conduct additional experiments that may help to clarify the situation.

Additional basic research on mechanisms may ultimately convert an uncertain finding to one clearly recognized as a sentinel of disease. It is also possible that increased body burdens may ultimately lead to clinical disease, perhaps after a long latent period. Thus, even in doubtful cases, some effort should be made in the publication of findings on the effects of environmental or occupational pollutants to explain their possible significance to human health.

EXPOSURE ASSESSMENT

An extremely important factor in the determination of any adverse health effect from exposure to an environmental or occupational pollutant is the accurate determination of the exposure related to the observed effect, whether in a toxicological or a human study. The exposure can usually be assessed and controlled quite accurately for animal toxicological or human clinical studies. It is quite a different matter, however, to assess the exposure to the same degree of confidence for human epidemiological studies in the

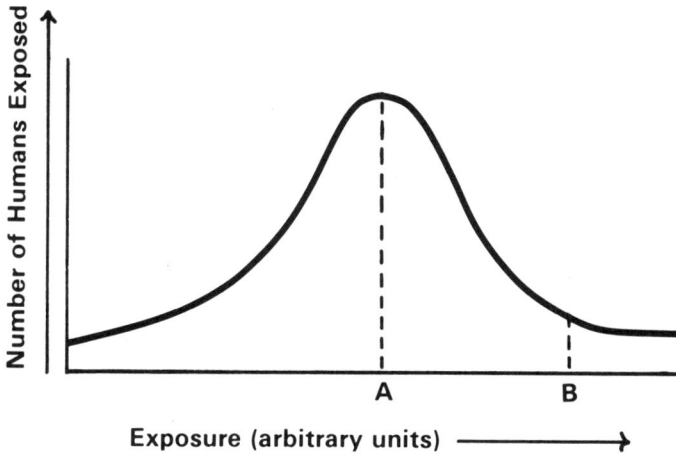

Figure 2. Hypothetical relationship between numbers of humans exposed and exposure to an environmental pollant.

general population. Figure 2 is a hypothetical graph relating exposure to number of subjects exposed for a typical field situation. In the idealized normal distribution graph depicted in Figure 2, the mean exposure would be A, so that about 50% of exposed humans would be exposed to levels higher than, and lower than, A. In particular, a finite number of people would be exposed to levels equal to, or greater than, B. In a field situation, the population exposure distribution function would probably not be normally distributed as in Figure 2. Therefore, it is necessary to determine experimentally the shape of this distribution at each study location.

In most epidemiological studies there are insufficient exposure data to construct an actual graph corresponding to Figure 2. Normally, an average value of exposure is used with a range around that average often given. That same exposure is then assumed to represent all human subjects in a defined region, which is usually relatively large geographically. What is really needed are individual exposure levels for each subject, which can then be related to the observed effects for the same individuals. With human activity patterns varying as they do, the ideal way to measure each subject's exposure is to have a personal exposure meter, which is attached to each subject wherever he goes. With study populations of large size, however, this is usually impractical. In large study populations, personal exposure meters on a representative sample of the population will be extremely helpful in estimating individual exposure.

Another approach to the problem would be to construct isopleths of constant exposure from fixed monitoring stations and then calculate the exposure of individuals from their documented activity patterns. This would require both indoor and outdoor monitoring stations. With this approach, it is extremely desirable to use personal exposure meters on a representative sample to assure quality control over the estimations. The indoor pollution situation is particularly difficult to get at any other way because of the wide differences in air exchange, insulation, indoor sources of pollution, etc.

Another complicating factor is the establishment of an appropriate averaging time for the exposure measurements. Since the correct averaging time for establishing the strongest correlation with observed health effects cannot be known in advance, the safest procedure is to take continuous measurements. From these one can construct at will measurements for various averaging times. Unfortunately, this is not always practical because of cost or lack of sufficient numbers of required monitoring instruments, among others.

The principal point is that if we wish to improve our epidemiological data base to serve as a better foundation for standards, we must devise better methods and procedures for documenting individual exposures to a higher level of confidence than in the past.

Where the pollutant of concern is held at a constant level for the exposure period, many toxicological or clinical studies may not give the same effects as would the same average exposure level delivered with peaks and valleys as they normally occur in field situations. Additional experiments should be conducted to examine this. Additional basic research into mechanisms might be extremely helpful in pinpointing the appropriate averaging time to use for exposure in relation to each observed effect. Different averaging times may have to be used with different effects.

RISK ASSESSMENT

The goal in risk assessment is relatively easy to state but extremely difficult to achieve. For each environmental or occupational pollutant, we would like to know the probability of an individual suffering a defined adverse effect when exposed to a defined pollutant in a field situation. In general, not all the necessary information is available to answer this at a stated level of confidence—not even for a single toxic pollutant. All the variability of individual human response is involved in attempting to answer the question in the previous paragraph. Also, the definition of an adverse effect may not be adequately established for the pollutant of concern. Furthermore, the spectrum of exposures actually existing in the field further complicates the problem. If one determines an "average" risk, it may be completely different from a more realistic "maximum reasonable" risk, where a more sensitive individual is exposed to a higher than average exposure level. Depending on the input assumptions, it is possible to obtain estimated risks that differ by orders of magnitude.

What is the answer then? How does one proceed? First the population to be protected must be specified. Then an acceptable level of adverse effects in a large population of the individuals to be protected must be specified. Once this has been done then we at least know what additional information provide more guidance in the development of critical series of experiments may have to be collected to finish the risk assessment. If it is not already available, the exposure distribution must be determined or estimated for reasonably realistic field conditions. Variations of response in the population to be protected must be determined or estimated. Agreement must be reached on the adverse effect of importance we are trying to prevent. With these results we can then proceed to develop an answer to the question of risk assessment. Without this information, a risk assessment can be accomplished only if assumptions are made to substitute for the missing data.

We have not yet discussed the situation in which no data are available from humans, but we do have data in one or more species of experimental animals, so a risk assessment is required. In this instance an assumption must be made about which animal model(s) should be used as an analogue for human response and about likely human exposure distributions. With these assumptions one can proceed.

IMPORTANCE OF MEASUREMENT AND MONITORING

The importance of measurement and monitoring to all of the above cannot be overemphasized. This also implies quality assurance as unknown confidence levels on measurements will always result without adequate quality assurance. This is true for the measurement both of biological endpoints and exposure levels.

In general, some independent method of cross-checking must be accepted and implemented to ensure the quality of the data. This implies that all collaborating researchers must agree on some "reference" measurement method. It also implies that one laboratory or agency must be accepted as the "referee" or "quality assurance" group to oversee the entire program and prepare reference samples as required for cross-checking analysis. The cross-check samples should be identified and treated by all collaborating laboratories as if they were *bona fide* experimental samples. Such quality control procedures are just as important for the measurement of biological endpoints as for chemical analyses. This is all simply normal operating procedure in good analytical laboratories and research programs. Conclusions from research experiments cannot be better than the data on which they are based. Many careful researchers allocate 10–20% of their total budget to quality assurance, which is money well spent.

CONCLUSIONS

I have attempted to establish a framework for the discussion of a wide variety of toxic environmental and occupational pollutants, which are rarely present in isolation. This means that it is important to study and assess combination effects of pollutants commonly found together. With nitorgen oxides in nature one commonly finds oxidants, CO, particulates, unburned hydrocarbons and sulfur oxides. Thus, adequate knowledge of the adverse effects of NO_x on human health must incorporate an assessment of any possible combination effects of NO_x with any of its usually concomitant pollutants. Such knowledge is particularly important to the interpretation of epidemiological studies, where many pollutants are often present simultaneously.

In addition to the need for additional investigations on combination effects of nitorgen oxides with other pollutants, more research is required to establish the effect on the biological endpoint being studied of varying the shape of the NO_2 exposure delivery curves as a function of time, e.g., experiments in which the average exposure is held constant but the time rate of delivery is altered in different ways.

More basic research is required about mechanisms of action of NO_2 in the development of adverse health effects; this, in turn, will be helpful in designing experiments aimed at formulating a more complete risk assessment of NO_x than is now possible.

Additional investigations are needed to develop improved methods of individual exposure assessment for NO_x along lines discussed previously. These investigations should include adequate consideration of indoor environments.

Continued efforts should be made to integrate toxicological, clinical and epidemiological studies on the health effects of NO_x into a coherent, mutually reinforcing research program.

SECTION I

Measurement and Monitoring of Nitrogen Oxides

CHAPTER 3

CRITIQUE OF MEASUREMENT TECHNIQUES FOR AMBIENT NITROGEN OXIDES

Bernard E. Saltzman

Department of Environmental Health
University of Cincinnati
Cincinnati, Ohio 45267

INTRODUCTION

The National Air Quality Standard for nitrogen dioxide, one of the six established in 1971, is an annual arithmetic mean of 0.05 ppm (100 $\mu g/m^3$). It is based on the effect of increasing acute and chronic respiratory disease. Nitric oxide is only of indirect interest as a major reactant in photochemical smog formation, and no standard has been set for it. Although it can form methemoglobin in the blood and nitrogen dioxide by slow reaction with oxygen, it is not regarded as a health hazard at ambient levels. This chapter focuses on the troubled evolution of regulatory methods for measurement of ambient nitrogen dioxide.

Part of the difficulty has been due to the complex chemistry of the nitrogen oxides: N_2O, NO, NO_2, N_2O_3, N_2O_4, N_2O_5 and HNO_2, and HNO_3 vapor may exist in ambient air. Numerous interactions occur between these forms at varying rates. NO and NO_2 are stable free radicals and typically participate in rapid simultaneous reactions resulting in nonintegral stoichiometries. It also has been extremely difficult to prepare accurately known low concentrations of NO_2 because of its high reactivity with impurities, other pollutants, the walls of the system, etc. This chapter briefly reviews common manual and continuous methods, calibration procedures, collaborative tests and the results of a routine study in which EPA-approved methods were carefully applied not to evaluate methodology, but to determine environmental levels.

METHODS EVALUATION PROCEDURES

The procedures that have been followed in standardizing air pollution methods for ambient nitrogen dioxide have been somewhat different than those that have been followed previously. Usually it has been expected that a proposed analytical method should be published in full detail in a scientific journal with supporting evidence and be confirmed, if possible, by independent investigators. However, these regulatory methods have been described mostly in the *Federal Register* and in U.S. Environmental Protection Agency (EPA) bulletins.

It has been customary for an independent referee to test the method's ruggedness, accuracy and suitability for use in his laboratory. The regulatory methods for ambient NO_2 have generally been developed and tested by EPA in its own or contract laboratories.

Finally, collaborative tests have usually been organized by professional societies. The purpose is not only to evaluate the method, but also the instructions in the procedure and the capability of the average chemist engaged in this work for obtaining acceptable accuracy. These collaborative tests have been conducted, however, under government-supervised and -financed contracts.

The final decision has generally been made by a consensus of interested professionals and users; however, these methods have been selected by the EPA administrator after invitations for comments on proposals published in the *Federal Register*. The results have been used to establish reference and equivalent methods suitable for use in regulatory activities.

The changes in the previous standardization practices resulted from the extreme pressures on EPA from Congress and the public to establish standards and requisite analytical methodologies within short deadlines. Publication in the *Federal Register* was necessary, of course, for legal purposes, and apparently time was lacking for presentation or publication in scientific journals. The EPA administrator preferred to make his own decisions regarding the analytical methods to be used in carrying out his legal responsibilities. However, these procedures have caused problems and dissent from some users of the methods.

MANUAL ANALYTICAL METHODS FOR NITROGEN DIOXIDE

Table I [1-13] lists manual colorimetric methods for determining ambient nitrogen dioxide. Among the factors that must be considered with such methods are the absorption efficiency of the gaseous compound; the method of standardization (for instance, generally with nitrite solution); the stoichiometries of the color production with nitrite ion and with gaseous NO_2 after its absorption; and the overall conversion efficiency measured as moles of nitrite ion producing the same color as one mole of gaseous nitrogen dioxide in the air samples. Also important are the stabilities of the reagents and collected samples on aeration during sampling and on storage before analysis. Frequent sources of error with manual methods are improper flow calibration, air leakage in the sampling system, malfunction of the flow control, leakage of the absorbing solutions or of the collected samples during transport, impure reagents or water, and improper calibration and analysis.

The first method listed in Table I was developed by Saltzman [1] in 1954 and has been widely used as a reference method (as well as in modified forms as a reagent for continuous analyzers). Although very reproducible, there has long been controversy about the stoichiometric factor for this method. It should be noted that this procedure has been used with sampling periods from 10 to 30 min, but not for 24 hr. Because longer storage intervals were desired for collected samples, the second procedure was developed by Jacobs and Hochheiser in 1958 and used with a sampling time of only 40 min. It was never demonstrated to be valid with known air samples. The ambient air analysis data presented originally [2] showed an inverse relationship between SO_2 and NO_2, suggesting interference. Table I lists numerous substantial modifications made in the absorbing reagents, the absorbing apparatus and the sampling time. These were used extensively without first demonstrating the validity of the results. This culminated in the Federal Reference Method (FRM), which was published in the *Federal Register* in 1971 and caused the debacle in 1973 when EPA withdrew it [12] because of variable stoichiometry and interference of nitric oxide.

It is instructive to examine some of the defects of the FRM method as shown in Figure 1, which has been recalculated from data presented by Hauser and Shy [14]. This shows colors obtained from known gaseous concentrations sampled for 24 hr by this and the Chattanooga Method, exhibiting an excessive scatter demonstrating the poor accuracy. The latter method was the one used for an extensive epidemiological study on which the standard was based and was similar to that listed in Table I for Meadows and Stalker. The yield of these methods can be calculated as the amount of color (the ordinate) divided by the amount of NO_2 gas absorbed (the abscissa); this is the slope of a line connecting any point on the gaseous calibration curve to the origin. The yield is variable, since one line is curved and the other does not pass through the origin. When such a method is used to sample fluctuating concentrations, the yield becomes indeterminate. The variability is due to irreversible losses and cannot be corrected by calculations. Thus, the amount of color does not give an accurate time-weighted average of the air concentrations for the sampling period. Subsequently, air concentrations for this epidemiological study, which formed the basic justification for the air quality standard, were recalculated from data of continuous analyzers using modified Saltzman reagent.

Later methods employed various additives to stabilize and increase the yield of nitrite. The effects were hypothesized [9] as follows:

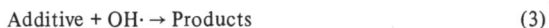

$$NO_2 + OH^- \rightarrow NO_2^- + OH\cdot \qquad (1)$$

$$NO_2 + OH\cdot \rightarrow HNO_3 \qquad (2)$$

$$\text{Additive} + OH\cdot \rightarrow \text{Products} \qquad (3)$$

Table I. Manual Methods

Name	Reference	Volume and Composition of Absorbing Reagent	Absorbing Apparatus
Griess–Saltzman	1	10 ml: 0.5% sulfanilic acid, 0.002% NEDA,[a] 14% AcOH (+1% acetone optional to reduce effect of SO_2).	Fritted bubbler (60 μm max. pore diameter).
Jacobs and Hochheiser (J–H)	2	30–35 ml: 0.1 N NaOH, 0.2% v/v butyl alcohol.	Course-fritted glass bubbler in sequential sampler.
Perry and Tabor (NASN)	3,4	20 ml: same as J–H.	Polypropylene test tube containing glass inlet tube with orifice 0.36 ± 0.05 mm. Thermostated at 35°C.
Meadows and Stalker (Alabama Study)	5	35 ml: 0.1 N NaOH (in each of 2 bubblers).	Two NASN bubblers in series.
1971 Federal Reference Method (FRM)	6	50 ml: 0.1 N NaOH.	Membrane filter leading to polypropylene test tube with fritted glass inlet of porosity B (70–100 μm max. pore diameter).
Christie et al.	7	5 ml: 0.025 N NaOH, 0.1% $NaAsO_2$, 0.75% sulfanilic acid.	Glass bubbler with open tube inlet.
Nash	8	4 ml: 0.1 N NaOH, 0.05% guaiacol (2-methoxy phenol).	Glass Arnold Tube (with open tube inlet).
Huygen and Steerman	9	15 ml: 0.1 N NaOH, 0.02% R-salt (2-naphthol-3,6-disulfonic acid disodium salt), 0.1% triethanolamine (in each of two bubblers).	Two bubblers in series, each with coarse glass frit (90–150 μm) of 6.5 cm² area.

(Continued)

for Ambient NO_2

Sampling Rate and Time	Procedure for Analysis	Remarks
0.4 liter/min, 10–30 min; not used for 24-hr sampling.	Read color.	Widely used as reference method. Absorption efficiency >95%, overall conversion 72–75% (with some dissent).
1.3 liter/min, 40 min.	Add 1 drop 1% H_2O_2, 10 ml of 2% sulfanilamide in 5% v/v H_3PO_4; 1 ml of 0.1% NEDA.[a]	No validation or stoichiometry studies; 90% collected in first of two bubblers.
0.15–0.20 liter/min, 24 hr.	Autoanalyzer, using J–H reagents.	Absorption in first of 4 bubblers varied from 30–69%. Collected samples stable for 3 weeks. Assumed 50% overall conversion. Large mass of data available using this method.
0.35–0.50 liter/min, 24 hr.	Same as NASN.	Absorption in first of 6 bubblers was 19%, in first 2 was 36%. Overall conversion of 36% used, with assumption of 100% stoichiometry.
0.20 liter/min, 24 hr.	Replace water evaporated in bubbler with distilled water up to original mark. To 10 ml add 1 ml 0.024% H_2O_2, 10 ml of 2% sulfanilamide in 5% v/v H_3PO_4, 1.4 ml 0.1% NEDA.[a]	Overall conversion of 35% is used. Retracted in 1973 because of variable conversion and interference of NO.
0.12 liter/min, 1 min.	Add 3 ml of 0.02% NEDA,[a] 6% oxalic acid.	Overall conversion 94% at 6 ppm NO_2. For industrial hygiene use.
0.6 liter/min, 2 hr.	Add 4 ml of 0.2 N HCl, 0.5% sulfanilic acid, 1.5% glycine, 0.002% NEDA.[a]	Limited amount of H_2O_2 may be added to reduce SO_2 interference. Stoichiometric factor 0.73, absorption efficiency 99%.
1 liter/min, 24 hr.	Add to each absorber 15 ml of 0.4% sulfanilamide, 0.03% Cleves acid (8-aminonaphthyl-2-sulfonic acid), 0.12 N HCl. A drop of H_2O_2 may be added first to eliminate SO_2 interference.	Stoichiometric factor varied from 96% to 80% in range 48–2200 $\mu g/M^3$ NO_2. Absorption efficiency averaged 96%.

(Continued)

Table I.

Name	Reference	Volume and Composition of Absorbing Reagent	Absorbing Apparatus
Levaggi, Siu and Feldstein (triethanolamine)	10	50 ml: 1.5% triethanolamine, 0.3% v/v n-butanol.	Polypropylene tube with fritted glass inlet of 70–100 μm max. pore diameter.
EPA Arsenite Method	11,12	50 ml: 0.1 N NaOH.	Polypropylene tube containing glass inlet tube with orifice 0.3–0.8 mm i.d., 6 mm from bottom of tube.
TGS-ANSA	12,13	50 ml: 2% triethanolamine (=T), 0.05% o-methoxyphenol (=G, guaiacol), 0.025% $Na_2S_2O_5$ (=S, sulfite).	Same as above, except for 0.3–0.6 mm i.d. orifice.

[a]NEDA: N-(1-Naphthyl)-ethylene diamine dihydrochloride.
[b]ANSA: 8-anilino-1-naphthalene-sulfonic acid ammonium salt.

Reaction 1 is the desired reaction, producing nitrite with 100% stoichiometry; however, its by-product OH· (free radical) could consume 50% of the NO_2 by Reaction 2, which yields no color, and the overall yield of the two reactions is 50%. If the additive competes efficiently with Reaction 2 to consume the OH·, as shown in Reaction 3, higher yields may be obtained. If Reaction 1 is not possible because of the high free energy of OH free radical, the competing reactions may be the trimolecular reactions:

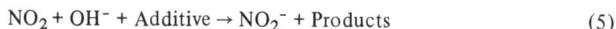

$$2NO_2 + OH^- \rightarrow HNO_3 + NO_2^- \tag{4}$$

$$NO_2 + OH^- + Additive \rightarrow NO_2^- + Products \tag{5}$$

The TGS-ANSA manual method and the arsenite method in manual and automated (Technician II) versions, were approved by EPA as "equivalent methods." [15] The TGS-ANSA method gives the highest yield (93%). This procedure for analysis utilizes ANSA because the sulfite in the absorbing reagent affects NEDA; however, the ANSA solution must be prepared fresh daily and be added within six minutes of the preceding step. Collected samples were said to be stable for three weeks at room temperature, but

Continued

Sampling Rate and Time	Procedure for Analysis	Remarks
0.15–0.20 liter/min, 24 hr.	Similar to J–H.	Overall conversion efficiency 76–92% over range 56–750 $\mu g/M^3NO_2$, collection efficiency 95–99%. Constant 0.85 factor recommended.
0.2 liter/min, 24 hr.	Same as FRM.	Overall conversion is 82% over range 9–750 $\mu g/M^3$ NO_2 concentration. Approved by EPA 12/77 as an equivalent method.
Same as above.	Replace water evaporated in bubbler with distilled water up to original mark. To 5 ml add 0.5 ml 0.024% H_2O_2, 2.7 ml sulfanilamide (2% in 4 N HCl), then 3 ml 0.1% ANSA[b] in CH_3OH (within 6 min. of preceding step).	Overall conversion of 93% is used. Approved by EPA 12/77 as an equivalent method.

supporting data have not yet been published. The arsenite method is more convenient, and collected samples were said to be stable for six weeks; however, the yield is somewhat lower (82%). Nitric oxide produces a 3–15% positive interference. The alkaline collection media suppress side reactions, which convert part of the NO_2 to NO, thus causing losses in acid reagents.

CONTINUOUS MONITORING METHODS FOR NITROGEN DIOXIDE

Continuous monitoring methods of two types—colorimetric and chemiluminescent—also have been tested by EPA. Current EPA performance specifications [16] require less than ±0.02 ppm drift in 24 hours for both zero and span readings, and a lower detection limit of 0.01 ppm. These are large fractions of the 0.05 ppm standard. There are no linearity specifications; however, readings must agree with the reference method within ±0.02 ppm in the range 0.02–0.2 ppm, and ±0.03 ppm in the range 0.25–0.35 ppm in a

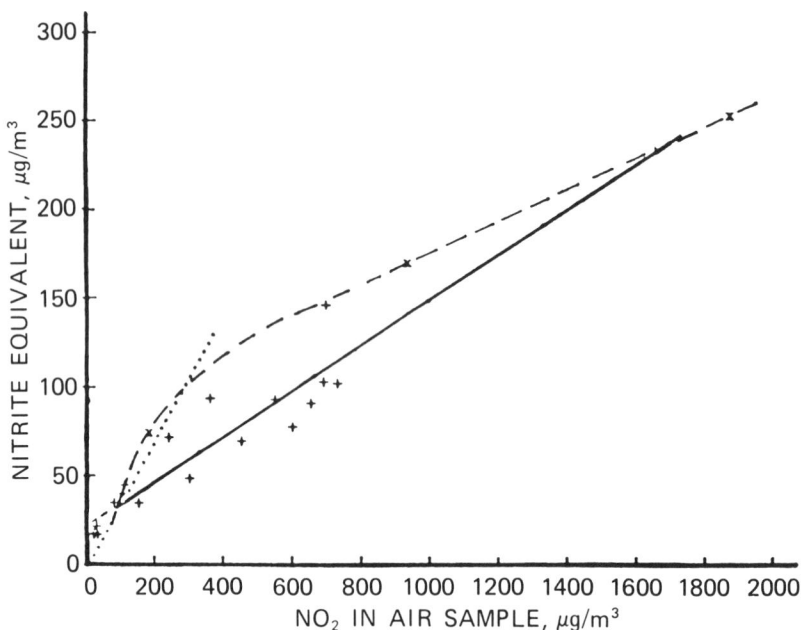

Figure 1. Overall conversion to nitrite in analysis of samples of NO_2 from permeation tubes, diluted with purified air. Legend: + = Federal Reference Method; x = Chattanooga Method; . . . . = theoretical relationship for 35% conversion.

7-day test. There are varying instrument designs and reagents, and the standard method requires the user to follow the manufacturer's directions. Interferent and test concentrations (in ppm) are: NH_3, 0.1; SO_2, 0.5; NO_2, 0.1; NO, 0.5; and water vapor, 20,000. Interference equivalent of each interferent is limited to ±0.02 ppm, and of their combined effect 0.04 ppm.

In the colorimetric type of monitor, metered air and liquid reagent flows are contacted in an absorbing column and the liquid then passes through a recording colorimeter. Much of the data has been collected using modified Griess-Saltzman reagent [17,18]: 0.5% sulfanilic acid, 5% acetic acid and 0.005% NEDA. Also used is Lyshkow [19] reagent: 0.15% sulfanilamide, 1.5% tartaric acid, 0.005% NEDA, 0.005% 2-naphthol 3,6 disulfonic acid, and 0.025% v/v Kodak photoflow. The manufacturer's directions vary, and differing instrument designs and reagents result in differing contact and response times, which affect overall conversion efficiencies and interferences. Static standards, produced by addition of standard nitrite solution to the reagent, may be used as reference points only. Accurate calibration requires dynamic standardization with known gaseous mixtures.

In the chemiluminescent monitor, adopted by EPA as the Federal Reference Principle (FRP) [16], a dual or cyclic design provides analyses of both

NO and NO_x, and NO_2 is determined as the difference. Nitric oxide is determined by mixing a metered sample air flow with a metered ozone flow (purified air irradiated by a controlled UV lamp). The light produced by their reaction is measured by a photomultiplier tube. The NO_x measurement is made similarly, after first passing the sample air stream through a heated converter, which reduces NO_2 to NO. Thermal converters (e.g., stainless steel at 600–800°C) decompose NO_2 into O_2 and NO, but may also oxidize ammonia to NO. Chemical converters (e.g., C, Mo, W, Pt) operate at lower temperatures to reduce NO_2 to NO, and may produce interferences from PAN and organic nitrogen compounds. They must be replaced periodically or reactivated with hydrogen. Converter efficiencies must be tested and maintained above 96%. In the cyclic design monitor, the sample air alternately passes through and bypasses the converter in a timed sequence. In the dual design there are separate sample airflow streams to two reaction chambers, which use either one time-shared photomultiplier tube or a matched pair. Several commercial monitors of the chemiluminescent type have been accepted [20] as "reference methods."

A complex gas-phase titration procedure was adopted by EPA [16] for calibration of monitoring instruments. Its publication required 31 pages of description and 40 pages of supporting explanations in a Technical Assistance Document [21]. In Alternative A, the primary standard is 50–100 ppm NO in nitrogen, with less than 1 ppm NO_2, in a gas cylinder. This must be traceable to a National Bureau of Standards Standard Reference Material, either a similar mixture or an NO_2 permeation tube. Whether this traceability is determined by the user or vendor and the precise procedures are not defined. A stable ozone generator, an ozone analyzer and an $NO/NO_2/NO_x$ analyzer are required for the titration procedure, and numerous calculations must be made to ensure correct operation of the flow system. In Alternative B, in addition to the above items, an NO_2 permeation device is required. This is used for the NO_2 span adjustment, and the NO cylinder is used for the NO and NO_x span adjustments. Either of the two working standards may be traceable to the NBS; the other must be intercompared and adjusted for consistency. This alternative makes determination of converter efficiency somewhat simpler. In the gas-phase titration procedure, increasing amounts of ozone are added to a constant concentration of nitric oxide in a flow system. The NO readings should decrease linearly, and the NO_2 readings should increase linearly. The NO_x readings should remain constant. If they drop then the converter efficiency is low.

COLLABORATIVE TESTS AND EVALUATIONS
OF NO_2 METHODS

A pioneer collaborative test of the Griess-Saltzman method (ASTM Designation D 1607) was conducted from 1971–1973 by the American Society for Testing and Materials at Los Angeles, Bloomington, Indiana and Manhattan, New York, with 8, 7 and 7 participating laboratories, respectively [22].

Ambient air was drawn by a fan at 150 scfm into a 3-in. aluminum pipe, past an aluminum flange fitting with 16 radial uniformly spaced sampling ports from which it was sampled. A 2-in. aluminum branch line carrying 3.5 scfm (100 liter/min) was spiked with NO_2 obtained by metering cylinder air over a calibrated permeation tube. This line carried a sampling fitting with eight ports. Each laboratory sampled unspiked and spiked air, generally for 1 hr, through an 8-mm o.d. Teflon®* tube using the standard bubbler containing the reagent with the optional acetone added to retard fading caused by sulfur dioxide. Most used both a glass rotameter and a calibrated dry test meter to measure sample volume, and placed the rotameter upstream from the bubbler. All the operators followed the specific instructions in the Test Method. Data from one laboratory at the Manhattan site were rejected as outlying. Results are given in Table II. A total of 704 measurements were made over a range of 10–400 $\mu g/m^3$ in ambient and spiked air. Nitrite solutions were used as the standards and an overall conversion efficiency of 72% was used in the calculations. The average bias reported for all three studies was +18%; thus, if 85% had been used for overall efficiency, the average bias would have been eliminated. The biases did not appear to depend on concentration, but differed in the three cities. It was suggested that this was possibly because of an interference in the ambient air. Since the coefficients of variation depended on the concentration, the table lists values at a "relevant concentration" selected to be typical and close to comparative values obtained in later studies of the other listed methods. This method was accepted as sensitive, accurate and precise, with minor revisions.

Four collaborative studies were conducted for EPA at Midwest Research Institute (MRI) in Kansas City during 1974–1975 under a contract [23-27]. Ambient air was drawn by an outside blower through 2-in. aluminum tubing into a baffled stainless steel chamber, then through two 1-in. aluminum lines through stainless steel venturi meters. Below these, only Teflon contacted the air. In each line the air passed through an NO_2 bleed-in unit, a diffuser, an equilibration section and a 45-port sampling manifold radially symmetrically arranged. A spike of NO_2 was added to one line by passing cylinder nitrogen gas over a set of permeation tubes in a thermostatted water bath. The permeation tubes were calibrated by the National Bureau of Standards and checked by EPA in North Carolina. The air streams also were monitored with a Bendix NO_x chemiluminescent analyzer. Effects of specific interfering gases were not examined.

The first study was of the manual arsenite method [23,24]. Ten collaborating laboratories prepared reagents and equipment at home, collected samples at the Kansas City site of ambient air unspiked and spiked, and then returned home to complete the analyses. Nitrite solutions were used as the standards, with an 82% overall conversion factor. Results of two laboratories were rejected, and the remainder are given in Table II. The −3% average

---

*Registered trademark of the E. I. du Pont de Nemours and Co., Inc., Wilmington, Delaware.

Table II. Collaborative Tests of Methods for Ambient NO_2

	Griess–Saltzman	Manual Arsenite	Manual TGS-ANSA	Continuous Colorimetric	Continuous Chemiluminescent
Reference	22	23,24	25	26	27
Detection Limit ($\mu g/m^3$)	3	8	15	19	22
Average Bias (%)	+18[a]	–3	–5	+6	–5
Coefficient of Variation (%)					
One lab	4	4	4	6	6
Between labs	10	4	5	12	13
Combined	11	6	6	13	14
Relevant concentration ($\mu g/m^3$)	180	184	180	168	163
Outlying Labs	1/22	2/10	1/10	1/10	2/10

[a]Average for three locations. Tests at Los Angeles, Bloomington, Indiana and Manhattan, New York showed +11, –11 and +35%, respectively.

bias that was found could have been eliminated if 80% were used in the calculation instead of 82%. However, the average bias for each of the eight nonoutlying laboratories ranged from −13.6 to +3.5%. In an earlier laboratory study [11], an absorption efficiency of 91.1% was reported and a stoichiometric factor of 0.72 was used, giving an overall conversion efficiency of 66%. A later laboratory study [28] determined that the overall conversion efficiency was 82%. Thus, there may be a variation in this value under different conditions. In the other laboratory studies [11,28] a small positive interference from nitric oxide and a negative interference from carbon dioxide were found. It is strange that an adequate test for the specific interference of sulfur dioxide, a very common air pollutant, has never been reported for this method.

The second EPA collaborative study [25] was of the manual TGS-ANSA method, conducted in the same way as the first. The ten collaborating laboratories prepared reagents at home, collected samples at the test site and shipped them home for analysis. Nitrite solutions were used for the calibration, and 93% overall conversion efficiency was assumed for the calculations. Data from one outlying laboratory were rejected. The cause of erroneous results was found to be contaminated methanol. Results (Table II) showed an average bias of −5%, which could have been eliminated if an 88% conversion factor had been used. The average bias of the nine nonoutlying laboratories ranged from +4 to 13%. Only a single paper has been published [13] describing this method (together with four other methods). The claim of no interferences was supported by very few data, some of which showed effects of −3% to +5% on recovery of 100 μg NO_2/m^3. No data were given on losses after storage of collected samples.

The third EPA collaborative study [26] was of the "continuous colorimetric" method, conducted with the same sample generation system. The procedure specified *either* the modified Saltzman [17] or the Lyshkow [19] reagent. No particular instrument was specified. Any could be used, provided they met the stated specifications: noise 0.005 ppm, lower detectable limit 0.01 ppm, 24-hr zero and span drifts ±0.02 ppm, lag time 20 min, 95% rise and fall times 15 min. The manufacturer's operating instructions were to be observed. The ten collaborating laboratories performed dynamic calibrations at home. A permeation tube method was described in the procedure and others were mentioned.

During the tests at MRI, static calibration checks were performed twice daily by adding nitrite solution to the absorbing reagent and adjusting the reading to the reference value established during the dynamic calibration. On each of four days a different spike level was sampled by all ten laboratories for 14 hr. Then half sampled the unspiked air for 3.5 hr, while the other half continued sampling spiked air. Finally, the first group sampled spiked air for 3.5 hr, while the second sampled unspiked air. Only half the laboratories achieved stable results, which are summarized in Table II. Because some collaborators showed biases as great as +80% at some levels, the comment was made that the method was difficult to use, although reliable in some hands.

The report did not indicate what reagents or instruments were used, how calibrations were carried out, or what the instrument specifications actually were and how they were checked. It should be kept in mind that a collaborative test evaluates the combination of three major factors: the merits of the method, the clarity of the instructions, and the abilities of the participants to execute them. Since the procedure did not specify a specific method, the proper interpretation of the poor results must be not to blame the "continuous colorimetric method," but rather the instructions and the execution. Obviously, in the cases of the five unsuccessful laboratories, the instruments either could not have met the required specifications for drift or the specifications were inadequate. The range of average bias for individual collaborators was −8 to +31%.

This suggests that an accuracy requirement should be added to the instrument specifications. No tests were made of interferences of any specific pollutants. One would expect substantial differences in interferences in different instruments, depending on the absorbing column design, the absorbing reagent used, and the liquid flow time between entering the absorber and leaving the spectrophotometer cell.

The fourth EPA collaborative study [27] was of the continuous chemiluminescent method. The procedure did not specify any particular instrument, but required the same specifications described above for the continuous colorimetric instruments, except that the specified lag time was 0.5 min and 95% rise and fall times 1.0 min. Calibration was required monthly and was conducted by the collaborators at their home laboratories. The standard for this complex procedure was an ultraviolet ozone generator calibrated by the neutral buffered potassium iodide method [6]. This was used to assay a secondary source, a gas cylinder containing 50−100 ppm NO in nitrogen by the gas-phase titration procedure. The nitrogen dioxide content of the cylinder was determined by the TGS-ANSA method. The converter efficiency was then determined and required to be at least 90%. (The current reference procedure requires 96% efficiency and uses the NO cylinder as a primary standard.) The sampling system and schedule were the same as those used for the continuous colorimetric tests. The instruments were zeroed and spanned twice daily. On each of the four test days a different spike level was used. Results are given in Table II. The overall average bias was −5%, and the range of average bias for individual collaboratores was −18 to +8%. One collaborator had biases that were unstable as a function of NO_2 level. Unfortunately, the report presented no information on what instruments were used, their actual specifications, how they were determined or the exact calibration procedures used. No comments of collaborators were given.

A number of evaluations of these methods were made in individual laboratories. All four methods were compared [29] with strict qualityy control procedures and were found to be in close agreement. Standards were based on permeation tubes, and for the chemiluminescent procedure an ultraviolet ozone generator analyzed by the neutral buffered potassium iodide method. The nitrite solution static colorimetric calibration recommended by Technicon showed only 94% of the value obtained by dynamic calibration. Manual bubbler orifice tolerances were made closer (0.35 ± 0.05 mm) than in the official procedures listed in Table I.

In a study of the continuous colorimetric method [30], two different reagents were used in three Technicon IV instruments calibrated dynamically with permeation tubes. Their overall conversion efficiencies were 0.87 for the Saltzman reagent, and 0.63 and 0.55 for the Lyshkow reagent, whereas the static calibration procedures recommended by Technicon used 0.72 as the factor. In the latter test, 32% of the NO_2 was not absorbed in the solution-air contact coils. Although both reagents produced 97% of the ultimate color in 3 minutes by manual procedures, 15 minutes was required in the Technicon instrument, indicating a large dead volume in the liquid plumbing. It was concluded by the investigators that dynamic calibration was essential. One could also conclude that different reagents and instruments must be considered individually, rather than interchangeably as "the continuous colorimetric method." An earlier study [31] in Los Angeles on modified Saltzman reagent showed that dynamic calibration of each instrument was essential to get its overall conversion factor, which remained constant over an 8-month period. The factors for various instruments ranged from 0.77 to 0.92.

Two dynamic calibration procedures were compared [32] for calibration of continuous chemiluminescent analyzers. The gas-phase titration technique (as described previously for the collaborative test) was found to agree within 1% with calibration by permeation tubes, and to be rugged. However, it was pointed out that the converter efficiency must be verified and the NO cylinder accurately assayed for NO_2 as well as NO.

The chemiluminescent NO_x method was shown [33] to give 92–103% response to ethyl nitrite, ethyl nitrate, n-propyl nitrate and peroxyacetyl nitrate. Of greater concern was the substantial, but nonlinear, response to nitric acid vapor, since this compound, ammonium nitrate, and other derivatives could be present in polluted atmospheres. These studies were made with both molybdenum and carbon converters. Another study [34] showed substantial interferences from hydrochloric acid vapor, chlorine, phosgene and trichloroacetyl chloride, which increased with the temperature of the carbon converter. It was hypothesized that the halogen products can produce an excited oxygen molecule that chemiluminesces. A ferrous sulfate converter eliminated this interference, but may not have completely converted the NO_2 to NO. These studies suggest the necessity of a more careful, complete study of interferences for specific converters and specific instruments.

COMPARATIVE RESULTS IN A MONITORING STUDY

It is instructive to compare the results between the official EPA reference method and the official arsenite equivalent method, as obtained simultaneously at the Red Cross Building in Boston in a careful environmental study [35] that focused not on analytical methods evaluation, but on making the measurements over a 2.5-month period in 1976. The chemiluminescent analyzer, TECO 14D NO/NO$_x$ analyzer, had a dual chamber and single detector. Automatic daily baseline and span checks were performed using an ML 8500 NO$_2$ calibrator (which contained an NO$_2$ permeation cylinder in a precision temperature bath). Manual checks of the zero point were made daily, and field calibrations and electronic checks made weekly. Monthly multipoint NO$_2$ calibrations were performed to ensure highest accuracy. The arsenite procedure also was in strict conformance with EPA requirements and quality control procedures. Five EPA blind control samples were analyzed in duplicate, with an average deviation of only −1%. Five calibration curves were prepared at intervals, all well within EPA recommended tolerances. A total of 43 standard and control samples were run on 17 days. Duplicate bubbler samples were found to be in close agreement.

In spite of all these precautions, the arsenite values were consistently higher than the chemiluminescent values, as shown in Figure 2. The data for August and September appeared to fall into one group, the equation of which was determined by the method of least squares as:

$$Y = 1.066\ X + 15.7 \qquad \text{Corr. Coeff.} = 0.851$$

For October, the relationship was:

$$Y = 1.087\ X + 2.1 \qquad \text{Corr. Coeff.} = 0.889$$

Figure 3 shows the relationships of the differences with chemiluminescent NO measurements made by the same instrument. For August and September combined:

$$Y = 0.21\ X + 11.7 \qquad \text{Corr. Coeff.} = 0.686$$

For October:

$$Y = -0.01\ X + 4.5 \qquad \text{Corr. Coeff.} = 0.055$$

If a positive interference of nitric oxide was the major cause of high arsenite results, one would expect a line through the origin with a positive slope. However, the October line has a negative slope, and both have substantial intercepts and scatter compared to the 50-ppb NAQS. The data suggest a change in operating procedures in October. Apparently, substantial differences may occur between these methods in practice.

DISCUSSION

From the standpoint of determination of compliance with the NAQS standard, 0.05-ppm annual arithmetic mean, the arsenite method appears to be a good choice because of its low cost. Table II shows it to be the most accurate in the collaborative tests. EPA preferred the chemiluminescent reference principle because it provided additional information on

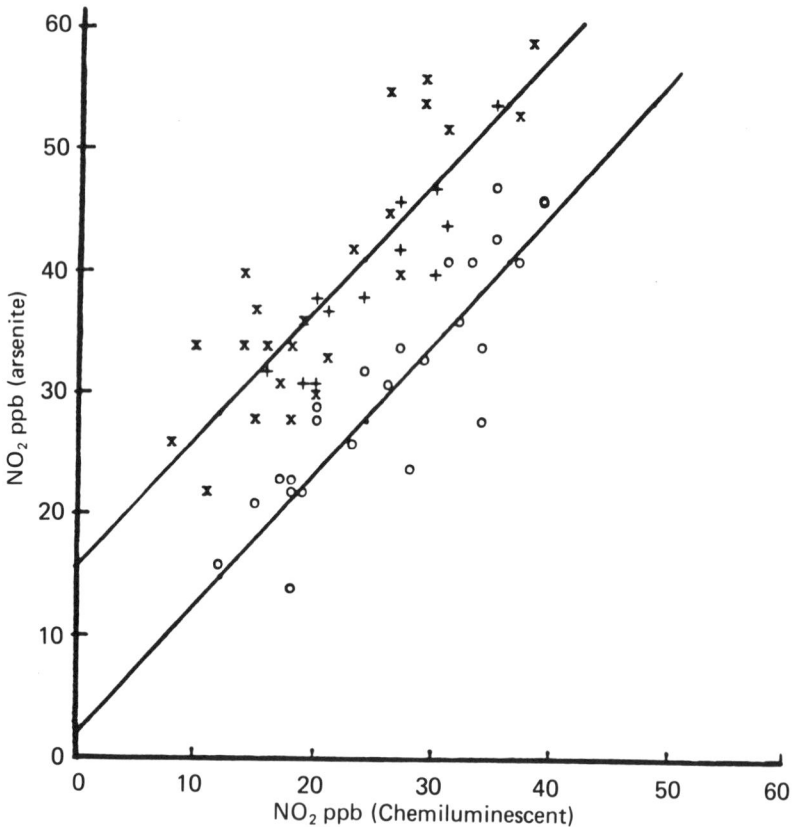

Figure 2. Discrepancies between arsenite and chemiluminescent measurements of NO_2 in Boston, 1976. Legend: + = August; x = September; o = October.

short-term levels, which were useful for studying emission and dispersion patterns, pollution episodes and control strategies. This is regarded as a principle, rather than a method to allow acceptance of new instrumental designs when they are determined by EPA to meet the specifications. The specifications have evolved since the collaborative tests. The primary standard is no longer a gravimetrically calibrated permeation tube or an ultraviolet ozone generator calibrated by the neutral buffered potassium iodide method, both within the capabilities of a user. The stoichiometry of the NBKI method became controversial, and it was dropped. Recently, EPA proposed [36] a gas-phase UV absorption procedure as the reference method for ozone, and a boric acid potassium iodide procedure as a transfer standard method. If adopted, these could be applied to the NO_x calibration. However, the primary standard is now either an $NO-N_2$ cylinder mixture or an NO_2 permeation tube prepared by the National Bureau of Standards. No specification

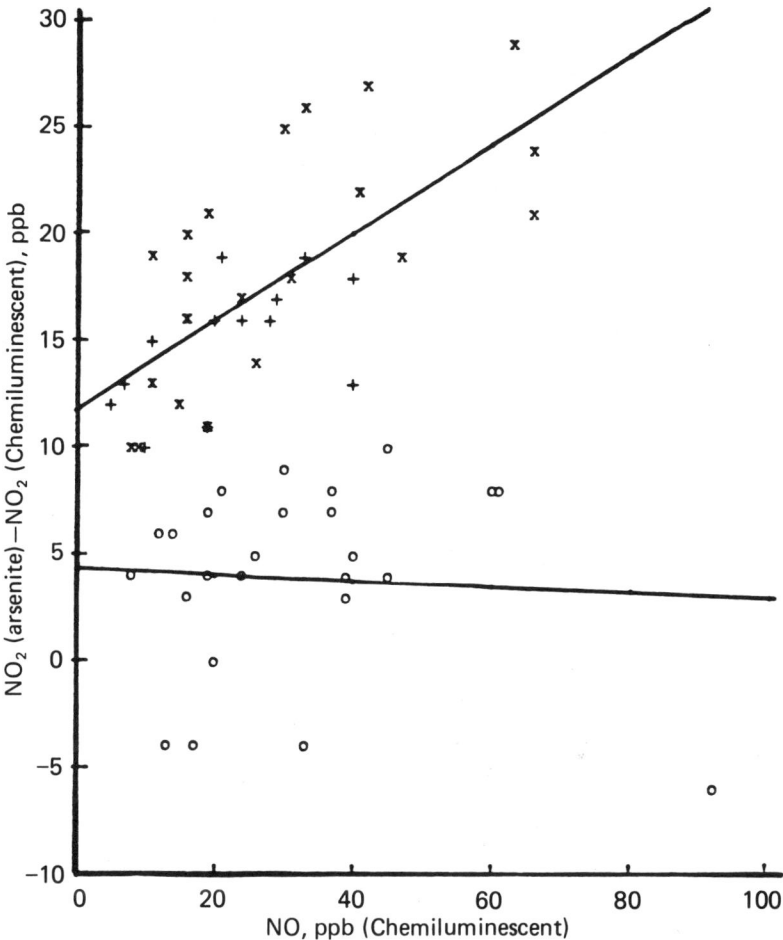

Figure 3. Relationship of nitric oxide concentration to differences between the NO_2 measurements in Boston, 1976. Legend: + = August; x = September; o = October.

is given as to how NBS is to calibrate them. Since both can deteriorate, there may be some problems for remote users. A reference method need neither be a working method nor a continuous 24-hour method, but rather a simple precise procedure suitable for checking other methods. Since a dynamic calibration system for NO_2 does not include interfering substances, any precise manual method could be used as a primary standard, after agreement on its overall conversion efficiency.

This chapter has commented on some of the inadequacies of the collaborative testing which were done under strong administrative pressures. Testing of monitoring instruments operated according to their manufacturers' directions clearly could give bad results in some cases, but these should not have been interpreted as discrediting the long used "colorimetric method." Effects of interferences were not considered in adequate detail. For the manual methods, sulfur dioxide interferences should be considered again, and also their abilities to obtain accurate time-weighted averages of fluctuating concentrations. New information indicates that interferences of nitric acid vapor, nitrates and halogenated compounds require more study for the chemiluminescent method. The ability of the average operator to accurately carry out the complex directions for the calibration procedure is also a question, and simpler methods could possibly yield more accurate results in practice. Further comparative field studies would clarify these problems. Currently allowable errors are a large fraction of the NAQS.

For an epidemiological study, the high cost of monitoring instruments allows measurements at only a few sites. This may result in large statistical sampling errors. Nitric oxide concentrations are not needed. If a short-term standard for NO_2 is to be established, it should be within the capabilities of current analytical methodology Thus, a 1-hr average standard (e.g., that for oxidant) not to be exceeded more than once a year requires that the monitoring instrument be operating 8759 of the 8760 hours in a year to demonstrate compliance if low values are obtained (although a few high values can show noncompliance). In practice, operating time may be 75-95%. If a short-term standard is not to be established, simpler analytical methods may be adequate to determine the annual average. Future studies should consider that people spend 80% of their time indoors, and young children and the chronically ill are indoors even more. In any case, simple inexpensive dosimetric devices, even if less accurate, would provide a valuable overall picture at many more indoor and outdoor sites.

REFERENCES

1. Saltzman, B. E. "Colorimetric Microdetermination of Nitrogen Dioxide in the Atmosphere," *Anal. Chem.* 26:1949-1955 (1954).
2. Jacobs, M. B., and S. Hochheiser. "Continuous Sampling and Ultramicro-determination of Nitrogen Dioxide in Air," *Anal. Chem.* 30:426-428 (1958).
3. Perry, W. H., and E. C. Tabor. "National Air Sampling Network Measurement of SO_2 and NO_2," *Arch. Environ. Health* 4:254-264 (1962).
4. Tabor, E. C., and C. C. Golden. "Results of Five Years' Operation of the National Gas Sampling Network," *J. Air Poll. Control Assoc.* 15:7-11 (1965).
5. Meadows, F. L., and W. W. Stalker. "The Evaluation of Efficiency and Variability of Sampling for Atmospheric Nitrogen Dioxide," *J. Am. Ind. Hyg. Assoc.* 27:559-566 (1966).
6. Environmental Protection Agency. "National Primary and Secondary Ambient Air Quality Standards," *Federal Register* 36:8186-8201 (April 30, 1971).

7. Christie, A. A., R. G. Lidzey and D. W. Radford. "Field Methods for Determination of Nitrogen Dioxide in Air," *Analyst* 95:519–524 (1970).

8. Nash, T. "An Efficient Absorbing Reagent for Nitrogen Dioxide," *Atmos. Environ.* 4:661–665 (1970).

9. Huygen, C., and P. H. Steerman. "The Determination of Nitrogen Dioxide in Air after Absorption in a Modified Alkaline Solution," *Atmos. Environ.* 5:887–889 (1971).

10. Levaggi, D. A., W. Siu and M. Feldstein. "A New Method for Measuring Average 24-Hour Nitrogen Dioxide Concentrations in the Atmosphere," *J. Air Poll. Control Assoc.* 23:30–33 (1973).

11. Merryman, E. L., C. W. Spicer and A. Levy. "Evaluation of Arsenite–Modified Jacobs–Hochheiser Procedure,"*Environ. Sci. Technol.* 7:1056–1059 (1973).

12. Environmental Protection Agency, "Reference Method for Determination of Nitrogen Dioxide," *Federal Register* 38:15174–15185 (June 8, 1973).

13. Mulik, J., R. Fuerst, M. Guyer, J. Meeker and E. Sawicki. "Development and Optimization of Twenty-Four Hour Manual Methods for the Collection and Colorimetric Analyses of Atmospheric NO_2," *Int. J. Environ. Anal. Chem.* 3:333–348 (1974).

14. Hauser, T. R., and C. M. Shy. "Position Paper: NO_x Measurement," *Environ. Sci. Technol.* 6:890–894 (1972).

15. Environmental Protection Agency. "Ambient Air Monitoring Equivalent Method Designations, Sodium Arsenite and TGS-ANSA Manual Methods for NO_2," *Federal Register* 42:62971 (December 14, 1977). Analytical procedures are available from EPA as: No. EQN-1277-026 (Manual Arsenite), No. EQN-1277-027 (Automated Arsenite) and No. EQN-1277-028 (TGS-ANSA).

16. Environmental Protection Agency. "Nitrogen Dioxide Measurement Principle and Calibration Procedure," *Federal Register* 41:52686 (December 1, 1976).

17. Saltzman, B. E. "Modified Nitrogen Dioxide Reagent for Recording Air Analyzers," *Anal. Chem.* 32:135–136 (1960).

18. Thomas, M. D., J. A. MacLeod, R. C. Robbins, R. C. Goettelman, R. W. Eldridge and L. H. Rogers. "Automatic Apparatus for Determination of Nitric Oxide and Nitrogen Dioxide in the Atmosphere," *Anal. Chem.* 28:1810–1816 (1956).

19. Lyshkow, N. A. "A Rapid and Sensitive Colorimetric Reagent for Nitrogen Dioxide in Air," *J. Air Poll. Control Assoc.* 15:481–484 (1965).

20. Environmental Protection Agency. "Ambient Air Monitoring and Equivalent Methods: Reference Method Designation," *Federal Register* 42: 37434–37435 (July 21, 1977); 42:46574 (September 16, 1977).

21. Ellis, E. C. "Technical Assistance Document for the Chemiluminescence Measurement of Nitrogen Dioxide," EPA-600/4-75-003, Environmental Protection Agency, Washington, D.C. (1975).

22. Foster, J. F., and G. H. Beatty. "Interlaboratory Cooperative Study of the Precision and Accuracy of the Measurement of Nitrogen Dioxide Content in the Atmosphere Using ASTM Method D1607," ASTM Pub. No. 05-055000-17, Philadelphia (1974).

23. Constant, P. C., Jr., M. C. Sharp and G. W. Scheil. *Collaborative Testing of Methods for Measurements of NO_2 in Ambient Air*, Volume I—*Report of Testing*, EPA-650/4-74-019-a (Washington, D.C.: Environmental Protection Agency, 1974).

24. Margeson, J. H., J. C. Suggs, P. C. Constant, Jr., M. C. Sharp and G. W. Scheil. "Collaborative Testing of a Manual Sodium Arsenite Method for Measurement of Nitrogen Dioxide in Ambient Air," *Environ. Sci. Technol.* 12:294–297 (1978).
25. Constant, P. C., Jr., M. C. Sharp and G. W. Scheil. "Collaborative Test of the TGS-ANSA Method for Measurement of Nitrogen Dioxide in Ambient Air," EPA-650/4-74-046, Washington, D.C. (1974).
26. Constant, P. C., Jr., M. C. Sharp and G. W. Scheil. "Collaborative Test of the Continuous Colorimetric Method for Measurement of Nitrogen Dioxide in Ambient Air," EPA-650/4-75-011, Washington, D.C. (1975).
27. Constant, P. C., Jr., M. C. Sharp and G. W. Scheil. "Collaborative Test of the Chemiluminescent Method for Measurement of NO_2 in Ambient Air," EPA-650/4-75-013, Washington, D.C. (1975).
28. Margeson, J. H., M. E. Beard and J. C. Suggs. "Evaluation of Sodium Arsenite Method for Measurement of NO_2 in Ambient Air," *J. Air Poll. Control Assoc.* 27:553–556 (1977).
29. Purdue, L. J., G. G. Akland and E. C. Tabor. "Comparison of Methods for Determination of Nitrogen Dioxide in Ambient Air," EPA-650/4-75-023, Research Triangle Park, NC (1975).
30. Margeson, J. H., and R. G. Fuerst. "Evaluation of a Continuous Colorimetric Method for Measurement of Nitrogen Dioxide in Ambient Air," EPA-650/4-75-022, Research Triangle Park, NC (1975).
31. Higuchi, J. E., A. Hsu and R. D. MacPhee. "Dynamic Calibrations of Saltzman Type Analyzers Using NO_2 in Air Mixtures," *J. Air Poll. Control Assoc.* 26:136–138 (1976).
32. Ellis, E. C., and J. H. Margeson. "Evaluation of Gas Phase Titration Technique as Used for Calibration of Nitrogen Dioxide Chemiluminescence Analyzers," EPA-650/4-75-021, Research Triangle Park, NC (1975).
33. Winer, A. M., J. W. Peters, J. P. Smith and J. N. Pitts, Jr. "Response of Commercial Chemiluminescent $NO-NO_2$ Analyzers to Other Nitrogen-Containing Compounds," *Environ. Sci. Technol.* 8:1118–1121 (1974).
34. Joshi, S. B., and J. J. Bufalini. "Halocarbon Interferences in Chemiluminescent Measurements of NO_2," *Environ. Sci. Technol.* 12:597–599 (1978).
35. Hilfiker, R. C., D. Muldoon and R. D. Andrews. "A Program for Short-Term Monitoring of TSP and NO_2 at Kenmore Square," Document P-2588, Environmental Research and Technology, Inc., Concord, MA (1977).
36. Environmental Protection Agency. "Photochemical Oxidants; Measurement of Ozone in the Atmosphere," *Federal Register* 43:26962–26985 (June 22, 1978).

CHAPTER 4

REVIEW OF U.S. ENVIRONMENTAL PROTECTION AGENCY NO$_2$ MONITORING METHODOLOGY REQUIREMENTS

Larry J. Purdue and Thomas R. Hauser

Environmental Monitoring and Support Laboratory
Environmental Protection Agency
Research Triangle Park, North Carolina 27711

INTRODUCTION

This chapter reviews the United States Environmental Protection Agency's (EPA) methodology requirements for measuring nitrogen dioxide (NO$_2$) in the ambient air from 1971, when the U.S. national ambient air quality standard for NO$_2$ was promulgated, to the present. The original NO$_2$ reference method, rationale for both its selection and subsequent replacement, and the current methodology requirements under the new ambient air monitoring reference and equivalent method regulations introduced in 1975, are reviewed.

BACKGROUND

The United States Clean Air Act was amended in 1967 to require that air quality criteria be developed and issued for those pollutants that may be harmful to health and welfare. These criteria were to serve as the basis for national ambient air quality standards; therefore, they had to accurately reflect the latest scientific information on the kind and extent of all identifiable effects on public health and welfare that might be expected from a pollutant in the ambient air. The Clean Air Act was further amended in 1970 to require the Administrator of EPA to establish such standards for pollutants that meet certain criteria relating to ubiquity, origin from a multiplicity of sources, and ability to cause adverse effects on health or welfare. Accordingly, on April 30, 1971, the Administrator promulgated national ambient air quality primary and secondary standards for six pollutants, including NO_2[1].

Concurrent with the promulgation of the ambient air quality standards, an analytical method was promulgated to measure each of the six pollutants. These methods were called "reference methods." Detailed descriptions of the reference methods prescribed for each of the six pollutants were published in appendices to the same part of the regulation that specified the standards. The air quality standard for each pollutant and its reference method are now codified in Title 40, Chapter I of the *Code of Federal Regulations*, Part 50 (40 CFR Part 50) [2].

On August 14, 1971, "Requirements for Preparation, Adoption and Submittal of Implementation Plans" were promulgated [3], which required each of the states to adopt and submit to the Administrator of EPA a plan providing for the implementation, maintenance and enforcement of the national ambient air quality standards. As a part of this plan, each state was required to establish an air quality surveillance system with specified minimum requirements for the number of air quality monitoring sites and sampling frequency. Known as State Implementation Plan (SIP) requirements, this regulation provides the basis for the requirement that pollutants for which air quality standards have been set be measured by the specified "reference method," or by a method demonstrated to be equivalent to it.

The six pollutants for which standards were set, the level of the standard for each pollutant, and the reference method measurement principle as they were promulgated in 1971, are shown in Table I.

National primary ambient air quality standards define the levels of air quality that the Administrator judges are necessary to protect the public health with an adequate margin of safety. National secondary ambient air quality standards define levels of air quality that the Administrator judges necessary to protect the public welfare from known or anticipated adverse effects of a pollutant.

The national primary and secondary ambient air quality standard for NO_2 was set as, and is still, an annual arithmetic mean of 100 $\mu g/m^3$ (0.05 ppm). This standard is currently under review by EPA, and serious consideration is being given to the promulgation of a short-term (1- or 3-hr) NO_2 standard, which will either replace or be included with the existing standard. The SIP requirements for air surveillance systems for determining compliance with the NO_2 standard specified a minimum of one 24-hr sample every 14 days (equivalent to 26 random samples per year) for each monitoring site to determine a valid annual arithmetic mean. A minimum of 3 air quality monitoring sites was required for areas with populations less than 100,000; 4 + 0.6 per 100,000 population for areas between 100,000 and 1,000,000; and 10 for areas above 1,000,000 population. These SIP requirements are currently being revised along with revisions to the requirements for ambient air quality monitoring for all the pollutants for which standards have been set [4].

Table I. National Ambient Air Quality Standards (40 CFR Part 50)

| Pollutant | Standard [$\mu g/m^3$ (ppm)] | | Reference Method Measurement Principle |
	Primary	Secondary	
Sulfur Dioxide	80 (0.03)[a] 365 (0.14)[b]	1300 (0.5)[c]	Spectrophotometric (pararosaniline)
Particulate Matter	75[d] 260[b]	60[d] 150[b]	High volume (gravimetric)
Carbon Monoxide	10×10^3 (9)[e] 40×10^3 (35)[f]	Same as primary Same as primary	Nondispersive infrared spectrometry
Photochemical Oxidants (Ozone)	160 (0.08)[f]	Same as primary	Chemiluminescence
Hydrocarbons	160 (0.24)[g]	Same as primary	Gas chromatography, flame ionization detection
Nitrogen Dioxide	100 (0.05)[a]	Same as primary	Spectrophotometric (Jacobs–Hochheiser)

[a] Annual arithmetic mean.
[b] Maximum 24-hour concentration, not to be exceeded more than once per year.
[c] Maximum 3-hour concentration, not to be exceeded more than once per year.
[d] Annual geometric mean.
[e] Maximum 8-hour concentration, not to be exceeded more than once per year.
[f] Maximum 1-hour concentration, not to be exceeded more than once per year.
[g] Maximum 3-hour concentration (6–9 A.M.), not to be exceeded more than once per year; for use only as a guide in devising implementation plans to achieve the ozone standard.

GRIESS–SALTZMAN METHOD

Prior to the promulgation of the air quality standards, methods based on the colorimetric Griess–Saltzman reaction [5] were deemed the most suitable for NO_2 measurements. The Griess–Saltzman method was adopted in 1969 as a tentative method by the Intersociety Committee on Manual of Methods for Ambient Air Sampling and Analysis [6]. The method is a manual method based on the reaction of NO_2 with diazotizing-coupling reagents to form a deeply colored azo dye. Air is sampled through a fritted bubbler into a solution containing sulfanilic acid, N-(1-naphthyl)-ethylenediamine dihydrochloride, and acetic acid (Griess–Saltzman reagent) for periods $\leqslant$30 min. Color development is complete within 15 minutes, and the intensity of the color produced, which is proportional to the amount of NO_2 sampled, is measured at a wavelength of 550 nm within 1 hr of the completion of sampling. High ratios of sulfur dioxide (SO_2) to NO_2 (about 30:1) can cause bleaching and result in erroneously low values. Interference from other oxides of nitrogen or ozone (O_3) is negligible at concentrations found in polluted air. Peroxyacetylnitrate (PAN) can give a response of up to 35% of an equivalent molar concentration of NO_2, but in ordinary ambient air, PAN concentrations are too low to cause significant error.

The Griess–Saltzman method is generally calibrated statically using dilute solutions of sodium nitrite as calibration standards. Saltzman reported that, under laboratory conditions, 0.72 mol of nitrite produces the same color as 1 mol of NO_2 gas. He incorporated this factor into the calculations to determine the ambient NO_2 concentration. This stoichiometric factor has been the source of considerable controversy. Values ranging from 0.5–1.00 have been reported [7]. The method can be dynamically calibrated using accurately known concentrations of NO_2 gas and thus eliminate the stoichiometric factor from the calculations. Concentrations of NO_2 in the ambient air from 40 to 1500 $\mu g/m^3$ (0.02–0.75 ppm) can be measured for time periods up to 30 minutes using this method.

ORIGINAL EPA REFERENCE METHOD

The original EPA reference method for NO_2, which was published in the *Federal Register* on April 30, 1971, along with the air quality standard, was a manual 24-hr integrated method based on a modification of the Griess–Saltzman method. The Griess–Saltzman method cannot be used successfully when the delay between sample collection and color measurement is more than one hour or when sampling periods longer than 30 min are required. Based on the work of Jacobs and Hochheiser [8], this method was modified by EPA scientists to allow for NO_2 sampling periods of 24 hr and analysis delay times of at least two weeks. This modified Griess–Saltzman method, which was commonly known as the Jacobs–Hochheiser method, was selected as the original EPA reference method in 1971. This method involves the collection of NO_2 by bubbling ambient air for a 24-hr period through an aqueous 0.1 N sodium hydroxide solution to form a stable solution of sodium nitrite. The nitrite ion formed in the absorbing reagent is then reacted with sulfanilamide and N-(1-naphthyl)-ethylenediamine dihydrochloride in acid media to form an azo dye, which is measured colorimetrically at 540 nm. The method is calibrated using standard nitrite solutions in accordance with directions published along with the reference method.

Deficiencies of the Original Reference Method

On June 14, 1972, the Administrator of EPA published a notice in the *Federal Register* [9] indicating that the original reference method for measuring NO_2 was suspected of being unreliable. Laboratory testing and air quality measurements made over a period of several months at many locations resulted in the identification of apparent deficiencies with this method. Accordingly, the Administrator announced that the method would be reevaluated.

On June 8, 1973, the results of EPA's evaluation [10] of these problems were published in the *Federal Register* [11]. The reference method was evaluated utilizing a newly developed NO_2 generation dilution system based on permeation devices [12] to generate accurately known dynamic NO_2 standards in air. Nitrogen dioxide–air mixtures of varying concentrations were generated, and the mixtures were sampled and analyzed by the reference method. The overall collection efficiency of the reference method was then determined by comparing the amount of NO_2 found versus the known amount of NO_2 generated. The results of this investigation (Figure 1) demonstrate that the collection efficiency of the reference method varied nonlinearly with NO_2 concentrations.

The reference method as originally published contained a correction factor for the overall efficiency of the method, which historically had been determined to be 35% [13]. Figure 1 shows that at low concentrations of NO_2 (20–50 $\mu g/m^3$), where collection efficiencies are much higher than 35%, the reported concentrations of NO_2 will be much higher than the actual ambient level. When the ambient level of NO_2 is about 100 $\mu g/m^3$, a 35% collection efficiency factor is approximately valid; but at higher atmospheric concentrations, the application of a 35% correction factor will underestimate the actual NO_2 concentration in the air sampled.

In addition to the variable efficiency problem, the reference method was found to be subject to an interference caused by nitric oxide (NO) in the ambient air. A positive interference results when NO and NO_2 are present simultaneously in the ambient air.

Because of these problems, EPA concluded that this method could no longer serve as the reference method and announced its intention, following completion of additional studies, to propose amendments to Appendix F of 40 CFR Part 50 that would supersede the original reference method for NO_2.

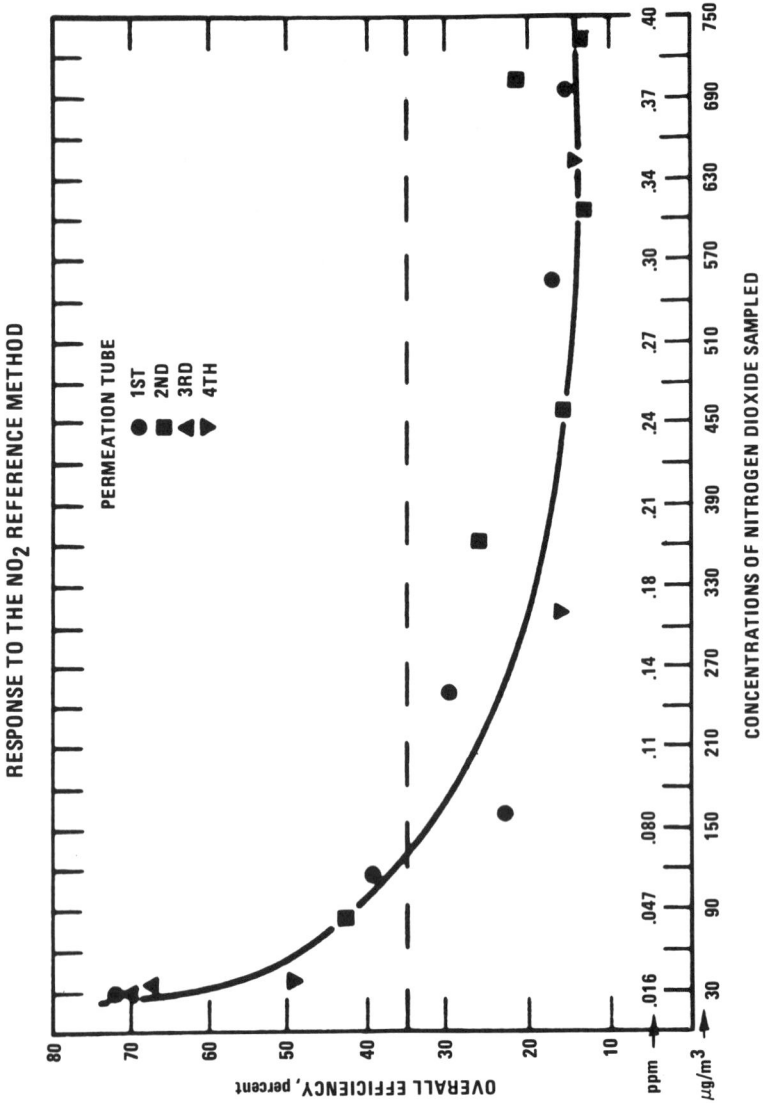

Figure 1. Collection efficiency of the former Federal Reference Method for nitrogen dioxide measurement.

TECHNICAL EVALUATION OF ALTERNATIVE METHODS

Three candidate methods for measurement of NO_2 were described in the June 8, 1973, *Federal Register* notice, which indicated that EPA would conduct complete evaluations and collaborative testing of these and other possible methods to select the best replacement. Accordingly, EPA evaluated three manual methods and two automated methods (analyzers) based on different measurement principles. The manual methods evaluated were sodium arsenite, triethanolamine-guiacol-sulfite (TGS) and triethanolamine (TEA). The automated methods evaluated were analyzers based on the continuous colorimetric and chemiluminescence measurement principles. Although the TGS and TEA methods were not identified as candidate methods in the June 8, 1973, notice, subsequent improvements in these methods and further consideration suggested that they should also be tested.

Two types of tests were used in the evaluations—method standardization tests and interrelatability tests. The method standardization tests included single investigator evaluations and collaborative testing. Information from developers, experienced users and the technical literature was reviewed to establish the best technical description of the methods. Evaluation tests, including statistical ruggedness test, when appropriate, were used to identify critical variables affecting the method or analyzer performance. The technical or procedural description and specifications for the methods were then revised to minimize adverse effects of these critical variables. Finally, comprehensive interlaboratory collaborative tests were conducted to determine the performance of the methods in actual use under typical onsite conditions by typical users to obtain statistical estimates of bias and precision. An interrelatability test was used to determine the intra- and intermethod comparability of the various methods and analyzers under carefully controlled conditions. Comprehensive reports of the results of the standardization tests for each of the five methods and the interrelatability tests were published in EPA's Environmental Monitoring series [14–24]. These results are summarized below.

Method Standardization Tests

Sodium Arsenite Method

The sodium arsenite method is a 24-hr integrated manual method similar to the original reference method. Nitrogen dioxide is collected by bubbling air through an orifice-type bubbler into a sodium hydroxide—sodium arsenite solution to form a stable solution of sodium nitrite. The nitrite ion produced during sampling is reacted with phosphoric acid, sulfanilamide, and N-(1-naphthyl)-ethylene-diamine dihydrochloride to form an azo dye and then determined colorimetrically.

The single investigator evaluations [14,16] indicated the following:

1. The method has a relatively constant collection efficiency of 82% over the range of 20–700 μg NO_2/m^3.

2. Nitric oxide (NO) is a positive interferent and carbon dioxide (CO_2) is a negative interferent. (Over the ranges of NO and CO_2 that are likely to occur in the ambient air, the effect of these two interferents is minimal.)
3. The sampling flowrate must be maintained constant within the range of 180–220 cm^3/min to obtain constant collection efficiency for NO_2.

The collaborative test [15] indicated the following:

1. The average bias of this method is approximately 3% and is independent of concentration.
2. The within laboratory standard deviation is 8 μg NO_2/m^3 and is independent of concentration over the range of 50–300 μg NO_2/m^3.
3. The between laboratory standard deviation is 11 μg NO_2/m^3 and is independent of concentration over the range of 50–300 μg NO_2/m^3.
4. The lower detectable limit is approximately 9 μg NO_2/m^3.

Triethanolamine-Guiacol-Sulfite (TGS) Method

The TGS method is a 24-hr integrated manual method. Nitrogen dioxide is collected by bubbling air through an orifice-type bubbler into a solution of triethanolamine, o-methoxyphenol (guiacol) and sodium metabisulfite. The nitrite ion produced during sampling is determined colorimetrically by reacting the exposed absorbing reagent with sulfanilamide and 8-amino-1-naphthalenesulfonic acid ammonium salt (ANSA).

The single investigator evaluations [17] indicated the following:

1. The method has a relatively constant collection efficiency of 92% over the range of 20–700 μg NO_2/m^3.
2. The analysis procedure requires careful attention to the method instructions.
3. The color developed in the analytical scheme depends on the time interval between the addition of reagents.
4. Nitric oxide and CO_2 are not significant interferents.

The collaborative test [18] indicated the following:

1. The average bias is approximately 5% and is independent of concentration.
2. The within laboratory standard deviation is 7 μg NO_2/m^3 and is independent of concentration over the range of 50–300 μg NO_2/m^3.
3. The between laboratory standard deviation is 12 μg NO_2/m^3 and is independent of concentration over the range of 50–300 μg NO_2/m^3.
4. The lower detectable limit is approximately 15 μg NO_2/m^3.

Triethanolamine (TEA) Method

The TEA method is a 24-hr integrated manual method. Nitrogen dioxide is collected by passing ambient air through a 3% aqueous triethanolamine solution. The nitrite ion produced is measured colorimetrically via a diazotization-coupling reaction similar to the original reference method. A bubble-dispersing frit is required to achieve high constant collection efficiency (80% over the range tested). The use of a fritted bubbler involves several serious disadvantages in comparison to using an orifice-type bubbler, which is used in the sodium arsenite and TGS methods. For this reason, the TEA method was not considered suitable to replace the existing reference method and was not evaluated further [19].

Continuous Colorimetric Measurement Principle

This measurement principle is based on the Griess–Saltzman reaction. Nitrite ion is reacted with diazotizing-coupling reagents to form a deeply colored azo dye that is measured colorimetrically in a continuously automated system. Sample air is drawn through a gas–liquid contact column at an accurately determined flowrate countercurrent to a constant controlled flow of absorbing solution. Nitrogen dioxide in the ambient air is converted to nitrite ion on contact with the absorbing solution containing the diazotizing-coupling reagents. Sufficient delay time is allowed for full color development; then the colored solution is passed through a colorimeter, in which the absorbance is measured continuously at a wavelength of about 550 nm. The absorbance of the azo dye is directly proportional to the concentration of NO_2 absorbed. Analyzers based on this measurement principle can be calibrated statically using nitrite ion solution, or dynamically, using known NO_2 in air mixtures.

The single investigator evaluation [20] of an analyzer based on this measurement principle indicated the following:

1. The analyzer response time is long (approximately 15 minutes), which is typical for colorimetric analyzers.
2. The manufacturer's recommended static calibration procedure is not acceptable; dynamic calibration is required.
3. Ozone is a significant negative interferent to the analyzers tested.

The collaborative test [21] indicated the following:

1. The average bias ranged from +3 to +15% over the range 50–300 $\mu g\ NO_2/m^3$.
2. The within laboratory relative standard deviation is approximately 6% of the concentration over the range of 50–300 $\mu g\ NO_2/m^3$.
3. The between laboratory relative standard deviation is approximately 12% of the concentration over the above range.
4. The lower detectable limit is approximately 19 $\mu g\ NO_2/m^3$.

Continuous Chemiluminescence Measurement Principle

Atmospheric concentrations of NO_2 are measured indirectly by photometrically measuring the light intensity at wavelengths greater than 600 nm, resulting from the chemiluminescent reaction of NO with O_3. Nitrogen dioxide is first quantitatively reduced to NO by means of a thermal converter. Nitric oxide, which commonly exists in ambient air together with NO_2, passes through the converter unchanged, causing a resultant total NO_x concentration equal to NO + NO_2. A sample of the input air is also measured without having passed through the converter. This latter NO measurement is subtracted from the former measurement (NO + NO_2) to yield the final NO_2 measurement. The NO and NO + NO_2 measurements are made concurrently with dual systems, or cyclically with the same system with less than 1-min cycle times. Dynamic calibration is required using NO_2 and NO calibration standards. Stable NO calibration standards are available in high-pressure cylinders. Nitrogen dioxide calibration standards can be generated utilizing a gas-phase titration (GPT) technique or utilizing NO_2 permeation devices.

The chemiluminescence measurement principle for NO_2 measurements was thoroughly evaluated in the course of its development [25–29]. These studies provided the basis for considering this measurement principle as a candidate for the reference method for NO_2. The results of these studies and a single investigator evaluation of the GPT calibration technique [22] indicated the following:

1. Analyzer response time is fast (less than 1 minute).
2. Certain organic nitrogen compounds are positive interferents but are a minor problem because of their typically low occurrence in ambient air at concentrations high enough to interfere.
3. The GPT calibration procedure is relatively complex, but is reliable in experienced hands.
4. The GPT calibration procedure is insensitive to normal variations in critical parameters such as O_3 and NO flowrates, dilution flowrates, reaction times and reactant concentrations.

The collaborative test [23] indicated the following:

1. The average bias is approximately 5% and is independent of concentration.
2. The within laboratory relative standard deviation is approximately 6% of the concentration over the range of 50–300 μg NO_2/m^3.
3. The between laboratory relative standard deviation is approximately 14% of the concentration over the above range.
4. The lower detectable limit is approximately 10 μg NO_2/m^3.

Interrelatability Tests

In addition to the method standardization tests, an interrelatability test was conducted [24] to determine the intra- and intermethod comparability of four of the five candidate methods (sodium arsenite, TGS and the chemiluminescence and colorimetric automated methods). The four candidates were compared by simultaneously measuring NO_2 in the ambient air with each method under a variety of carefully controlled conditions. Quadruplicate samples were taken for the manual methods, and duplicate analyzers were used for each of the two continuous methods. The various test conditions were designed to include: (1) tests to investigate comparability of the methods over NO_2 concentration ranges expected to occur in ambient air; (2) tests for comparability under various patterns of NO_2 concentration fluctuations; and (3) special tests for possible interference from O_3. The results of the comparability tests from the first phase of this study are shown in Table II, and the statistical analysis of the method differences is shown in Table III.

The primary observations resulting from the statistical analysis of the results from the interrelatability tests are presented below.

Intramethod Comparisons

1. The overall agreement within each method throughout all phases of the study was good, as characterized by a mean difference in results from analyzer or bubbler pairs of no worse than 7.5 μg NO_2/m^3.

Table II. NO$_2$ Methods Comparison Data (22-hr average), μg NO$_2$/m^3

	Automated Methods				Manual Methods							
	Chemilu- minescence		Colorimetric		Sodium Arsenite				TGS			
					Sampler A		Sampler B		Sampler A		Sampler B	
Day	CM-1	CM-2	C-1	C-2	ARS-1	ARS-2	ARS-3	ARS-4	TGS-1	TGS-2	TGS-3	TGS-4
1	173	165	164	164	172	169	180	178	164	160	161	156
2	88	86	81	75	96	94	87	95	92	84	83	81
3	73	73	66	62	80	68	75	73	64	50	73	62
4	148	152	147	143	146	155	144	150	141	138	136	146
5	273	276	274	267	273	262	269	270	274	275	281	271
6	58	58	49	47	57	64	66	65	61	63	64	68
7	290	288	286	284	275	281	286	286	279	282	282	276
8	128	132	147	130	133	137	143	133	128	138	126	126
9	102	100	103	98	106	107	109	106	95	100	117	115
10	107	109	103	98	129	127	122	126	203	112	108	108
11	175	179	173	169	184	191	195	198	172	175	187	161
12	73	62	94	68	81	75	74	74	74	72	70	74
13	128	126	128	132	132	133	141	144	122	122	124	119
14	53	49	53	43	52	48	50	53	49	49	46	45
15	235	233	252	237	227	220	233	221	219	215	215	226
16	147	145	—	—	157	163	160	147	159	162	137	134
17	164	165	148	145	156	154	158	161	141	150	144	147
18	263	265	261	259	251	249	240	254	240	245	233	247
19	150	150	169	147	140	141	143	140	134	138	138	133
20	169	147	179	—	170	167	196	187	164	163	160	174
21	192	194	197	182	197	193	198	196	173	174	176	189
22	17	13	15	8	17	17	17	18	20	21	19	18

Table III. Statistical Analysis of Method Differences ($\mu g/m^3$)

Comparison	Pairs	Mean[a]	Standard Deviation[b]	95% C. I.[c] Lower	95% C. I.[c] Upper	Correlation Coefficient[d]
Intramethod						
Chemil/Chemil	22	1.3	5.6	−1.1	3.8	0.999
Color/Color	20	7.5	7.5	3.8	11.3	0.995
ARS/ARS (A)	22	0.6	5.6	−1.9	+3.8	0.997
ARS/ARS (B)	22	0.2	5.6	−1.9	1.9	0.996
TGS/TGS (A)	20	−0.9	3.8	−1.9	0.8	0.999
TGS/TGS (B)	22	0.2	9.4	−3.8	+3.8	0.992
ARS (A)/ARS (B)	22	−2.6	5.6	−5.6	+0.1	0.997
TGS (A)/TGS (B)	20	0.6	7.5	−3.8	+3.8	0.996
Intermethod						
Chemil/Color	20	3.8	7.5	0.0	+7.5	0.994
Chemil/ARS	22	−1.9	9.4	−5.6	+1.9	0.991
Chemil/TGS	20	5.6	9.4	+0.9	9.4	0.990
Color/ARS	20	−3.8	11.3	−9.4	1.5	0.989
Color/TGS	18	3.8	11.3	−1.9	9.4	0.985
ARS/TGS	20	7.5	7.5	+3.8	11.3	0.994

[a]Signed difference.
[b]Standard deviation.
[c]95% confidence interval of mean difference.
[d]Correlation coefficient between paired values in calculating mean difference.

2. Intracomparison data correlations were better than 0.95, indicating that each method responded similarly to changes in concentration of NO_2.
3. There was a small (<10%), although statistically significant, bias between the data generated by the two colorimetric analyzers.
4. Fifty percent of the absolute differences between duplicate measurements for any of the four methods were less than 10 μg NO_2/m^3 for each of the four methods.
5. The highest maximum difference was 26 μg NO_2/m^3, which occurred in the comparison of the two continuous colorimetric analyzers.

Intermethod Comparisons

1. The overall agreement among NO_2 concentrations measured by the different methods or principles was impressive, as reflected by mean differences of less than 10 μg NO_2/m^3.
2. All intercomparison data correlations were better than 0.95.
3. Fifty percent of the absolute differences between average NO_2 concentrations measured by each of the four measurement systems were less than 10 μg NO_2/m^3.

The most important conclusion drawn from the results of the interrelatability study is that all four methods, when used by skilled technicians under carefully controlled conditions, can produce data that are in remarkably good agreement. The effect on the results produced by variable NO_2

concentration fluctuations during a sampling period was relatively minor. The results of special O_3 interferent tests verified suspected interference of O_3 to the automated colorimetric analyzers. Significant negative interference was found at NO_2 concentrations of 75 and 100 $\mu g/m^3$ in combination with 353 and 667 μg O_3/m^3. However, at a low O_3 concentration (100 $\mu g/m^3$), no effect was detected. The effect of high O_3 concentrations on the performance of the other methods was not significant.

The performance of all methods and analyzers was monitored closely throughout this study. The performance of the chemiluminescence analyzers was better than that of the colorimetric analyzers, with respect to zero drift, span drift, response times and overall operation. Zero drift and span drift data for these analyzers are shown in Table IV. Of the two manual procedures, the sodium arsenite method gave the best overall performance. The good agreement between the ambient data generated by the continuous analyzers and the independently calibrated manual methods validated the GPT calibration procedure used for the continuous analyzers. It was also an indication that the methods tested can produce reasonably accurate NO_2 data, with the exception of potential O_3 interference in the continuous colorimetric analyzers.

Based on consideration of the results of the methods' standardization activities and the interrelatability tests, the sodium arsenite method, the TGS method and analyzers based on the chemiluminescence measurement principle were all considered to be excellent candidates for the replacement

Table IV. Zero Drift and Span Drift Data

	Analyzer[a]			
	CM-1	CM-2	C-1	C-2
Zero Drift[b]				
Number of recordings	35	34	34	34
Minimum value	−33	−10	−17	−5
Maximum value	30	28	38	19
Mean	0	3	10	3
Standard deviation	8	7	12	4
Span Drift[c]				
Number of recordings	33	35	34	34
Minimum value	−5.4	−4.6	−10.7	−6.5
Maximum value	4.8	6.3	13.1	18.7
Mean	−0.20	0.24	0.23	1.02
Standard deviation	2.20	2.02	4.95	4.38

[a]CM-1 and CM-2 are chemiluminescence instruments; C-1 and C-2 are colorimetric instruments.
[b]Zero drift = adjusted zero − unadjusted zero (μg NO_2/m^3).
[c]Span drift = $\dfrac{\text{adjusted span} - \text{unadjusted span}}{\text{adjusted span}} \times 100$, at 80 ± 10 % of full scale.

of the original reference method. The TEA method was eliminated as a possible candidate because of the requirement for a fritted disperser for maintaining a high constant collection efficiency. Analyzers based on the colorimetric measurement principle were eliminated on the basis of poorer performance relative to the chemiluminescence analyzers (e.g., slow response time, unquantifiable bias and possible O_3 interference).

RATIONALE FOR SELECTION OF THE CHEMILUMINESCENCE MEASUREMENT PRINCIPLE

On March 17, 1976, EPA published a proposal in the *Federal Register* [30] to amend Appendix F of 40 CFR Part 50 that would replace the original manual reference method for NO_2 with the continuous chemiluminescence measurement principle and a specified calibration procedure. After consideration of public comments, EPA promulgated this amendment on December 1, 1976 [31]. The rationale for selecting the chemiluminescence measurement principle was based on the results from the technical evaluations and on the capability of chemiluminescence analyzers for continuous NO_2 measurements.

As noted above, on the basis of the technical evaluations, consideration was limited to the sodium arsenite and TGS methods and analyzers based on the continuous chemiluminescence principle. Quantitative performance indicators such as collection efficiency, bias, precision, interferences and intra- and intermethod correlations were used to assess the relative technical merits of each method. Based on these indicators, any of the three candidate methods would have been adequate for measuring 24-hr average NO_2 concentrations to determine compliance with the national ambient air quality standard. However, EPA recognized the increasing need for short-term (hourly average) data: (1) to analyze diurnal patterns; (2) to study the relationship between emission sources and receptor or monitoring sites; (3) to develop effective and efficient control strategies for both NO_2 and photochemical oxidants; and (4) to provide data during air pollution episodes. In addition, there was some probability, based on current health data, that EPA would be considering a new short term NO_2 standard, which would require hourly averages rather than 24-hr measurements. Reference methods based on the continuous chemiluminescence principle would meet these needs, whereas the 24-hr manual methods could not. The selection of the chemiluminescence measurement principle for reference methods for NO_2 was based primarily on this consideration. EPA is now developing a national ambient air quality short-term standard for NO_2 based on the 1977 amendments to the Clean Air Act. These amendments require the Administrator of EPA to promulgate a short-term standard for NO_2 concentrations over a period of no more than three hours, unless he finds no significant evidence that such a standard is necessary to protect the public health. Because this action was anticipated in 1976 when the reference method principle was selected, there will be no need to amend EPA's NO_2 monitoring methodology requirements when and if a new short-term standard is promulgated.

REFERENCE AND EQUIVALENT METHODS

On February 18, 1975, EPA promulgated 40 CFR Part 53, entitled "Ambient Air Monitoring Reference and Equivalent Method" [32], which established procedures governing the designation of reference and equivalent methods for the measurement of SO_2, O_3 and CO. On December 1, 1976, concurrent with the promulgation of the new reference measurement principle, the scope of 40 CFR Part 53 was extended to include NO_2. Consistent with the basic definitions and policies established in 40 CFR Part 53, the original NO_2 reference method has been replaced by a "measurement principle and calibration procedure." Appendix F now specifies only the measurement principle and associated calibration procedures on which reference methods for NO_2 must be based. As with the reference methods for CO and photochemical oxidants, analyzers based on the specified measurement principle and calibration procedure are designated as reference methods if they meet the performance specifications and other requirements set forth in 40 CFR Part 53, in accordance with the test procedures set forth in Part 53. The required performance specifications that continuous chemiluminescence NO_2 analyzers must meet are shown in Table V.

Since Appendix F specifies only the measurement principle and calibration procedure applicable to NO_2 reference methods, any commercially available

Table V. Performance Specifications for NO_2 Automated Methods

Performance Parameter	NO_2
1. Range (ppm)	0–0.5
2. Noise (ppm)	0.005
3. Lower Detectable Limit (ppm)	0.01
4. Interference Equivalent	
Each interferant (ppm)	±0.02
Total interferant	0.04
5. Zero Drift, 12- and 24-hr (ppm)	±0.02
6. Span Drift, 24-hr	
20% of upper range limit (%)	±20.0
80% of upper range limit (%)	±5.0
7. Lag Time (min)	20
8. Rise Time (min)	15
9. Fall Time (min)	15
10. Precision	
20% of upper range limit (ppm)	0.02
80% of upper range limit (ppm)	0.03

NO_2 analyzer (model) based on the specified measurement principle is a candidate for designation as an NO_2 reference method. Therefore, a number of different analyzers can be designated as reference methods for NO_2.

As noted above, the "Ambient Air Reference and Equivalent Methods" regulations also established procedures for governing the designation of equivalent methods. Equivalent methods are methods based on measurement principles different from the reference method, which have been demonstrated to meet the requirements of 40 CFR Part 53. Two kinds of equivalent methods are possible—manual and automated (continuous monitoring analyzers). Candidate automated methods may be designated as equivalent methods if they meet the performance specifications listed in Table V and demonstrate a consistent relationship to the reference method during side-by-side ambient monitoring. Candidate manual methods need only demonstrate a consistent relationship to the reference method to be designated as equivalent methods. Table VI shows the test specifications that must be met to demonstrate a consistent relationship with the reference method.

In specifying that NO_2 reference methods be based on the chemiluminescence measurement principle, EPA was not necessarily advocating or encouraging increased use of chemiluminescence analyzers. The technical evaluation of the sodium arsenite and TGS manual methods indicated that both methods had good performance and would likely be more economical to use than automated methods. Accordingly, EPA tested and formally designated these two manual methods as equivalent methods, which allows state and local control agencies to use these methods for determining compliance with the ambient air quality standard for NO_2. However, these methods are 24-hr integrated methods and, therefore, are not applicable for shorter term integrating times or for continuous monitoring.

NO_2 Reference Methods

As noted earlier, the measurement principle specified for reference methods for NO_2 involves the conversion of NO_2 to NO and the subsequent measurement of NO using the chemiluminescence gas-phase reaction of NO with O_3. Analyzers using this measurement principle measure NO_2 indirectly after conversion to NO by means of a thermal converter. Virtually all ambient

Table VI. Consistent Relationship Test Specifications for NO_2

Concentration Range (ppm NO_2)	Maximum Discrepancy Specification (ppm)
Low, 0.02–0.08	0.02
Medium, 0.10–0.20	0.02
High, 0.25–0.35	0.03

chemiluminescence NO_2 analyzers are also capable of measuring ambient concentrations of NO and NO_x (NO_2 + NO) in addition to the NO_2 measurement. Figure 2 shows a schematic design of a typical automated NO_2 analyzer.

Interferents

Chemiluminescence analyzers will respond to other nitrogen-containing compounds, such as peroxyacetylnitrate (PAN), which are reduced to NO in the thermal converter. Atmospheric concentrations of these potential interferences are generally low relative to NO_2, and valid measurements are normally obtained. In certain geographical areas where the concentration of these potential interferents is known or suspected to be high relative to NO_2, the use of an equivalent method for the measurement of NO_2 is recommended.

Sampling Considerations

The use of integrating flasks on the sample inlet line of chemiluminescence analyzers is strongly discouraged. The sample residence time between the sampling point and the analyzer should be kept to a minimum to avoid erroneous NO_2 measurements resulting from the reaction of ambient levels of NO and O_3 in the sampling system.

The use of particulate filters on the sample inlet line is optional and left to the discretion of the user or manufacturer. Use of the filter should depend on the analyzer's susceptibility to interference, malfunction or damage due to particulates. Filters should be changed frequently because particulate matter concentrated on a filter may cause erroneous NO_2 measurements.

Calibration

Calibration of analyzers based on the chemiluminescence measurement principle requires standard atmospheres of NO_2 and NO. Two calibration procedures are prescribed in Appendix F of 40 CFR Part 50. One procedure generates NO_2 standards by gas-phase titration of NO with O_3. This calibration technique is based on the rapid gas-phase reaction between NO and O_3 to produce stoichiometric quantities of NO_2 in accordance with the following equation:

$$NO + O_3 \rightarrow NO_2 + O_2$$

The quantitative nature of this reaction is such that when the NO concentration is known, the concentration of NO_2 can be determined. Ozone is added to excess NO in a dynamic calibration system, and the NO channel of the chemiluminescence $NO/NO_x/NO_2$ analyzer is calibrated and used as an indicator of changes in NO concentration. On addition of O_3, the decrease in NO concentration observed on the calibrated NO channel is equivalent to the concentration of NO_2 produced. The amount of NO_2 generated may be varied

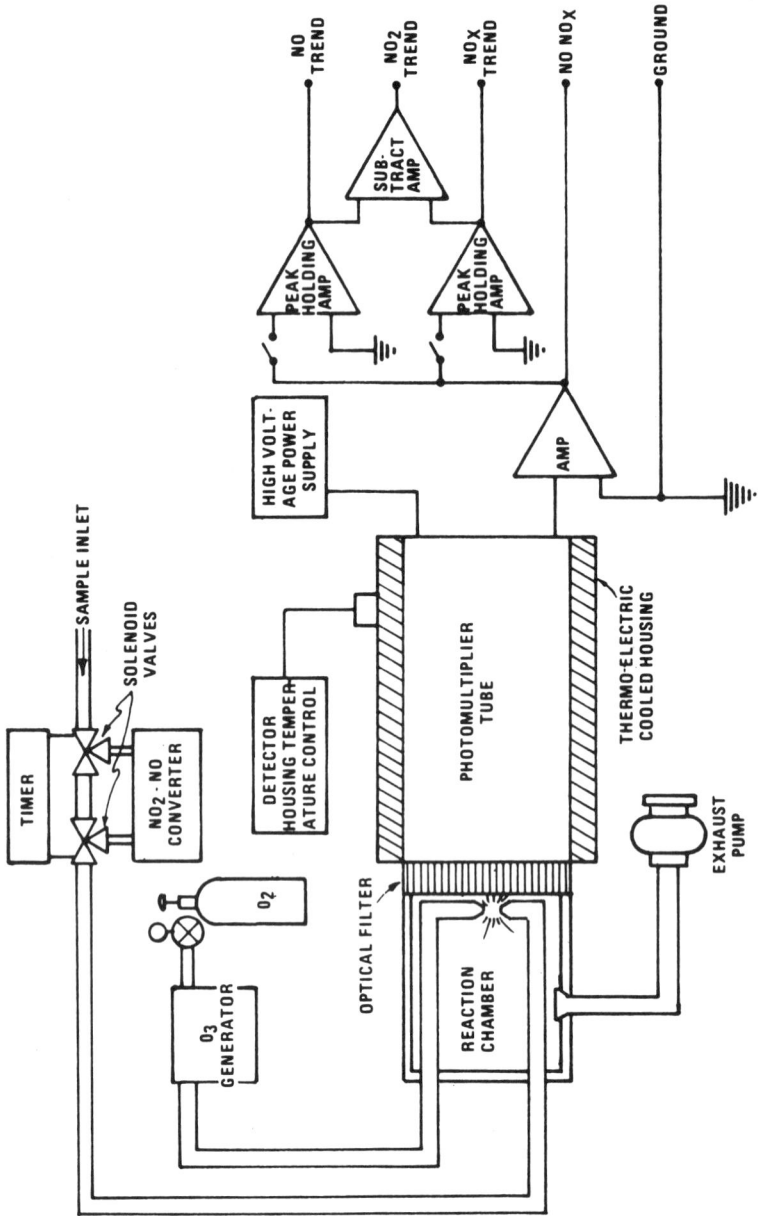

Figure 2. Automated NO, NO$_2$, NO$_X$ chemiluminescence analyzer.

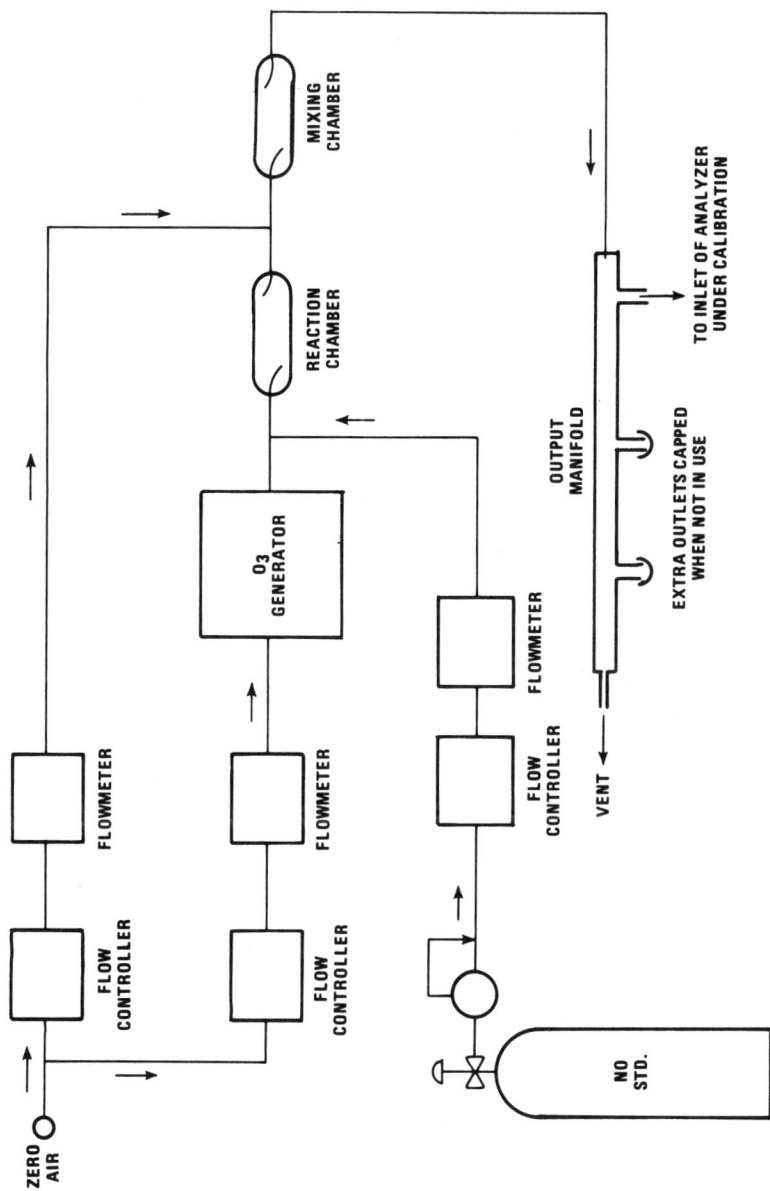

Figure 3. Schematic diagram of a typical GPT apparatus.

by adding variable amounts of O_3 from a stable uncalibrated O_3 generator. A schematic of a typical GPT calibration system is shown in Figure 3.

In the other calibration procedures, NO_2 standards are generated directly, utilizing an NO_2 permeation device. The permeation device emits NO_2 at a known constant rate when the temperature of the device is held constant ($\pm0.1°C$) and the device has been accurately calibrated at the temperature of use. The NO_2 emitted from the device is diluted with zero air to produce NO_2 concentrations suitable for calibration of the NO_2 channel of the analyzer. An NO concentration standard is used for calibration of the NO and NO_x channels of the analyzer. A typical system suitable for generating the required NO and NO_2 concentrations is shown in Figure 4.

Designated Reference and Equivalent Methods

Under the 40 CFR Part 53 regulations, analyzer manufacturers submit applications to EPA to have their analyzers considered for designation as reference or equivalent methods. After EPA determines that a candidate method satisfies the requirements for either a reference or equivalent method, notification of the designation is sent to the applicant and published in the *Federal Register*. Information, such as the designation number, method identification, options and operational limitations, and the source of the method (manufacturer), is provided in the *Federal Register* notice. In addition, EPA is required to maintain a current list of designated reference and equivalent methods and to provide the list to any person or group on request. A copy of the current list can be obtained by writing to the Environmental Monitoring and Support Laboratory, Department E (MD-77), U.S. Environmental Protection Agency, Research Triangle Park, NC 27711.

When an applicant's analyzer has been designated as a reference or equivalent method, the applicant must agree that the method offered for sale will be accompanied by a comprehensive operational or instructional manual. The applicant must also ensure that any such analyzer offered for sale will generate no unreasonable hazard to either the operators or the environment during normal use or when malfunctioning. In addition, an analyzer offered for sale must continue to meet the required performance specifications for a period of one year after delivery to the purchaser, provided the analyzer is maintained and operated by the purchaser in accordance with the operating manual supplied with the analyzer.

As of this date, EPA has designated seven NO_2 analyzer models as NO_2 reference methods. These analyzers are manufactured by six different commercial vendors. All analyzers sold by these vendors under their respective designation numbers are acceptable for use in determining compliance with EPA's National Ambient Air Quality Standard for NO_2. In addition to the seven reference methods that have been designated, three manual (24-hr integrated) equivalent methods have been designated. These methods are also acceptable for determining compliance with the air quality standard. Table VII gives all the NO_2 methods that have been designated as of February 22, 1979, and information pertinent to each method.

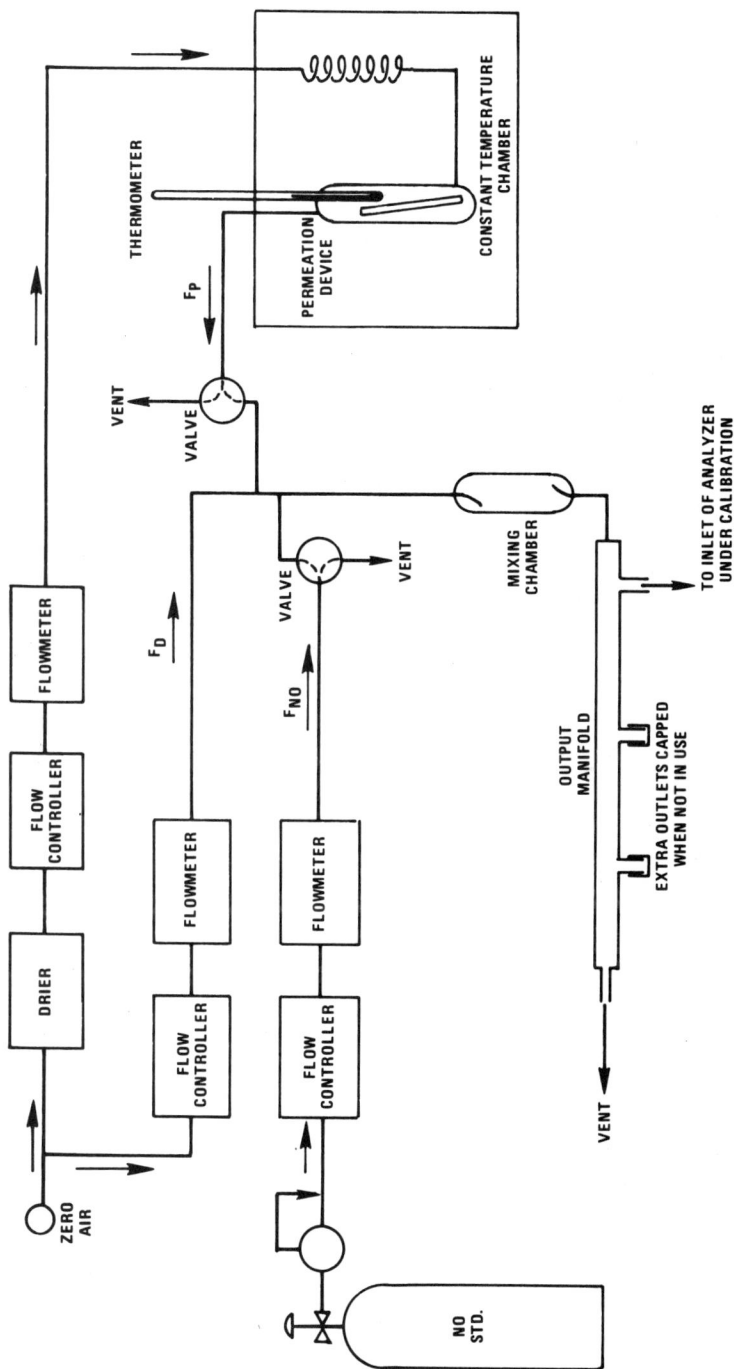

Figure 4. Schematic diagram of a typical calibration apparatus using an NO_2 permeation device.

Table VII. List of Designated NO$_2$ Methods

Designation Number	Identification	Source	Federal Register Vol.	Page	Notice Date
	Reference Methods—Automated				
RFNA-0677-021	"Monitor Labs Model 8440E Nitrogen Oxides Analyzer," operated on a 0–0.5 ppm range (position 2 of range switch) with a time constant setting of 20 sec and with or without any of the following options: TF – Sample particulate filter with TFE filter element VT – Zero/span valves and timer V – Zero/span valves FM – Flowmeters DO – Status outputs R – Rack mount	Monitor Labs, Inc. 10180 Scripps Ranch Blvd. San Diego, CA 92131	42	37434	7/21/77
RFNA-0777-022	"Bendix Model 8101-C Oxides of Nitrogen Analyzer," operated on a 0–0.5 ppm range with a Teflon sample filter (Bendix P/N 007163) installed on the sample inlet line	The Bendix Corp. Environment & Process Instruments Division P.O. Box 831 Lewisburg, WV 24901	42	37435	7/21/77
RFNA-0977-025	"CSI Model 1600 Oxides of Nitrogen Analyzer," operated on a 0–0.5 ppm range with a Teflon sample filter (CSI P/N M951-8023) installed on the sample inlet line, with or without any of the following options: 951-3053 Rack Mounting Kit 951-3054 Chassis Slide Kit 951-3066 Tilt Stand 951-8032 Local-Remote Ambient, Zero, Span Kit M951-0007 External Pump Unit M951-0008 Diagnostic Output	Columbia Scientific Industries 11950 Jollyville Rd. P.O. Box 9908 Austin, TX 78766	42	46574	9/16/77

RFNA-1078-031 "Meloy Model NA530R Nitrogen Oxides Analyzer," operated on the following ranges and time constant switch positions:

Range (ppm)	Time constant setting
0–0.1*	4
0–0.25*	3 or 4
0–0.5	2, 3 or 4
0–1.0	2, 3 or 4

Operation of the analyzer requires an external vacuum pump, either Meloy Option N-10 or an equivalent pump capable of maintaining a vacuum of 200 torr (200 inches mercury vacuum) or better at the pump connection at the specified sample and ozone–air flowrates of 1200 and 211 cm^3/min, respectively. The analyzer may be operated at temperatures between 10° and 40°C and at line voltages between 105 and 130 V, with or without any of the following options:

N-1A Automatic zero and span
N-2   Vacuum gauge
N-4   Digital panel meter
N-6   Remote control for zero and span
N-6B Remote zero/span control and status (pulse)
N-6C Remote zero/span control and status (timer)
N-9   Manual zero/span

N-10 Vacuum pump assembly (see alternate requirement above)
N-11 Auto ranging
N-14B Line transmitter
N-18 Rack mount conversion
N-18A Rack mount conversion

Meloy Labs, Inc.
Instruments & Systems Division
6715 Electronic Drive
Springfield, VA 22151

43 50734 10/31/78
44 8327 2/09/79

RFNA-0179-034 "Beckman Model 952-A NO/NO$_2$/NO$_X$ Analyzer," operated on the 0–0.5 ppm range with 5-µm Teflon sample filter (Beckman P/N 861072) supplied with the analyzer installed on the sample inlet line, and with or without the Remote Operation Option (Beckman Cat. No. 635539).

Beckman Instruments, Inc.
Process Instruments Division
2500 Harbor Blvd.
Fullerton, CA 92634

44 7806 2/09/79

RFNA-0179-035 "Thermo Electron Model 14 B/E Chemiluminescent NO/NO$_2$/NO$_X$ Analyzer," operated on the 0–0.5 ppm range and with or without the following options:
14-001 Teflon Particulate Filter
14-002 Voltage Divider Card

Thermo Electron Corp.
108 South St.
Hopkinton, MA 01748

44 7805 2/09/79

(Continued)

Table VII. (Continued)

Designation Number	Identification	Source	Federal Register Vol.	Page	Notice Date
Reference Methods—Automated					
RFNA-0279-037	"Thermo Electron Model 14 D/E Chemilumi-nescent NO/NO$_2$/NO$_X$ Analyzer," operated on the 0–0.5 ppm range and with or without any of the following options: 14-001 Teflon Particulate Filter 14-002 Voltage Divider Card	Thermo Electron Corp. 108 South St. Hopkinton, MA 01748	44	10429	2/20/79
Equivalent Methods—Manual					
EQN-1277-026	"Sodium Arsenite Method for the Deter-mination of Nitrogen Dioxide in the Atmosphere,"	Environmental Monitoring & Support Laboratories Department E (MD-77) U.S. Environmental Protection Agency Research Triangle Park, NC 27711	42	62971	12/14/77
EQN-1277-027	"Sodium Arsenite Method for the Deter-mination of Nitrogen Dioxide in the Atmosphere—Technicon II Automated Analysis System"	Environmental Monitoring & Support Laboratories Department E (MD-77) U.S. Environmental Protection Agency Research Triangle Park, NC 27711	42	62971	12/14/77
EQN-1277-028	"TGS-ANSA Method for the Determination of Nitrogen Dioxide in the Atmosphere"	Environmental Monitoring & Support Laboratories Department E (MD-77) U.S. Environmental Protection Agency Research Triangle Park, NC 27711	42	62971	12/14/77

REFERENCES

1. U.S. Environmental Protection Agency. "National Primary and Secondary Air Quality Standards," *Federal Register* 36(84):8186–8201 (April 30, 1971).
2. U.S. Environmental Protection Agency. "National Primary and Secondary Air Quality Standards," Title 40–Protection of Environment, Chapter I, *Code of Federal Regulations*, Part 50 (Washington, D.C.: U.S. Government Printing Office, 1978).
3. U.S. Environmental Protection Agency. "Requirements for Preparation, Adoption, and Submission of Implementation Plans," *Federal Register* 36(158):15486–15506 (August 14, 1971).
4. U.S. Environmental Protection Agency. "Air Quality Surveillance and Data Reporting," Proposed Regularity Revisions, *Federal Register* 43(152):34892–34934 (August 7, 1978).
5. Saltzman, B. E. "Colorimetric Microdetermination of Nitrogen Dioxide in the Atmosphere," *Anal. Chem.* 26:1949–1955 (1954).
6. Intersociety Committee on Manual Methods for Ambient Air Sampling and Analysis, Subcommittee 3. "Tentative Method of Analysis for Nitrogen Dioxide Content of the Atmosphere (Griess-Saltzman Reaction)," *Health Lab. Science* 6:106–113 (1969).
7. U.S. Environmental Protection Agency. *Air Quality Criteria for Nitrogen Dioxide*, (Washington, D.C.: U.S. Government Printing Office, 1971), AP-84, Chapter 5.
8. Jacobs, M. B., and S. Hochheiser. "Continuous Sampling and Ultramicrodetermination of Nitrogen Dioxide in Air," *Anal. Chem.* 30:426–428 (1958).
9. U.S. Environmental Protection Agency. "Notice of Proposed Rules," *Federal Register* 36:11826 (June 14, 1972).
10. Hauser, T. R., and C. M. Shy. "Position Paper: NO_x Measurement," *Environ. Sci. Technol.* 6(10):890–894 (1972).
11. U.S. Environmental Protection Agency. "Reference Method for Determination of Nitrogen Dioxide," Notice of Proposed Rules, *Federal Register* 38(110):15174–15180 (June 8, 1973).
12. Scaringelli, F. P., A. E. O'Keeffe, E. Rosenberg and J. P. Bell. "Preparation of Known Concentrations of Gases and Vapors with Permeation Devices Calibrated Gravimetrically," *Anal. Chem.* 42:871 (1970).
13. Purdue, L. J., J. E. Dudley, J. B. Clements and R. J. Thompson. "Studies of Air Sampling for Nitrogen Dioxide," *Environ. Sci. Technol.* 6:152 (1972).
14. U.S. Environmental Protection Agency. "An Evaluation of the Arsenite Procedure for Determination of NO_2 in Ambient Air," Environmental Monitoring Series, EPA-650/4-75-023 (June 1975).
15. U.S. Environmental Protection Agency. *Collaborative Testing of Methods for Measurements of NO_2 in Ambient Air*, Volume I, *Report of Testing*, Environmental Monitoring Series, EPA-650/4-74-019a (Washington, D.C.: U.S. Government Printing Office, 1974).
16. U.S. Environmental Protection Agency. "Evaluation of Effects of NO, CO_2, and Sampling Flow Rate on Arsenite Procedure for Measurement of NO_2 in Ambient Air," Environmental Monitoring Series, EPA-650/4-75-019 (1975).

17. U.S. Environmental Protection Agency. "An Evaluation of TGS-ANSA Procedure for Determination of NO_2 in Ambient Air," Environmental Monitoring Series, EPA-650/4-74-047 (1974).

18. U.S. Environmental Protection Agency. "Collaborative Test of the TGS-ANSA Method for Measurement of Nitrogen Dioxide in Ambient Air," Environmental Monitoring Series, EPA-650/4-74-046 (1974).

19. U.S. Environmental Protection Agency. "Evaluation of Triethanolamine Procedure for Determination of NO_2 in Ambient Air," Environmental Monitoring Series, EPA-650/4-74-031 (1974).

20. U.S. Environmental Protection Agency. "Evaluation of a Continuous Colorimetric Method for Measurement of NO_2 in Ambient Air," Environmental Monitoring Series, EPA-650/4-75-022 (1975).

21. U.S. Environmental Protection Agency. "Collaborative Test of the Continuous Colorimetric Method for Measurement of Nitrogen Dioxide in Ambient Air," Environmental Monitoring Series, EPA-650/4-75-011 (1975).

22. U.S. Environmental Protection Agency. "Evaluation of Gas Phase Titration Technique as Used for Calibration of NO_2 Chemiluminescence Analyzers," Environmental Monitoring Series, EPA-650/4-75-021 (1975).

23. U.S. Environmental Protection Agency. "Collaborative Test of the Continuous Chemiluminescence Method for Measurement of Nitrogen Dioxide in Ambient Air," Environmental Monitoring Series, EPA-650/4-75-013 (1975).

24. U.S. Environmental Protection Agency. "Comparison of Methods for Determination of NO_2 in Ambient Air," Environmental Monitoring Series, EPA-650/4-75-023 (1975).

25. Fontijn, A., A. J. Sabadell and R. J. Ronco. "Homogeneous Chemiluminescent Measurement of Nitric Oxide with Ozone," Anal. Chem. 41:575 (1970).

26. Stedman, D. J., E. E. Daby, F. Stuhl and H. Niki. "Analysis of Ozone and Nitric Oxide by a Chemiluminescent Method in Laboratory and Atmospheric Studies of Photochemical Smog," J. Air Poll. Control Assoc. 22:260 (1972).

27. Martin, B. E., J. A. Hodgeson and R. K. Stevens. "Detection of Nitric Oxide Chemiluminescence at Atmospheric Pressure," paper presented at 164th National ACS Meeting, New York City, August 1972.

28. Stevens, R. K., and J. A. Hodgeson. "Applications of Chemiluminescence Reactions to the Measurement of Air Pollutants," Anal. Chem. 45:443A (1973).

29. Winer, A. M., J. W. Peters, J. P. Smith and J. N. Pitts, Jr. "Response of Commercial Chemiluminescent $NO-NO_2$ Analyzers to Other Nitrogen-Containing Compounds," Environ. Sci. Technol. 8:1118 (1974).

30. U.S. Environmental Protection Agency. "Nitrogen Dioxide in the Atmosphere," Notice of Proposed Rules, Federal Register 41(53):11258–11266 (March 17, 1976).

31. U.S. Environmental Protection Agency. "Nitrogen Dioxide Measurement Principle and Calibration Procedure," Notice of Final Rulemaking, Federal Register 41(232):52686–52694 (December 1, 1976).

32. U.S. Environmental Protection Agency. "Ambient Air Monitoring Reference and Equivalent Methods," Federal Register 40(33):7042–7070 (February 18, 1975).

CHAPTER 5

ATMOSPHERIC INTERACTIONS OF NITROGEN OXIDES

James N. Pitts, Jr.
Statewide Air Pollution Research Center
and Department of Chemistry
University of California
Riverside, California 92521

INTRODUCTION

To establish the nature and magnitude of the health effects of nitrogen oxides (NO_x)—key factors in determining cost-effective strategies for their control—we need accurate dose–response functions for nitrogen dioxide (NO_2). Furthermore, we require reliable chemical, physical and biological information on a number of nitrogenous species formed as secondary pollutants in complex homogeneous and heterogeneous atmospheric interactions of NO_x. Some of these are already known to be toxic, mutagenic and/or carcinogenic in experimental animals. Others, as yet untested, may also present health hazards. Unfortunately, our data bases are too meager to make the kinds of reliable assessments necessary to protect public health in smog-impacted urban and suburban areas.

Congressional concern is evidenced in section 104(b) of the 1977 Amendments to the Clean Air Act, which states:

> Not later than six months after the date of the enactment of the Clean Air Act Amendments of 1977, the Administrator (of the EPA) shall revise and reissue criteria relating to concentrations of NO_2 over such a period (not more than three hours) as he deems appropriate. Such criteria shall include a discussion of nitric and nitrous acids, nitrites, nitrates, nitrosamines, and other carcinogenic and potentially carcinogenic derivatives of oxides of nitrogen.

Elucidation of the "dose" of these compounds to which individuals might be exposed in various types of air pollution episodes requires a detailed knowledge of their physical and chemical nature *at the point of impact on man.* Obtaining such information for "trace" nitrogenous species is particularly difficult for photochemical smog. Thus, not only are homogeneous gas-phase processes of NO_x involved, but also heterogeneous reactions are important. These include the formation of secondary nitrate aerosols and reactions on the surfaces of primary organic particulates. These chemical and physical transformations occur during transport of NO_x in the atmosphere for hours, days or even weeks, all in the presence of sunlight, oxygen, water and a spectrum of copollutants. Consequently, mixtures of gaseous and particulate pollutants characteristic of "real smog" are very complex; identification and determination of the "trace" nitrogenous species therein requires sophisticated instrumentation and analytical procedures.

For some years, research in our laboratory has dealt with this subject, both in homogeneous and heterogeneous systems. Recent results from our studies on atmospheric interactions of NO_x using a variety of environmental chambers and spectroscopic techniques for the detection of labile trace products are summarized in this chapter. Some environmental implications relevant to health effects are also discussed briefly.

SMOG CHAMBER STUDIES OF THE PRODUCTS OF REACTIONS OF NO_x WITH OLEFINS AND AROMATICS

Propylene-NO_x

The short scan times and large wave number range per scan of the interferometer makes long-path, Fourier transform-infrared (FT-IR) spectroscopy an ideal tool for identifying and obtaining time-concentration profiles of labile gaseous nitrogenous species formed in experiments on simulated photochemical air pollution [1]. For example, an Eocom interferometer, interfaced with an 85-m multiple-reflection cell in the SAPRC 5800-liter evacuable smog chamber (Figure 1) was used to obtain infrared spectra during an experiment in which a mixture of propylene, nitric oxide (NO) and NO_2 (at concentrations in the ppm range) in air at 9.4°C (48°F) was irradiated with a 24-kW solar simulator [2].

A high-resolution (0.125 cm^{-1}) spectrum from this run, shown in Figure 2, identifies and contains quantitative information about such species as N_2O_5, HNO_3, pernitric acid (HO_2NO_2), methyl nitrate (CH_3OHO_2) and PAN ($CH_3COO_2NO_2$). Also of importance when considering possible health effects of real photochemical oxidant, versus simply ozone in air in simulated atmospheres, are formic acid and formaldehyde, which are also products.

It was in similar FT-IR studies that the formation of pernitric acid was first discovered [3,4]. Formation of HO_2NO_2 by reaction of HO_2 with NO_2 in photochemical smog (as well as in the stratosphere) is plausible both thermodynamically and kinetically. However, while the rate of formation of HO_2NO_2 in such systems could be as large as that of peroxyacetyl nitrate (PAN), it is less stable thermally [5]. Thus, only low concentrations of

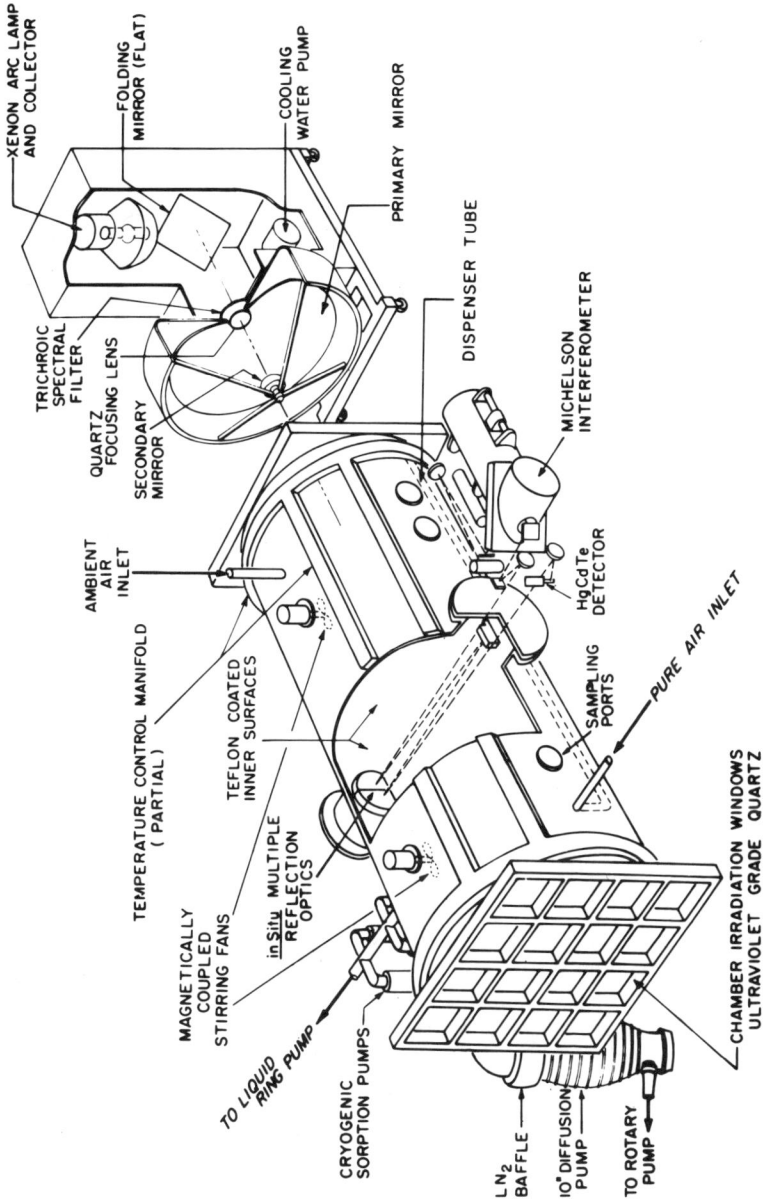

Figure 1. SAPRC 5800-liter evacuable chamber, 25-kW solar simulator facility and FT-IR spectrometer system.

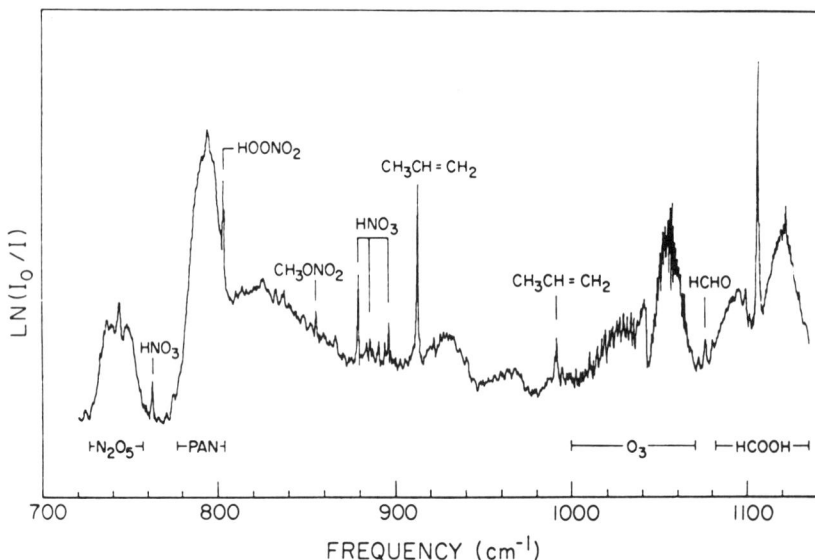

Figure 2. FT-IR spectrum (0.85-m pathlength; 0.125 cm^{-2} spectral resolution) of irradiated propylene–NO_x–air mixture at 48°F.

HO_2NO_2 (relative to PAN) should be expected under warm, ambient conditions. However, as the temperature is lowered, HO_2NO_2 becomes increasingly more stable. Thus, it may be present in the ppb range in photochemical smog on cool (e.g., 40–50°F), bright days. If so, its possible effects on plants and man warrant consideration.

Toluene-NO_x

Much attention has been paid to the role of toluene and other substituted benzenes in the production of ozone (O_3) in simulated photochemical smog; however, until recently, little was known about the products or mechanisms of formation of nitroaromatic compounds in this system. Recently it has been shown in several smog chamber studies [6-8] that irradiation of toluene–NO_x–air mixtures produced a variety of nitrogenous compounds. These include such species as nitrotoluene and benzylnitrate in the gas phase and both nitrocresols and hydroxynitrocresols in the particulate phase (Figure 3).

Little is known about either the atmospheric reactions or the health effects of these compounds. However, the possibility that some of these may be mutagenic and/or carcinogenic, or exhibit other forms of biological activity warrants investigation. In view of the increasingly high levels of

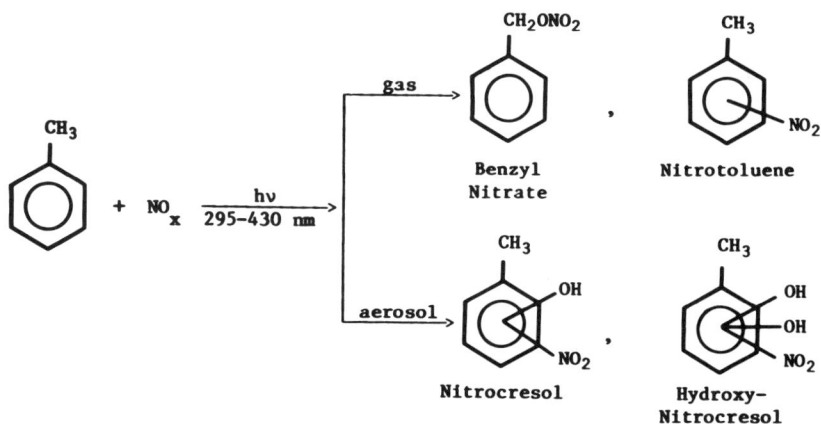

Figure 3. Aromatic products of toluene-NO_x photooxidations.

aromatics in gasoline and of NO_x in urban air basins, research on atmospheric transformations and related health effects in this system should have a high priority.

AMBIENT AIR STUDIES OF LABILE GASEOUS POLLUTANTS

A number of the species identified in smog chamber FT-IR spectroscopic studies of simulated atmospheres (e.g., Figure 2) have now been identified and measured in ambient photochemical smog. Thus, in collaboration with Dr. P. Hanst of the U.S. Environmental Protection Agency (EPA), who designed and furnished the original instrument, another long-path FT-IR facility (in addition to that associated with the evacuable chamber, vide supra) was established at SAPRC in the summer of 1976. The prime task was to employ this system to identify and measure trace species in the ppb-pphm concentration range in ambient atmospheres. The design and operation of this FT-IR facility and some of the initial results obtained with it are discussed in detail elsewhere [9].

In this system, a Digilab FT-IR spectrometer was interfaced with a multiple-reflection cell consisting of eight gold-coated mirrors with a 22.5-m base path (Figure 4). This cell has been routinely operated at total pathlengths of 1 km or greater, and detection limits of 2–10 ppb have been established for many species suspected or known to be present in photochemical smog.

The first spectroscopic detection of HNO_3 and $HCHO$ in ambient smog was achieved during the summer of 1976 and time–concentration profiles were obtained for these species as well as for formic acid, ammonia, ozone

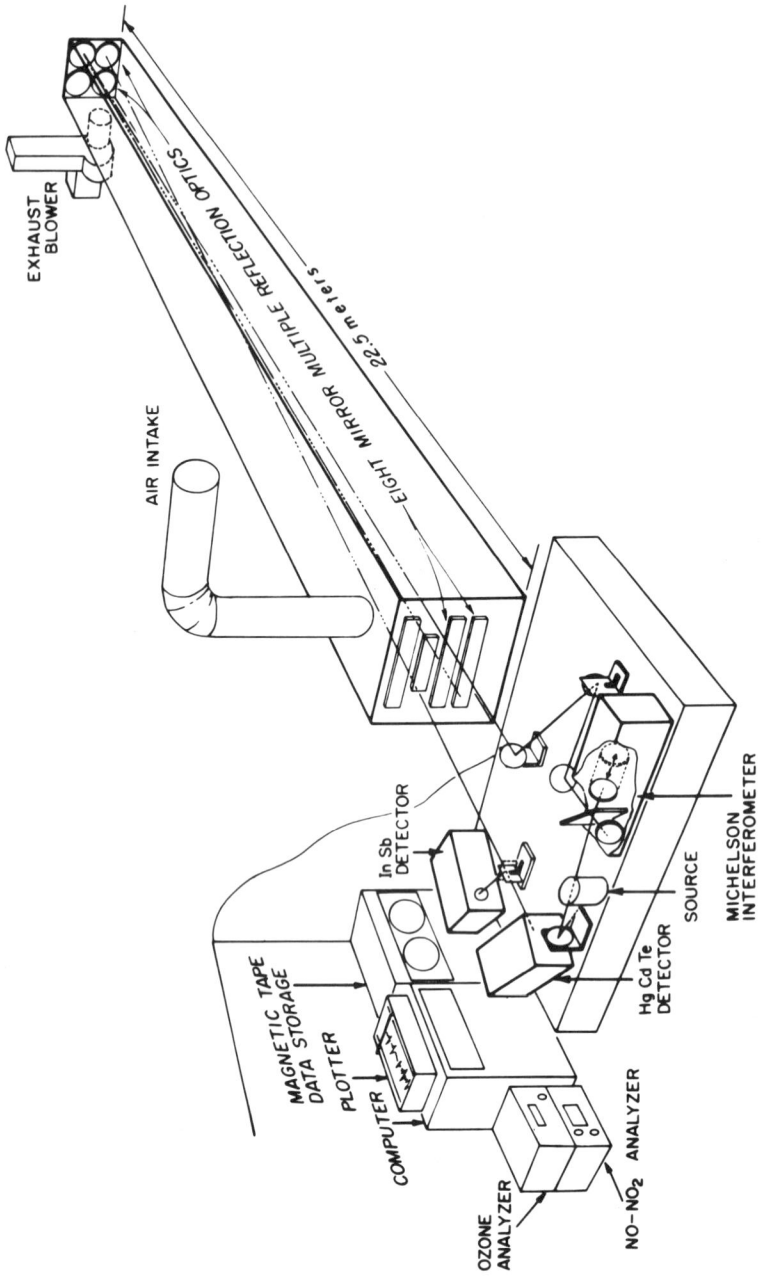

Figure 4. Kilometer pathlength multiple reflection infrared cell and FT-IR spectrometer.

and PAN [9]. Concentrations of a number of these trace species detected at 12:32 P.M. during a smog episode in Riverside, California, on August 12, 1977, as determined with this FT-IR facility operating at a pathlength of 900 m, are: HCOOH (10 ppb), HCHO (19 ppb), HNO_3 (16 ppb), NH_3 (15 ppb), PAN (7 ppb), and O_3 (181 ppb).

These studies were expanded in 1978. Thus, Figure 5 shows recent FT-IR spectra of nitric acid and formaldehyde which were measured by Dr. E. Tuazon concurrently with ozone, PAN, formic acid and ammonia at Claremont, California, on October 12 and 13, 1978. As seen in Figure 6, the pollutant levels were significantly higher than those experienced the previous year.

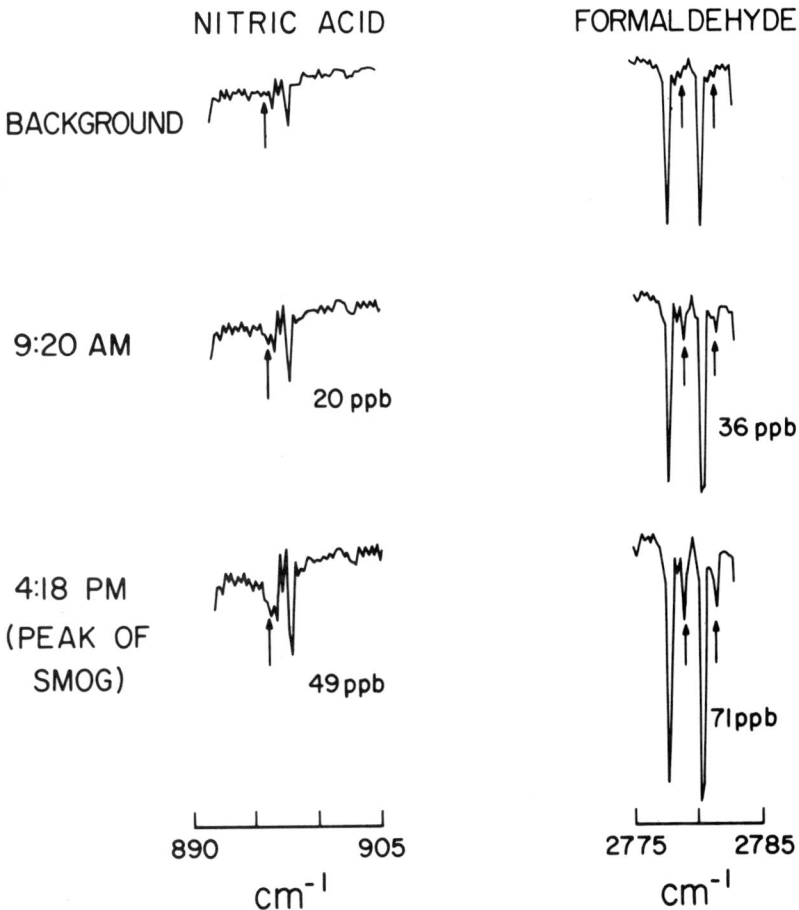

Figure 5. Infrared spectroscopic identification of nitric acid (HNO_3) and formaldehyde (HCHO) in photochemical smog, Claremont, CA, October 13, 1978.

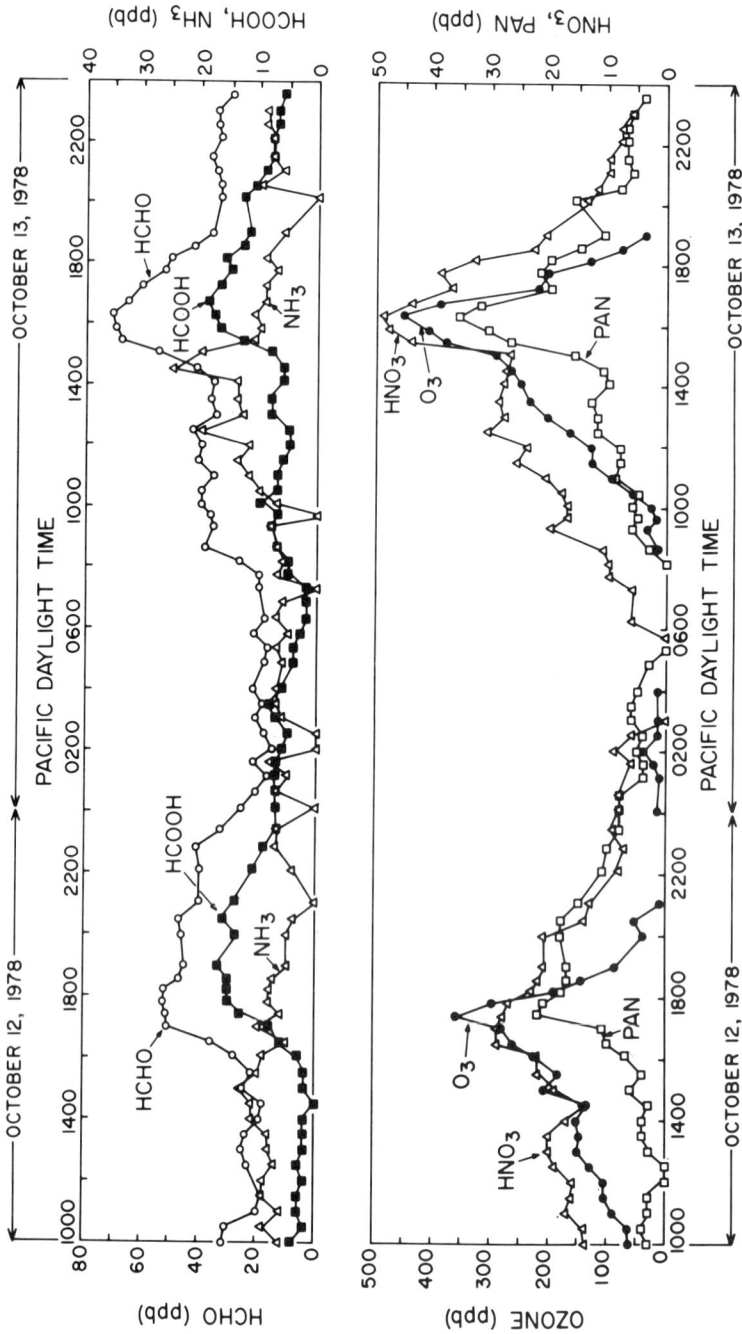

Figure 6. Concentrations of ozone and other toxic pollutants present in photochemical smog, i.e., photochemical oxidant, determined with 1-km pathlength, Fourier transform infrared spectroscopic system, Claremont, CA, October 1978.

Indeed one can see from these plots that on this particular occasion the aggregate of these real or potentially toxic species actually approached the levels of the ozone itself. For example, on the afternoon of October 13, when the ozone level was 0.12 ppm, the sum of formaldehyde, formic acid, nitric acid and PAN was 0.11 ppm.

Of particular interest in the Riverside study is that substantial levels of NH_3 and HNO_3 can coexist in ambient smog [9] as seen from the spectra taken at 12:32 P.M. (Figure 7). Clearly, much research lies ahead to understand better the obviously complex interactions in the heterogeneous system:

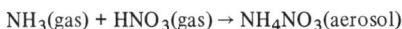

$$NH_3(gas) + HNO_3(gas) \rightarrow NH_4NO_3(aerosol)$$

Indeed, such an understanding may well be essential in determining the extent and causes of a so-called nitrate filter artifact that has been reported recently by several laboratories [10].

In this regard, during the Riverside study, measurements were also made of the concentrations of NH_4^+ and NO_3^- ions in 24-hr, high-volume particulate samples (washed and fired glass fiber filters). These were collected concurrently with the FT-IR analyses for gaseous NH_3 and HNO_3. Analysis of these data supports the hypothesis that particulate ammonium nitrate is in equilibrium with its precursors, NH_3 and HNO_3, and suggests that the ammonium

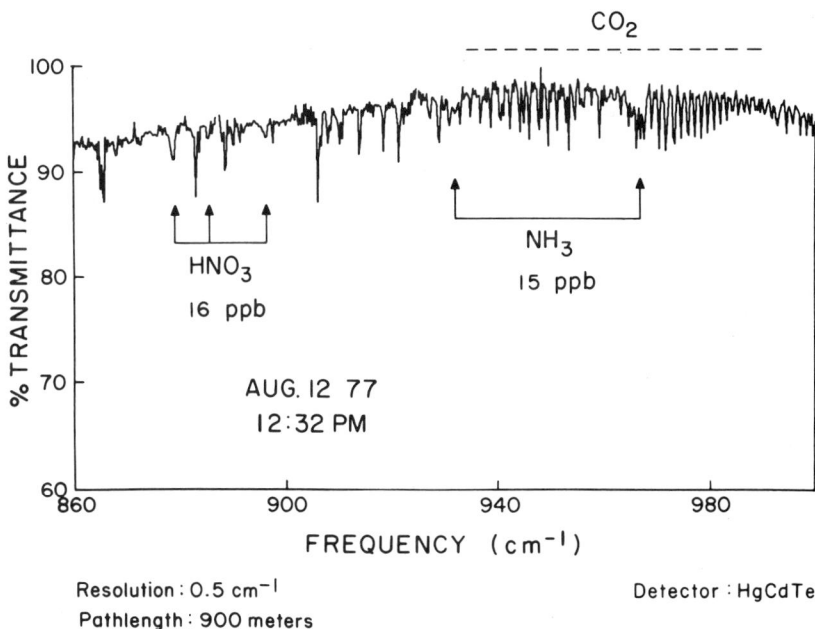

Figure 7. Riverside ambient air spectrum showing simultaneous presence of nitric acid and ammonia.

nitrate equilibrium constant derived from published thermochemical data gives a reasonable upper limit for the concentration product of gas-phase NH_3 ahd HNO_3 in the atmosphere [11].

FORMATION OF NITROSAMINES AND NITRAMINES FROM NO_x-AMINE INTERACTIONS IN SIMULATED ATMOSPHERES

We recently studied reactions of possible atmospheric *precursors* to diethylnitrosamine $(C_2H_5)_2NNO$, that is, mixtures of secondary or tertiary ethylamines with NO_x containing some HONO. Experiments were carried out both at sub-ppm levels in air in a 50-m^3 outdoor Teflon environmental chamber [12] and at ppm levels in air in the outdoor long-path (720 m) FT-IR facility described earlier [13].

In the presence of ambient levels of NO_x (300 ppb), both diethylamine (DEA) and triethylamine (TEA) readily form photochemical oxidants (e.g., O_3, PAN and aerosols). Additionally, small but significant amounts of $(C_2H_5)_2NNO$ are formed in the dark from DEA (~3% yield), but are destroyed in sunlight. In contrast, $(C_2H_5)_2NNO$ is formed on irradiation of TEA-NO_x mixtures containing some HONO, reaches a maximum (~1.8% yield), and then subsequently photodecomposes. Other gaseous photoproducts from both DEA-NO_x and TEA-NO_x mixtures include amounts of dialkylnitramines (R_2NNO_2) and a number of substituted amides.

The product spectrum (720 m) from irradiation of a DEA-NO_x-air mixture shows both $(C_2H_5)NNO_2$ and HONO (Figure 8). Additionally, $(CH_3)_2NNO_2$ was shown to be the "unknown major product" formed on irradiation of dimethylnitrosamine in air [14].

Finally, small amounts of acetamide were formed in the particulate phase from irradiation of both DEA-NO_x and TEA-NO_x mixtures. The health effects implications of these findings are discussed later in this chapter.

REACTION OF NO_x WITH POLYCYCLIC AROMATIC HYDROCARBONS

Discovery in these laboratories of the *direct* mutagenic activity [15], as determined by the Ames *Salmonella typhimurium* reversion microbiological assay [16] in the organic fractions of all ambient urban aerosols collected throughout southern California, led to our investigation of the reactions of benzo[a]pyrene (B[a]P) deposited on washed glass fiber filters with (a) ambient photochemical smog, and (b) with ppm levels and less of O_3 and NO_2 in simulated atmospheres.

A variety of derivatives of B[a]P were readily formed in both types of experiments. Furthermore, in contrast to B[a]P, which requires microsomal activation to produce mutagenic activity, *directly* mutagenic mononitroderivatives are readily formed on exposure of the B[a]P-coated filters to

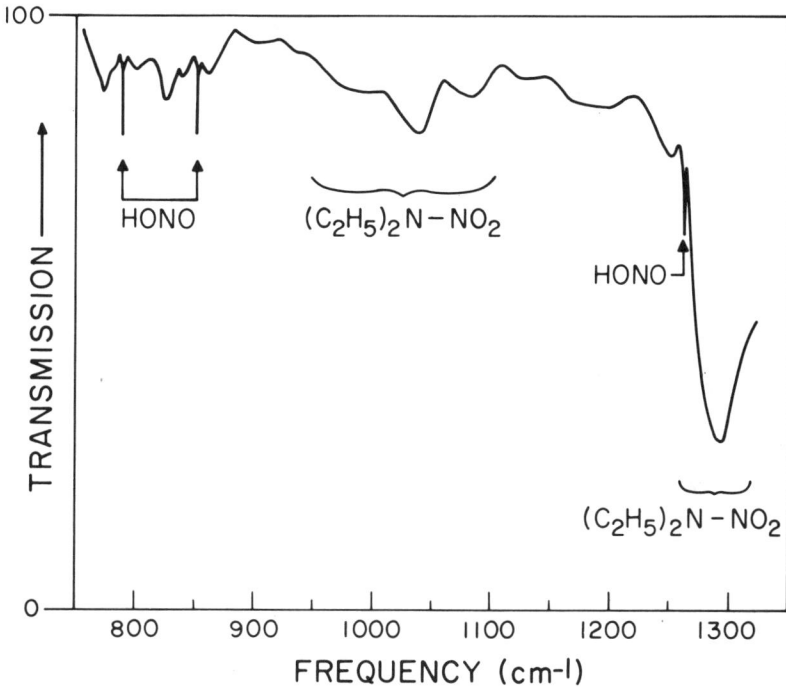

Figure 8. IR spectrum of diethylnitramine found on irradiation of a diethylamine-NO_x-HONO mixture in air, optical pathlength 720 m.

1 ppm of NO_2 (+ 10 ppb HNO_3) in air [17,18]. It was subsequently found that this nitration process occurred with only 0.25 ppm NO_2 and a trace of HNO_3 in air. The value of 0.25 ppm NO_2 for one hour is the California Air Quality Standard. It is frequently exceeded in the coastal and downtown regions of Los Angeles and its immediate environs (e.g., West Los Angeles and Pasadena). The absorption spectra of B[a]P, 6-nitro-B[a]P and a mixture of 1- and 3-isomers are shown in Figure 9.

It is also interesting that a directly mutagenic mononitroderivative, 3-nitroperylene, was formed when ppm levels of NO_2 (+ 10 ppb HNO_3) in air interacted with perylene similarly deposited on a glass fiber filter. This polycyclic aromatic hydrocarbon is an isomer of B[a]P, but a much weaker activable mutagen than B[a]P [19-22].

Such reactions with NO_2 and O_3 may, in part, account for the formation of the compounds (as yet unidentified) responsible for the "excess" carcinogenicity, i.e., over that which could be ascribed to other known carcinogenic polycyclic aromatic compounds actually measured in the samples, that has been observed in animals treated with organic extracts of the particulates collected from ambient smog. Indeed, we have shown recently that

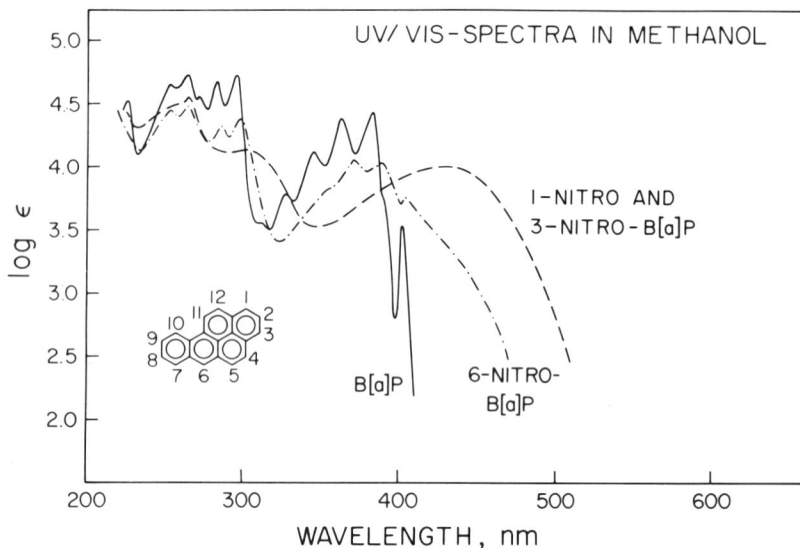

Figure 9. Ultraviolet-visible (UV/vis) spectra in methanol of 6-nitro-B[a]P, 1-nitro- and 3-nitro-B[a]P.

direct mutagens are readily formed in the B[a]P-O_3 experiments and that the half-life of B[a]P in these experiments is less than one hour, even at 0.1 ppm O_3 in air [23].

However, a complication is that reactions of B[a]P with NO_2 or O_3 may have occurred not only in the atmosphere, but also on the filters used to collect the ambient particulates. We are currently investigating in detail such "filter artifacts." This problem applies not only to our current experiments, but also to studies of the carcinogenicity of ambient particulate organic matter (POM) conducted in other laboratories over several decades [24,25].

ENVIRONMENTAL IMPLICATIONS

The health impact of photochemical air pollution results not just from NO_2 and O_3, but from the entire spectrum of gas-phase and particulate pollutants which, as we have seen, coexist in ambient smog with these major "criteria" pollutants. Thus, there is incontrovertible spectroscopic evidence for a host of potentially harmful compounds, including not only PAN, but species such as nitric acid, formic acid and formaldehyde. In addition, a number of other potentially toxic species have been identified in smog chamber experiments and may be present in ambient air (e.g., HO_2NO_2), especially under relatively cool conditions.

Although the presence of such compounds in ambient smog is now firmly established, little can be said today concerning their individual health effects, much less possible synergistic effects that may occur in a complex heterogeneous mixture of oxygenated and nitrogenous pollutants [26]. However, it should be recognized that the recent raising of the ozone/oxidant federal air quality standard from 0.08 to 0.12 ppm is tantamount to a de facto increase in the allowable levels of these potentially harmful pollutants [27] that coexist with O_3 in photochemical smog.

With respect to amine-NO_x interactions, both dimethylnitramine $[(CH_3)_2NNO_2]$ and acetamide $[CH_3CONH_2]$ are carcinogens in animals, although less potent than nitrosamines; the activity of diethylnitramine $[(C_2H_5)_2NNO_2]$ apparently has not been determined. Although not enough is known about actual concentrations of amines in ambient air, they are probably usually very low. Thus, the risk of forming significant amounts of nitrosamines or nitramines in urban air from their precursors seems correspondingly small. However, in situations in which sub-ppm concentrations of amines may be released (e.g., from certain industries) into urban atmospheres containing NO_x and HONO, within and immediately downwind from the facility, formation of significant amounts of nitrosamines in the dark— and nitramines and amides in sunlight—seems possible and should be kept in mind.

The ease with which B[a]P and perylene deposited on glass fiber filters react with ambient levels of NO_2 and O_3 has a number of implications. In preparing bioenvironmental impact statements or health effects criteria documents, investigators frequently concentrate on the emission inventories of *primary* particulate polycyclic organic matter (POM) and its associated PAH (such as B[a]P), serious errors in judgment can be made. As we have seen, in actuality many of the more reactive PAH in POM, such as the pentacyclic isomers B[a]P and perylene, may be substantially modified chemically, physically and biologically during (1) transport through polluted urban and suburban atmospheres; (2) sampling of the POM from the ambient air; (3) extraction of compounds from the POM; and (4) separation and analysis of the PAH and associated oxidized or nitrated species. Additionally, we should note that artifacts may well be generated in any or all of processes 2 through 4 when POM is collected directly from primary sources, e.g., the exhaust of diesel and spark ignition engines, fly ash from coal-fired power plants, etc.

In this context, consider the following scenario: It is known that in 1972, levels of B[a]P in ambient air in California's South Coast Air Basin were much lower than in most other major cities in the world [24]. One could conclude:

- that primary emissions from industry and motor vehicles were much less in the Los Angeles area; and
- since B[a]P levels were lower than in other urban areas, the risk of lung cancer that may be associated with B[a]P was correspondingly lower.

Actually, if the rates of reaction of B[a]P with such species as O_2 or NO_2 are as fast in photochemical smog as they are on the sampling filters, these processes could convert the B[a]P into species *not* detected by the analytical techniques commonly used in the 1950–1970 period (fluorescence). This could account for the low levels of B[a]P reported for Los Angeles, an observation made by the National Academy of Sciences Committee that investigated the POM question [24] and published its comprehensive report in 1972.

A second major point can be made about this scenario. A common assumption is that if the B[a]P levels are reduced, e.g., by a catalytic converter on a diesel engine operating in a mine, the health impact of the POM will necessarily be lessened, i.e., the POM will be safer to humans. This may or may not be the case, especially for an old and inefficient catalyst. Consider the possibility that as B[a]P reacts (and disappears from the system) new species are formed that are *more* mutagenic (or of a different type) and/or carcinogenic than the original B[a]P! If this is the case, the health impact may be worse than before treatment of the POM, either by an oxidizing catalyst or by oxidizing species in ambient smog. As we have seen, not only are the reactions of O_3 and NO_2 with B[a]P very fast on a conventional glass fiber filter, but they convert it from a promutagen into other compounds, several of which are direct mutagens. The possibility of such processes occurring in vivo with polycyclics deposited in the lungs should also be explored.

It is not yet known, nor will it be known until animal tests are conducted, whether certain products of nitrated or ozonized B[a]P are more or less carcinogenic (or carcinogenic at all) than B[a]P itself. At this point there is simply no information available on this important issue. All one can say is that, just because B[a]P levels are low in certain regions, one cannot conclude that the health impact of the POM is therefore necessarily decreased.

Indeed, if such atmospheric reactions of B[a]P and other carcinogenic PAH and hetero-PAH are significant in polluted urban atmospheres, and new, more potent mutagens and/or carcinogens are formed, the following anomaly reported by the epidemiologists Goldsmith and Friberg [28] and quoted in "Environmental Causes of Cancer" [29] may in part be explained:

> If urban pollution by benzo(a)pyrene makes an important contribution to the urban excess, lung cancer in the locations most polluted by this material should be highest, and when the agent decreases, lung cancer should do so as well. This has not been shown to occur.

In fact, our results suggest that there may be little reason to expect that such a correlation should exist in heavily polluted areas. In these regions, the B[a]P may be efficiently transformed into other compounds in the air or on the sampling filters. Thus, the ambient levels of B[a]P, as reported by air monitoring networks and utilized by epidemiologists and control officials estimating health effects, may be seriously misleading.

CONCLUSIONS

In light of the considerations we have discussed in this paper, which point to a variety of existing or newly discovered nitrogenous air pollutants that may pose real or potential health hazards, the following two quotes from the National Academy of Sciences document, "Nitrates: An Environmental Assessment" [26], seem somewhat puzzling and perhaps mutually inconsistent.

> Nitric acid vapor is an irritant, with a threshold limit value (TLV) for industrial exposure of 5 mg/m³. In one preliminary report, acute inhalation exposure to nitric acid mist at approximately seven times the TLV exposure produced osteosarcomas in four of 58 rats. Further investigation of this possible link between nitric acid and cancer is desirable. In general, the data base is both too limited and too ambiguous (because nitrates are almost always present with many other pollutants) to provide any definite indication of the health hazards of ambient levels of atmospheric nitrates.

> In the absence of more positive evidence of a risk to health, other effects of atmospheric nitrates, such as corrosion of materials and contribution to the acidity of precipitation, provide the primary impetus for controls and emissions of nitrogen oxides.

Clearly, a closer working relationship between atmospheric chemists and biological scientists working on health effects of air pollutants, as well as those responsible for control decisions, would be productive.

ACKNOWLEDGMENTS

The author is deeply indebted to the SAPRC-Chemistry-Biology team that participated in the research efforts discussed in this chapter. This work was funded by the National Science Foundation (RANN Grant No. ENV73-02904 and ASRA Grant No. ENV78-01004, Dr. R. D. Carrigan, project officer) and the U.S. Environmental Protection Agency (Grant No. EPA R804546-3, Dr. P. L. Hanst, project officer). We greatly appreciate their scientific and monetary support. The opinions in this chapter are those of the author and do not necessarily reflect the views and/or policies of the NSF, EPA or the University of California.

REFERENCES

1. Pitts, J. N., Jr., B. J. Finlayson-Pitts and A. M. Winer. "Optical Systems Unravel Smog Chemistry," *Environ. Sci. Technol.* 11(6):568–573 (1977).
2. Winer, A. M., R. A. Graham, G. J. Doyle and J. N. Pitts, Jr. "An Evacuable Environmental Chamber and Solar Simulator Facility for the Study of Atmospheric Photochemistry," *Adv. Environ. Sci. Technol.* (in press).
3. Niki, H., P. Maker, C. Savage and L. Breitenbach. "Fourier Transform IR Spectroscopic Observation of Pernitric Acid Formed via $HOO + NO_2 \rightarrow HOONO_2$," *Chem. Phys. Lett.* 45(3):564–566 (1977).
4. Hanst, P., and B. Gay. "Photochemical Reactions Among Formaldehyde, Chlorine and Nitrogen Dioxide in Air," 11(12):1105–1109 (1977).
5. Graham, R. A., A. M. Winer and J. N. Pitts, Jr. "Pressure and Temperature Dependence of the Unimolecular Decomposition of HO_2NO_2," *J. Chem. Phys.* 68:4505–4510 (1978).

6. O'Brien, J., P. Green and R. Dotz, "Interactions of Oxides of Nitrogen with Aromatic Hydrocarbons in Polluted Air," paper presented at American Chemical Society Meeting, Division of Environmental Chemistry, Anaheim, CA, March 13-17, 1978.

7. Grosjean, D., K. Van Cauwenberghe, D. Fitz and J. N. Pitts, Jr. "Photo-oxidation Products of Toluene-NO_x Mixtures Under Simulated Atmospheric Conditions," paper presented at the American Chemical Society Meeting, Division of Environmental Chemistry, Anaheim, CA, March 13-17, 1978.

8. Hoshino, M., H. Akimoto and M. Okuda, "Photochemical Oxidation of Benzene, Toluene and Ethylbenzene Initiated by OH Radicals in Gas-Phase," *Bull. Chem. Soc. Japan* 51:718-724 (1978).

9. Tuazon, E. C., R. A. Graham, A. M. Winer, R. R. Easton, J. N. Pitts, Jr. and P. L. Hanst. "A Kilometer Pathlength Fourier-Transform Infrared System for the Study of Trace Pollutants in Ambient and Synthetic Atmospheres," *Atmos. Environ.* 12:865-875 (1978).

10. *EPA Workshop on the Measurements of Atmospheric Nitrates*, Southern Pines, NC; and Attachment No. 1, *Handbook of Atmospheric Nitrates*, R. Stevens, Ed. (1978).

11. Doyle, G. J., E. C. Tuazon, T. M. Mischke, R. A. Graham, A. M. Winer and J. N. Pitts, Jr. "Simultaneous Concentrations of Ammonia and Nitric Acid in a Polluted Atmosphere and Their Equilibrium Relationship to Particulate Ammonium Nitrate," *Environ. Sci. Technol.*, in press.

12. Pitts, J. N., Jr., D. Grosjean, K. Van Cauwenberghe, J. P. Schmid and D. R. 'Fitz, "Photooxidation of Aliphatic Amines Under Simulated Atmospheric Conditions: Formation of Nitrosamines, Nitramines, Amides and Photochemical Oxidant," *Environ. Sci. Technol.* 12:946-953 (1978).

13. Tuazon, E. C., A. M. Winer, R. A. Graham, J. P. Schmid and J. N. Pitts, Jr. "Fourier Transform Infrared Detection of Nitramines in Irradiated Amine-NO_x Systems," *Environ. Sci. Technol.* 12:954-958 (1978).

14. Hanst, P., J. Spence and M. Miller. "Atmospheric Chemistry of N-Nitroso Dimethylamine," *Environ. Sci. Technol.* 11(4):403-405 (1977).

15. Pitts, J. N., Jr., D. Grosjean, T. M. Mischke, V. F. Simmon and D. Poole. "Mutagenic Activity of Airborne Particulate Organic Pollutants," *Toxicol. Lett.* 1:65-70 (1977).

16. Ames, B. N., J. McCann and E. Yamasaki. "Methods for Detecting Carcinogens and Mutagens with the *Salmonella*/Mammalian Microsome Mutagenicity Test," *Mutation Res.* 31:347-364 (1975).

17. Pitts, J. N., Jr., K. Van Cauwenberghe, D. Grosjean, J. Schmid, D. Fitz, W. Belser, Jr., G. B. Knudsen and P. M. Hynds. "Atmospheric Reactions of Polycyclic Aromatic Hydrocarbons: Facile Formation of Mutagenic Nitro Derivatives," *Science* 202:515-519 (1978).

18. Pitts, J. N., Jr. "Photochemical and Biological Implications of the Atmospheric Reactions of Amines and Benzo(a)pyrene," *Phil. Trans. Roy. Soc. London, A* 200:551-576 (1979).

19. Kaden, D. A., and W. G. Thilly. "Genetic Toxicology of Kerosene Soot," in *Proc. Workshop on Unregulated Diesel Emissions and their Potential Health Effects*, U.S. Department of Transportation, Washington, D.C. (1978).

20. Eisenstadt, E. Personal communication (1979).

21. Dickson, J. Personal communication (1979).

22. Belser, W. L., Jr., S. Shaffer, P. Hynds and J. N. Pitts, Jr. Unpublished results (1979).
23. Van Cauwenberghe, K., L. Van Vaeck and J. N. Pitts, Jr. Unpublished results (1979).
24. *Particulate Polycyclic Organic Matter* (Washington, D.C.: Printing and Publishing Office, National Academy of Sciences, 1977).
25. U.S. Environmental Protection Agency. "Health Assessment Document for Polycyclic Organic Matter," Washington, D.C. (1979).
26. *Nitrates: An Environmental Assessment* (Washington, D.C.: Printing and Publishing Office, National Academy of Sciences, 1978).
27. Pitts, J. N., Jr. Testimony presented to the Subcommittee on Natural Resources and the Environment of the Committee on Science and Technology of the U.S. House of Representatives, February 15, 1979.
28. Goldsmith, J. R., and L. T. Friberg. In: *Air Pollution*, Vol. II, A. Stern, Ed. (New York: Academic Press, Inc., 1977), p. 458.
29. Eisenbud, M. "Environmental Causes of Cancer," *Environment* 20:6–16 (1978).

CHAPTER 6

NITROGEN DIOXIDE MONITORING IN JAPAN

Michio Nakano
    Environment and Public Health Bureau
    City of Osaka
    Osaka, 541 Japan

INTRODUCTION

Air pollution in Japan has become increasingly serious because of increased industrialization and economic growth since 1960. Especially up until 1967, the pollution problem was intensifying because of increased consumption of fossil fuel.

Since 1967, however, the sulfur dioxide (SO_2) concentration levels have been decreased by strict legislation. As a result, the National Ambient Air Quality Standards for SO_2 (daily average concentration $\leqslant 0.04$ ppm) has been attained now in most areas, excluding some major cities.

In the summer of 1970 high school students in the outskirts of Tokyo were injured by photochemical smog. This episode brought national attention to the necessity for an immediate solution to pollution resulting from oxides of nitrogen.

In May 1973 the Environment Agency determined the National Ambient Air Quality Standard for nitrogen dioxide (NO_2) be set at a daily average concentration of 0.02 ppm. At the time, this was the strictest standard in the world. But, in 1978, according to the recommendation of the Central Advisory Committee on Environmental Pollution Control, the standard was readjusted to a daily average from 0.04 to 0.06 ppm.

The monitoring network for nitric oxide (NO), NO_2 and photochemical oxidants has been improved throughout the country. Local governments were required to monitor ambient levels of NO and NO_2 based on the National Air Pollution Control Law.

AIR MONITORING NETWORK FOR NITRIC OXIDE
AND NITROGEN DIOXIDE

Continuous monitoring for NO and NO_2 in the ambient air was first established in two major cities, Tokyo and Osaka, and was later followed by the other cities as shown in Table I.

As of 1977, there were 968 automatic monitoring stations functioning in 470 cities.

In Japan, measurements of ambient NO and NO_2 are conducted using automatic analyzers based on the modified Saltzman wet chemical method.

Selection of the sites for an air monitoring station is based on population density, industrial development and weather factors. Local governments select station locations to be representative of the air quality for an area in which people either work or live. Meteorological factors play a major role in determining the size of the area a station represents as well as the location of the station.

It is required by the Air Pollution Control Law and Cabinet Order of Environment Agency that the air monitoring network be established by the local governments as follows:

1. The monitoring network must be developed in habitable areas. For air monitoring mainly from stationary sources, the number of stations required is obtained by dividing the habitable area by 25 km^2 where the population density is more than 2600/km^2. If necessary more monitoring stations are expected to be installed in heavily polluted areas. Where the population density is no more than 2600/km^2 the value is determined by multiplying the value obtained above by the ratio of the population density in a given area to the average population density of habitable areas in Japan, which is 2600/km^2.

2. To monitor air polluted mainly by motor vehicle exhaust, monitoring is conducted adjacent to roads with heavy traffic volume or at intersections where residences or human activities are found (Table II).

3. Ambient air samples are taken at a height of 1.5–10 m above the ground, which is considered an average sample of ambient air that inhabitants breathe; however, in the case of tall buildings, sites above the 10 m level can be chosen.

Table I. Number of Ambient Air Monitoring Stations in Japan

Fiscal year[a]	1968	1969	1970	1971	1972	1973	1974	1975	1976	1977
Cities	8	12	13	44	112	192	303	385	417	470
Stations	8	17	20	68	176	329	582	772	859	968

[a]Fiscal year begins in April and ends in March.

Table II. Increase in the Number of Cities and Monitoring Stations
Primarily Testing Motor Vehicle Exhaust from 1971 to 1977

Fiscal year	1971	1972	1973	1974	1975	1976	1977
Cities	15	42	65	89	97	103	113
Stations	33	75	122	164	182	194	210

4. The air quality standard for NO_2 is based on the daily average concentration. If more than 4 hours of data per day were lacking it was considered invalid and discarded. Examination of long-term data (over a year) is necessary for consideration of pollution control legislation in any area concerned.

5. A modification of World Health Organization data estimation method is employed in Japan (WHO Health Criteria for Oxides of Nitrogen) [1] because of inaccuracies in the existing measuring methods, the inadequate measurement duration and daily fluctuations in the monitoring conditions.

The annual data is processed excluding 2% of the highest values among daily average (data for 7 days out of 365 days).

GENERAL ASPECTS OF NITROGEN DIOXIDE
CONCENTRATION IN JAPAN

Figure 1 shows the annual mean concentration of ambient NO_2 from six monitoring stations in Japan. Figure 2 depicts NO_2 concentrations in four

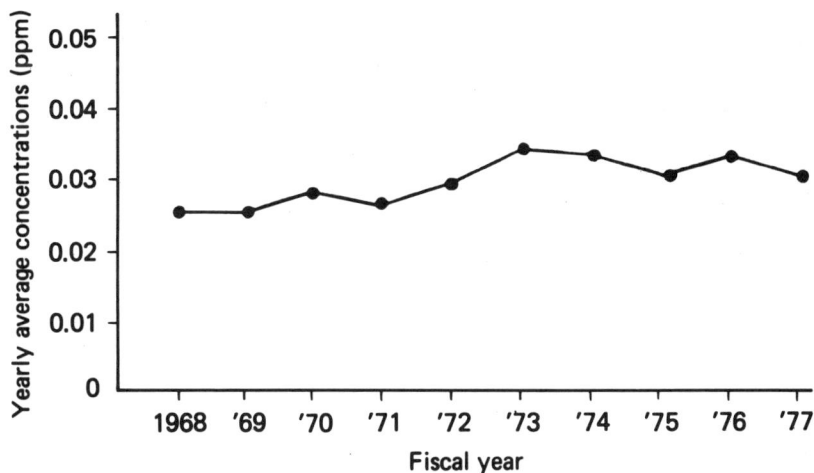

Figure 1. Trends of NO_2 concentration in the ambient air (yearly average from 6 monitoring stations).

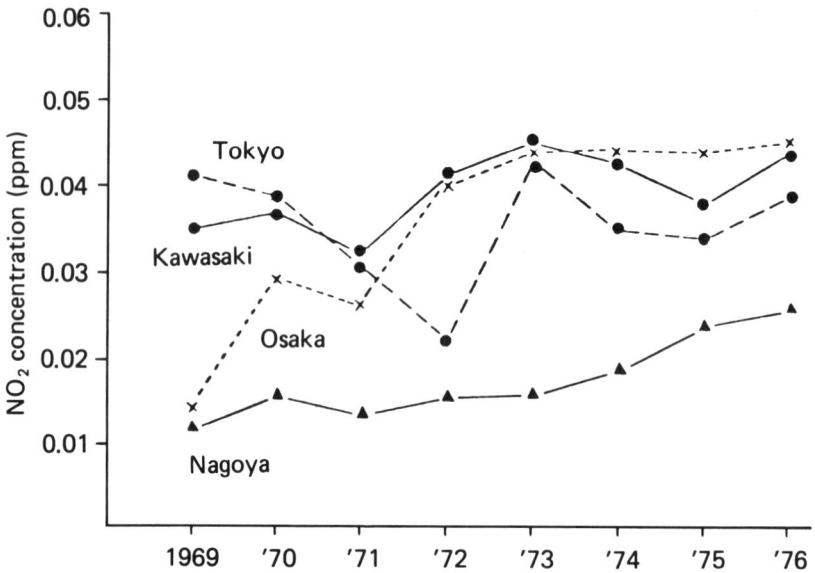

Figure 2. Annual change in NO₂ concentration in four major cities.

major cities. Figure 3 shows the annual mean NO_2 concentrations in the motor vehicle exhaust monitoring stations, and as expected this is somewhat higher than the average ambient NO_2 concentrations.

The NO_2 concentrations had been, on the whole, increasing before 1973 and then reached a plateau. This trend proved true among other cities in Japan.

Table III shows levels of NO_2 for monitoring stations in each level throughout the country in FY 1976. Analysis of these data indicates that about 70% of the monitoring stations which exceeded the level of the National Air Quality Standard are located in the large cities. The remaining monitoring stations which exceeded the level of National Air Quality Standard are located near the heavy traffic roads.

Analysis of 1976 data indicated that the ten highest concentrations (annual mean concentrations ranged from 0.047 to 0.060 ppm) were observed by stations which were located in large cities.

NO₂ MONITORING AND THE AMBIENT AIR QUALITY IN OSAKA

Since basic design and operations of monitoring stations are the same regardless of location or city, Osaka will be used to exemplify the quality of the monitoring system in Japan.

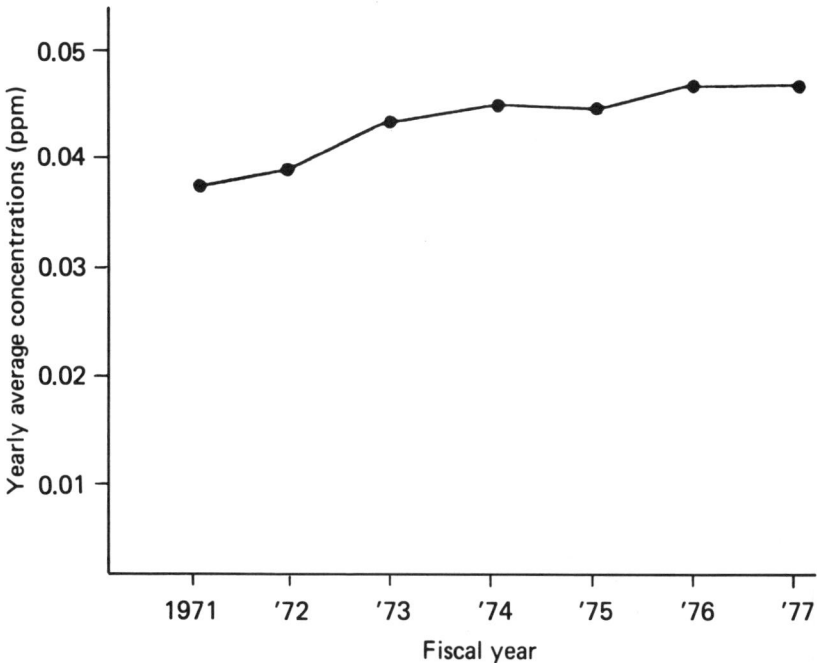

Figure 3. Trends of NO_2 concentrations near major roads (yearly average for 26 monitoring stations).

Table III. NO_2 Levels and Number of Monitoring Stations in 1976

NO_2 (pphm)	0.0–1.0	1.1–2.0	2.1–3.0	3.1–4.0	4.1–5.0	5.0
Stations	136 (17.0%)	301 (37.7%)	231 (28.9%)	101 (12.7%)	25 (3.2%)	4 (0.5%)

Osaka City, which faces Osaka Bay to the west, is located west of the Osaka Plain (an area of 1849 km^2), and is surrounded by mountains on three sides: north, east and south. The city itself covers an area of 208 km^2 and has a population of 2.78 million, plus 1.21 million daytime commuters.

There are 12 major monitoring stations in Osaka for ambient air quality and 15 stations for motor vehicle exhaust (Figure 4). These stations continuously provide data on air pollutant concentrations (Figure 5).

In Japan, NO_2 measurement is performed using automated systems based on modified Saltzman's method. Static calibration using sodium nitrite solution is employed, and the Saltzman factor has been determined to be 0.72.

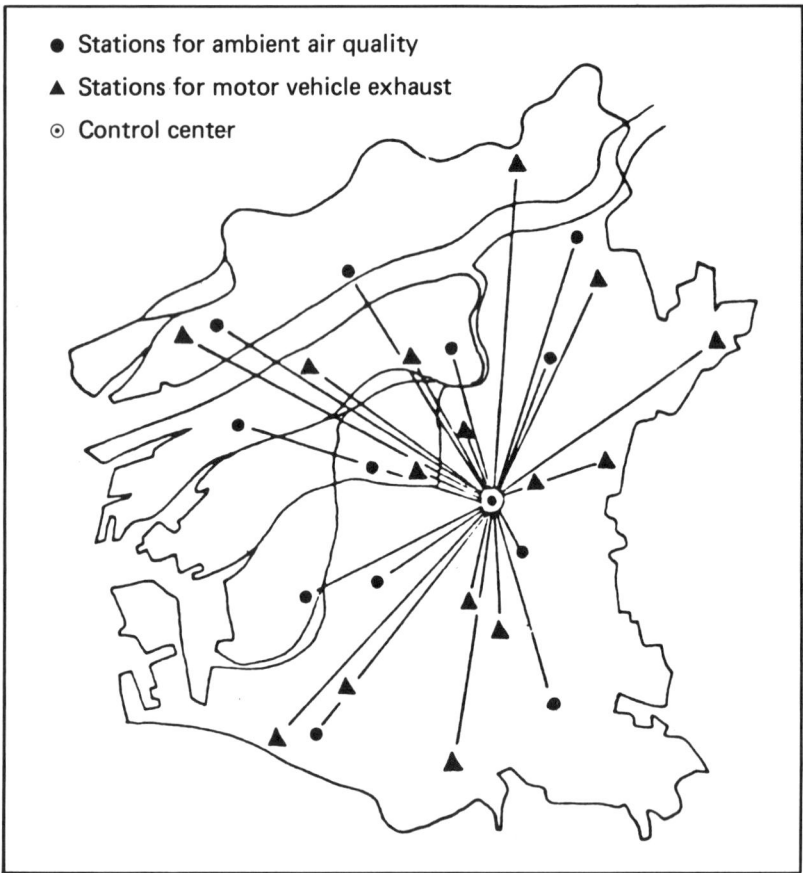

Figure 4. Map of monitoring network in Osaka City.

The variables in the Saltzman method for NO_x measurement have been, to some extent, made clear by numerous reports. They may be generally classified into two principal factors: (1) errors caused by the measuring principle and (2) errors due to mechanical malfunction peculiar to the analyzer itself. According to the original report, Saltzman reagent required a maximum color developing time of ten minutes depending on the type of cell used [2,3]. The Saltzman factor varies up to 1.0 depending on the type of bubbler.

Recent reexaminations of the Saltzman factor, using standard gas from a standard gas generator on the various types of existing automatic analyzers in Japan [4] resulted in a factor of 0.84 ± 0.03 over the range 0.02–0.35 ppm of NO_2, independent of NO_2 concentrations. Based on this result the previous

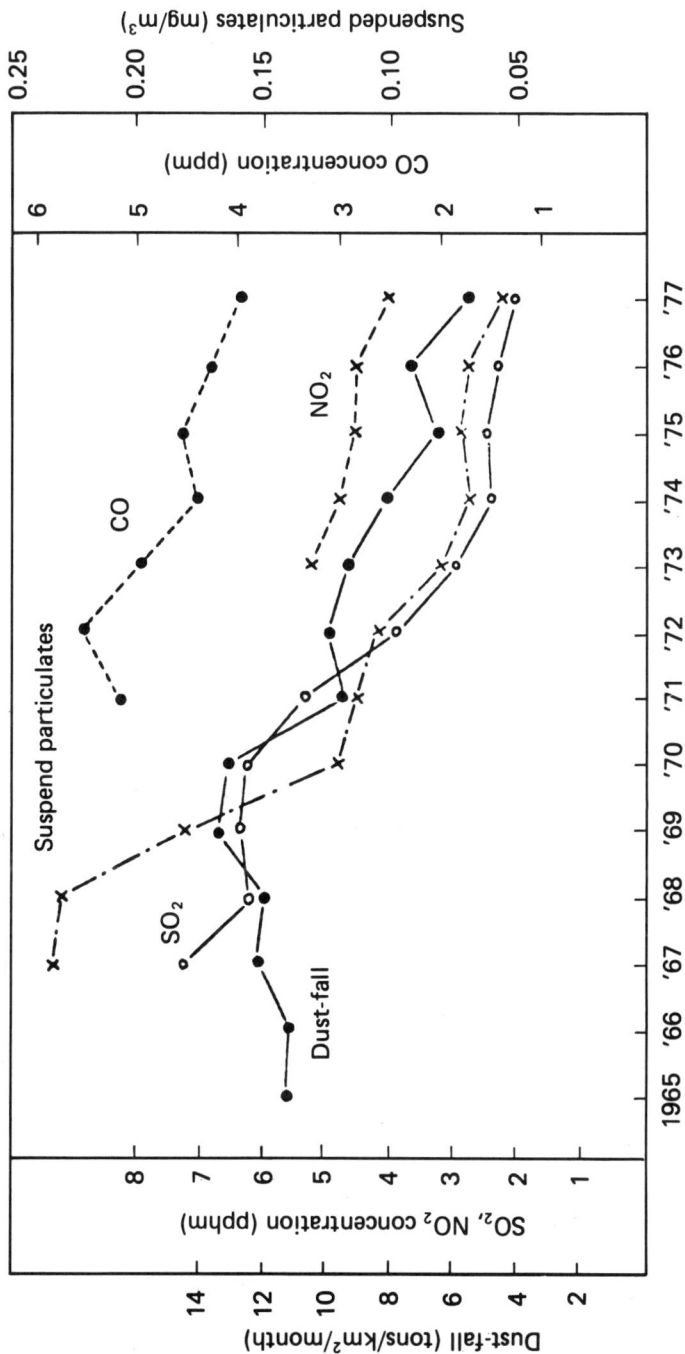

Figure 5. Selected air pollutant concentration patterns.

value of 0.72 as the Saltzman factor was replaced by the new value of 0.84 in July 1978 by the Environment Agency in Japan.

An absorption rate of more than approximately 95% of NO_2 can be expected for the automatic analyzers used in Japan. In the future, dynamic calibration procedures will be adopted in order to enable more accurate measurement.

DISTRIBUTION PATTERN OF NO_2 CONCENTRATIONS IN OSAKA

The concentration distribution as shown in Figure 6 was obtained by measurements at approximately 100 sites in Osaka City by the Triethanolamine Method [5]. It is useful in determining the general pattern of NO_2 concentrations. The data gathered by this method together with the data obtained by the use of the Saltzman method, were used in evaluating NO_2 concentration pattern in Osaka City.

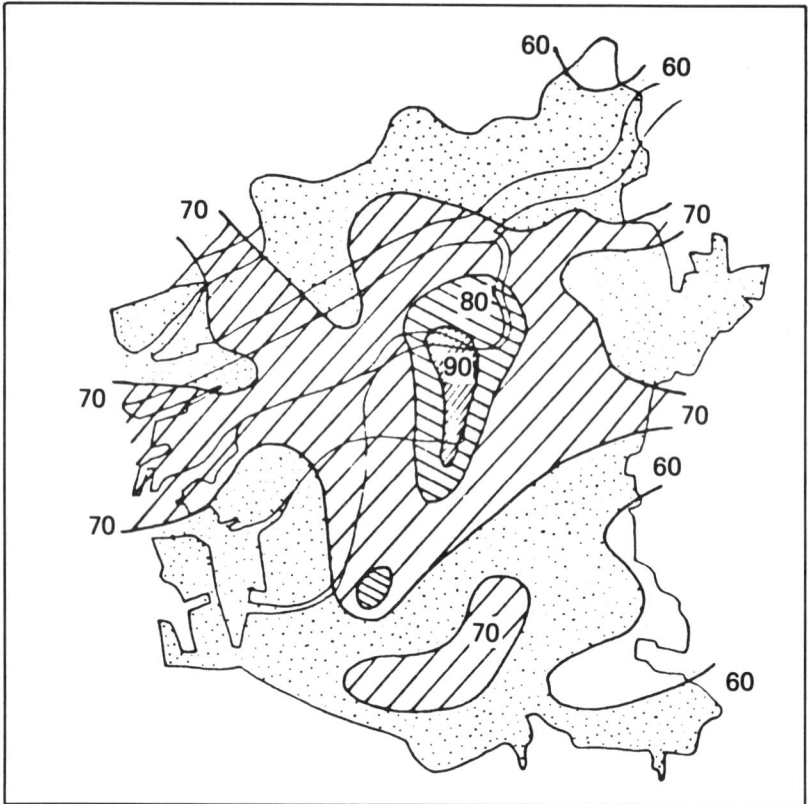

Figure 6. NO_2 distribution in Osaka City (μg/100 cm³/day) for fiscal year 1977.

PERFORMANCE EVALUATION OF THE MODIFIED SALTZMAN ANALYZER

Performance evaluation of modified Saltzman analyzer conducted by Japan Environment and Sanitary Center [4] is as follows:

1. The maximum error for absorber was less than ±4%.
2. The error of sampling air volume being less than ±3.3%.
3. The variable range of the static span calibration was less than 4% after one or two months and the fluctuation factor was within 1–4%.
4. The Saltzman factor was 0.855 ± 0.023; after adjustment of errors resulting from sampling air volume and others, the average Saltzman factor was 0.835 ± 0.0125.
5. The oxidation rate of NO was 0.731 ± 0.026.
6. The reduction rate of NO_2 was 0.077 ± 0.11.

STANDARDIZATION OF AMBIENT MONITORING DATA PROCESSING

To utilize ambient monitoring data for pollution control standardization of data processing [6] is important.

First, in the data gathering process (measurement, examination, tabulation), the standardization of the monitored data itself is necessary. Second, the specification for the magnetic tape for data-storage must be standardized. Due to lack of standardized procedure for data processing and storage, at the present time, the utility of data is rather limited.

Table IV. Annual Monitoring Data

Prefecture	
City	
Name of Monitoring Station	
Category of Area by Land Use	
Effective Monitoring Days	
Monitoring Hours	
Yearly Average Concentration	
Days When Air Quality Standard Was Exceeded	
Maximum Hourly Concentration	

Table V. Monthly Monitoring Data

City	Name of Monitoring Station	Items	1978 Fiscal Year											
			Apr.	May	Jun.	Jul.	Aug.	Sep.	Oct.	Nov.	Dec.	Jan.	Feb.	Mar.
		Effective monitoring days												
		Effective monitoring hours												
		Monthly average concentration (ppm)												
		Hours and days when air quality standard was exceeded												
		One-hour maxima concentrations (ppm)												
		Maximum of daily averages (ppm)												

STANDARDIZATION IN DATA GATHERING AND PROCESSING

A tentative guideline has been proposed by the Environment Agency for a standardized data gathering process, however, it is unsatisfactory, at this time. Therefore, the agency has been making an effort to complete a manual for operation and maintenance of continuous air pollution monitoring systems. The methods of air pollutants measurement are determined by the Japan Industrial Standards.

In order to determine the quality of monitoring data, errors are corrected by comparison with data recorded on sites against the data transmitted by telemetry. However, no standard procedure has yet been established for quality assurance.

Tables IV and V show the data tabulation formats mandated by the Environmental Agency for annual and monthly reports, respectively. The data from all the local governments are made public in the (Environment) Agency's annual report, "The Air Pollution of Japan."

REFERENCES

1. WHO. "Environmental Health Criteria for Oxide of Nitrogen," *Environmental Health Criteria* 4, World Health Organization, Geneva (1977).
2. Saltzman, B. E. "Colorimetric Microdetermination of NO_2 in the Atmosphere," *Anal. Chem.* 32:135 (1960).
3. Thomas, M. S. et al. "Automatic Apparatus for Determination of Nitric Oxide and Nitrogen Dioxide in the Atmosphere," *Anal. Chem.* 28:1810 (1956).
4. Nakano, K., N. Aihara, A. Ibusuki, M. Mori, S. Omichi, Y. Suzuki and H. Yamakawa. "Report of the Examination of the Fundamental Performance of the Saltzman Analyzer," Japan Environment and Sanitary Center, Environment Agency (1978).
5. U.S. EPA. "Summary of Technical Evaluations and Rationale for Selection of the Chemiluminescence Measurement Principle to Replace the Existing NO_2 Reference Method," (January 15, 1976).
6. Matsumoto, Y., M. Fujita, T. Oshima and J. Himeno. "Standardization of Ambient Monitoring Data Processing," *J. Environ. Poll. Control* 15(3): 69-80 (1979).

CHAPTER 7

REVIEW OF U.S. ENVIRONMENTAL PROTECTION AGENCY NO$_2$ MONITORING QUALITY ASSURANCE PROGRAM

John B. Clements
Chief, Quality Assurance Branch

Thomas R. Hauser
Director
Environmental Monitoring and Support Laboratory
U.S. Environmental Protection Agency
Research Triangle Park, North Carolina 27711

INTRODUCTION

The U.S. Environmental Protection Agency (EPA) has an active program to upgrade the quality of air pollution monitoring data collected in the United States. Regulated pollutants of the ambient air and in source emissions receive the major emphasis of the program. This quality assurance program for air pollution measurements is divided into two major subprograms. The first subprogram evaluates the methodology used for measuring regulated pollutants to determine if it can make the required measurements. A part of this subprogram corrects defects in measurement methodology whenever they are identified. The subprogram on methodology evaluation is discussed in Chapter 4 and elsewhere [1].

The other part of the program evaluates the performance of control agencies required to monitor for air pollutants in support of regulations and of monitoring projects that make air pollution measurements important to EPA for various reasons. Performance evaluation is determined using samples of materials that contain the appropriate pollutants in known amounts at several concentration levels. These samples are submitted in a blind fashion to the organization under evaluation, which then analyzes the samples; their results are compared with the true values for the samples. The term performance audit is used to describe this activity. A discussion of the EPA program for evaluating the performance of a variety of organizations that measure nitrogen dioxide (NO_2) follows.

AUDIT SAMPLES

Nitrogen dioxide, as is the case with a number of gaseous pollutants, can be measured on a time-integrated basis or as a continuous measurement. For each type of measurement, the most satisfactory audit sample would be an atmosphere containing known amounts of gaseous NO_2, whose concentration could be varied to cover the concentration range most likely to be encountered. For analyzers that measure NO_2 continuously, atmospheres of NO_2 are the only feasible audit samples. Time-integrated measurement methods generally collect ambient NO_2 in an absorbing reagent, and appropriate solutions can be used as audit samples.

NO_2 Atmosphere Audit Samples

Preparing and using an atmosphere containing subpart per million levels of NO_2 presents special problems. Compressed gaseous mixtures in cylinders have been used successfully, but the slow change in NO_2 concentration in these mixtures requires frequent analysis to be certain of the NO_2 concentration at the time of audit. Another method to obtain audit atmospheres of NO_2 takes advantage of the rapid and quantitative gas-phase reaction of nitric oxide (NO) with ozone (O_3) to produce NO_2, as shown in Equation 1 [2]:

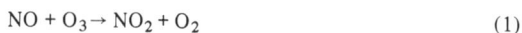

$$NO + O_3 \rightarrow NO_2 + O_2 \qquad (1)$$

In practice, atmospheres containing known amounts of NO_2 are continuously generated in a dynamic system, in which a gas stream containing a known concentration of NO is mixed with O_3 to give an atmosphere containing known concentrations of NO_2. This atmosphere is then used to audit a continuous NO_2 analyzer.

This procedure, known as gas-phase titration, requires a rather skilled operator and specialized equipment. For these reasons, gas-phase titration, as a basis for auditing NO_2 analyzers, is used only in special monitoring programs where the expense is justified. Prototype devices that can be shipped easily from point to point and simply operated are presently being evaluated for possible use in nationwide audits of control agencies using continuous NO_2 analyzers.

NO_2 Solution Audit Samples

For time-integrated samples, NO_2 is generally collected by passing a measured amount of ambient air through a solution that selectively removes the NO_2 and converts it to nitrite ion. The concentration of nitrite ion is determined by conventional conversion to an azo dye and colorimetric determination. Aqueous solutions of sodium nitrite are convenient audit samples for the analytical portion of time-integrated measurements of ambient NO_2.

AUDITS OF NO_2 MEASUREMENTS

Nationwide Audits of Time-Integrated Measurements

Introduction

Since 1972, EPA has conducted a performance audit program of the various monitoring groups throughout the United States and several foreign countries. Auditing the analytical performance of the measurement of NO_2 by time-integrated methodology has been a part of this program. The performance audit program is one part of an overall quality assurance program for NO_2 measurements, and the results can give quite meaningful information on the overall quality of NO_2 measurements.

The most important purpose of the performance audit program is to provide a monitoring organization a means to rapidly evaluate their analyses themselves. Performance in the analysis of audit samples when the concentrations are unknown to the analyst is a very useful means of identifying problems and determining the corrective action needed. Rapid self-evaluation and correction of errors is one of the fundamental precepts of EPA's quality assurance program. The results of the performance audit program also provide EPA with the continuing index of the validity of monitoring data used in making important decisions.

Program Coordination

Participants are selected by the Regional Quality Control Coordinators in each of EPA's ten regions throughout the United States. Additional participants are obtained through individual request and through notification in the *Journal of the Air Pollution Control Association.*

After the audit roster is completed for a particular survey, instructional materials and audit samples are mailed to participants, who are allowed 5–6 weeks to analyze the samples and return their results. The results are entered into a computerized data handling system, which calculates statistics. Individual reports are returned to participants showing the acceptable ranges for each sample as well as the values they reported.

To determine whether an incorrect result is due to inherent problems in the laboratory procedure or whether the error is a one-time phenomenon, recheck samples are sent to laboratories whose results do not meet certain criteria. Thus, by having a second chance to analyze a set of samples, real deficiencies are distinguished from one-time problems.

Statistical Approach

Acceptance criteria, which can be used to judge the performance of participants, are an important, yet vexing, requirement of the program. The program has developed acceptance criteria that the participants themselves can use to judge their own performance. The criteria must be used with judgment because it is unlikely that any one set of criteria can serve the needs of all data users. The system used in this program was developed from the results of analyses by experienced laboratories, by the variability in results caused by the samples themselves and by the variability expected between laboratories during a survey.

Two acceptable ranges have been defined. The Sample Range contains the variability attributable to the sample material. In each survey, this range is based on the results of repeated analyses of several samples from each concentration range by one laboratory. These ranges are goals for each participant. It is not reasonable to expect all participants to fall within these ranges on any given analysis. However, falling within these ranges repeatedly indicates a laboratory with excellent precision and accuracy in the analysis for NO_2. Falling outside the sample range does not indicate a problem.

A second acceptable range, the Target Range, has been constructed to include between-laboratory variability as well as sample variability. A participant falling outside the sample range but within the Target Range should feel comfortable with his results. If, however, the results fall outside the Target Range, the accuracy of the analysis should be suspect. The Target Range is based in part on past performance of all participants. For a particular survey, an average percentage difference between the reported and true results for all past surveys is determined and used to calculate the Target Range for a survey.

The method of evaluating the results of a survey is under review, and changes are likely in the means of judging performance.

Results

Results from four performance audits during 1976 and 1977 are presented to show the type of information this program produces [3,4]. Nitrogen dioxide audit number 0676 began in June 1976: out of 119 sample sets requested by participants, 100 sets of data were returned for a response rate of 84%. This and corresponding information for audits begun in December 1976, June 1977 and December 1977, are shown in Table I. Table I also indicates the monitoring agency-type distribution.

Methods used to analyze samples can be grouped into five broad categories, as defined by the participants. Table II lists the analytical methods used and the number of participants using the method. Participants tend to define the analytical method used in very general terms. A laboratory reporting the use of the manual sodium arsenite procedure as the method of choice may have used that procedure with various modifications. Thus, Table II should be interpreted to indicate the number of laboratories that used the procedure of approximating the method specified in the Table.

Table I. NO$_2$ Agency Distribution

	Foreign	EPA	State	Local	Private	Total	% Return
NO$_2$ Audit–0676							
Agencies Requesting Samples	2	10	45	55	7	119	
Agencies Returning Data	2	8	38	50	2	100	84.0
NO$_2$ Audit–1276							
Agencies Requesting Samples	2	11	47	57	13	130	
Agencies Returning Data	1	8	41	50	10	110	84.6
NO$_2$ Audit–0677							
Agencies Requesting Samples	1	9	41	60	12	123	
Agencies Returning Data	1	6	37	44	7	95	77.2
NO$_2$ Audit–1277							
Agencies Requesting Samples	1	9	41	58	15	124	
Agencies Returning Data	1	5	37	40	7	90	72.6

AVERAGE: 79.6%

Table II. NO$_2$ Analytical Methods

	Number Agencies Using Method			
Method	Audit 0676	Audit 1276	Audit 0677	Audit 1277
Saltzman–Manual	5	3	4	4
Sodium Arsenite–Manual	65	72	62	56
Sodium Arsenite–Automatic	24	28	23	24
TGS–ANSA–Manual	1	3	2	2
Other	4	4	4	4

Table III. NO$_2$ Sample and Target Ranges (%)

	Conc. 1	Conc. 2	Conc. 3	Conc. 4	Conc. 5
Sample Range	±10	±10	±10	±10	±10
Target Range	±20	±20	±20	±20	±20

Acceptable Range. As described above, two performance ranges were used as one means of judging performance. The Sample Ranges and Target Ranges for NO$_2$ in the five concentration ranges used in these audits are shown in Table III in the order of increasing concentration. The percentages given are the allowable differences from true values. Because only a few audits had been conducted previous to the one reported here, considerable judgment has been used in constructing the limits shown in Table III. As more audits are conducted and more data become available, the ranges will be refined.

Data Summary. Using the Target Ranges as one means of eliminating non-representative data, the arbitrary decision was made that any laboratory not reporting at least one value within the Target Ranges would be considered an outlier. The Target Ranges are sufficiently broad that unless the analyses are totally out of control, at least one value would fall within the ranges of the laboratory's performance similar to most of the study population. The remainder of this chapter deals with all reported results except those values identified as outliers using the above test.

Table IV is a frequency distribution of the percent differences between the reported value and the true value for each sample concentration for each audit. Table IV is very useful for laboratories trying to determine their performance relative to other participants. For example, in the 1276 audit, only 10% of the laboratories reporting results for concentration 1 had a percent difference of 0.8% or less, while 50% of the laboratories reported a percent

Table IV. NO_2 Absolute Percent Difference

	No.	Minimum	10%	30%	50%	70%	90%	Maximum	Mean
				NO_2 Audit–0676					
Conc. 1	91	0.40	1.2	4.8	8.7	12.7	28.2	78.2	12.7
Conc. 2	92	0.00	0.5	1.7	2.4	4.8	11.4	287.4	7.8
Conc. 3	91	0.00	1.0	1.7	3.2	5.1	11.4	104.8	6.2
Conc. 4	91	0.12	0.6	1.6	2.5	4.4	8.4	45.3	4.5
Conc. 5	90	0.10	0.5	1.6	2.4	4.4	15.3	86.4	6.7
All Samples	455	0.00	0.7	1.7	3.2	6.5	15.1	287.4	7.7
				NO_2 Audit–1276					
Conc. 1	106	0.0	0.8	5.0	8.3	16.7	30.8	80.0	15.3
Conc. 2	105	0.6	0.6	5.0	8.8	11.6	22.6	81.8	11.4
Conc. 3	108	0.0	0.9	2.4	4.9	7.4	16.6	32.2	7.1
Conc. 4	108	0.2	0.8	3.0	6.0	8.2	14.2	40.5	7.3
Conc. 5	107	0.0	0.5	1.7	3.0	6.3	13.8	34.1	5.4
All Samples	534	0.0	0.8	2.8	6.1	10.1	22.5	81.8	9.3
				NO_2 Audit–0677					
Conc. 1	88	0.0	0.8	5.9	9.2	14.3	24.4	77.3	12.7
Conc. 2	91	0.0	0.8	3.4	5.5	9.2	16.4	82.4	8.5
Conc. 3	86	0.4	1.2	2.5	4.9	7.8	17.2	76.2	7.8
Conc. 4	92	0.0	0.5	2.1	3.4	6.9	15.1	74.9	7.8
Conc. 5	90	0.0	0.4	2.0	3.8	5.9	12.2	75.7	5.8
All Samples	448	0.0	0.8	2.9	5.0	9.2	17.6	82.4	8.5
				NO_2 Audit–1277					
Conc. 1	84	0.0	0.6	2.2	5.0	7.7	22.1	58.6	9.0
Conc. 2	86	0.3	1.3	3.5	4.5	7.4	15.8	35.7	7.1
Conc. 3	87	0.2	1.8	3.4	4.8	6.8	12.5	19.4	6.0
Conc. 4	86	0.0	1.0	2.2	3.4	5.4	11.9	20.6	5.0
Conc. 5	88	0.1	0.9	1.8	3.1	5.2	9.2	96.1	5.3
All Samples	431	0.0	1.0	2.6	4.2	6.6	13.7	96.1	6.5

difference of 8.3% or less for the same concentration. The mean percent difference for "All Samples" is a most useful statistic for comparing one audit with another. Thus, Table IV shows that this value ranges from 6.5% in audit 1277 to 9.3% in audit 1276; there seems to be no pronounced trend. This discovery indicates that the level of performance in the analysis of NO_2 is stable and that in general, accuracy of about 8–10% can be expected. However, there is a dependence on concentration levels, which must be carefully considered.

Analytical Method Summary. Table V was developed to determine if a particular analytical method produces unduly imprecise or biased results. The results show that all methods estimate true values quite adequately and that no unusually large imprecision is present.

Table V. Mean and Standard Deviation of

Method	Concentration 1		Concentration 2	
	Mean	Standard Deviation	Mean	Standard Deviation
				NO$_2$ Audit
Saltzman–Manual	0.214	0.045	0.411	0.007
Sodium Arsenite–Manual	0.236	0.040	0.426	0.157
Sodium Arsenite–Automatic	0.228	0.028	0.405	0.017
TGS–ANSA–Manual				
Other	0.228	0.121	0.423	0.033
True Value		0.25		0.41
				NO$_2$ Audit
Saltzman–Manual	0.11	0.02	0.16	0.03
Sodium Arsenite–Manual	0.12	0.03	0.17	0.04
Sodium Arsenite–Automatic	0.12	0.02	0.16	0.05
TGS–ANSA–Manual	0.14	0.06	0.21	0.04
Other	0.12	0.01	0.17	0.01
True Value		0.12		0.18
				NO$_2$ Audit
Saltzman–Manual	0.13	0.00	0.25	0.00
Sodium Arsenite–Manual	0.12	0.02	0.25	0.04
Sodium Arsenite–Automatic	0.13	0.01	0.24	0.02
TGS–ANSA–Manual	0.15	0.02	0.26	0.02
Other	0.12	0.02	0.26	0.02
True Value		0.12		0.24
				NO$_2$ Audit
Saltzman–Manual	0.19	0.02	0.33	0.02
Sodium Arsenite–Manual	0.19	0.03	0.32	0.03
Sodium Arsenite–Automatic	0.18	0.01	0.32	0.01
TGS–ANSA–Manual	0.20	0.00	0.32	0.02
Other	0.22	0.03	0.34	0.02
True Value		0.18		0.31

Special Audits of Continuous Measurements

There are monitoring programs that measure NO_2 continuously and are of enough importance to justify periodic audits of the performance of the analyzers. Portable devices that can generate atmospheres of NO_2 using

NO_2 Results by Analytical Method ($\mu g/ml$)

Concentration 3		Concentration 4		Concentration 5	
Mean	Standard Deviation	Mean	Standard Deviation	Mean	Standard Deviation
−0676					
0.408	0.014	0.807	0.033	1.00	0.033
0.415	0.066	0.828	0.076	1.01	0.181
0.403	0.017	0.810	0.032	1.00	0.064
0.425	0.041	0.810	0.023	0.96	0.075
	0.41		0.83		1.02
−1276					
0.34	0.01	0.49	0.04	0.81	0.05
0.32	0.04	0.51	0.06	0.85	0.08
0.32	0.02	0.52	0.03	0.86	0.04
0.35	0.05	0.54	0.04	0.89	0.04
0.31	0.01	0.54	0.01	0.86	0.01
	0.33		0.54		0.87
−0677					
0.25	0.00	0.38	0.00	0.68	0.02
0.25	0.04	0.39	0.06	0.73	0.08
0.24	0.02	0.38	0.03	0.70	0.06
0.25	0.02	0.52	0.19	0.69	0.01
0.26	0.01	0.40	0.04	0.72	0.02
	0.24		0.38		0.69
−1277					
0.52	0.04	0.52	0.02	0.87	0.04
0.52	0.03	0.52	0.03	0.89	0.05
0.52	0.02	0.52	0.02	0.84	0.18
0.52	0.01	0.52	0.00	0.89	0.02
0.53	0.03	0.53	0.03	0.91	0.04
	0.52		0.50		0.87

gas-phase titration are used for these special audits. Because of the complexity of the equipment, a team is required at the monitoring site to perform the audit. All field gas-phase titration audit devices are compared with a laboratory standard apparatus, which is never used for field work. These comparisons are made before and after each audit to ensure that the audit device in fact delivered known NO_2 concentrations.

To conduct this audit, normal monitoring is temporarily interrupted and atmospheres of NO_2 at zero and several upscale levels, generally five or six, are produced by the gas-phase titration apparatus. The monitoring instrument is challenged at each concentration level, and the value indicated by the analyzer is recorded. Statistical analysis of the audit concentrations and analyzer responses is then made to determine how well the analyzer was performing at the time of the audit.

Determination of Performance

Linear regression analysis is used to determine the slope and intercept of the plot of analyzer response vs audit concentration. Perfect agreement between analyzer response and audit concentration would result in a straight line with unity slope and zero intercept. The slope of the line is the measure used to determine overall analytical accuracy, provided the intercept is small compared to the concentrations being analyzed.

Criteria for acceptable performance, as a result of these audits, can only be made in some arbitrary fashion. Each monitoring program is unique, and the requirements, with respect to data quality, differ from one to another. However, there is now enough experience to show that differences of more than 15% between the slope and unity indicate problems, and some corrective action should be taken. Likewise, close agreement between the slope and unity indicates good to excellent performance, and the following arbitrary criteria are used to describe the results of these audits:

- excellent—slope of the regression of analyzer response on audit concentration is $0.95 \leqslant$ slope $\leqslant 1.05$;
- satisfactory—slope of the regression of analyzer response on audit concentration is $0.85 \leqslant$ slope < 0.95 or $1.05 <$ slope $\leqslant 1.115$; and
- unsatisfactory—slope of the regression of analyzer response on audit concentration is slope <0.85 or slope >1.15.

RESULTS

Field performance audits for NO_2 analyzers in a number of important monitoring projects have been conducted and the results reported to the respective program managers for these projects. The results from one project will be given in some detail to show that an audit program does have an impact on improving data quality.

EPA is concerned about the impact on the environment of present and proposed energy developments. The impact on air quality is one of the major concerns with respect to these energy developments, and many air quality monitoring data are being gathered to determine this impact. It is very important that air quality data obtained from diverse monitoring programs be comparable, and a large-scale quality assurance program has been established for this purpose. The program is being conducted for EPA under

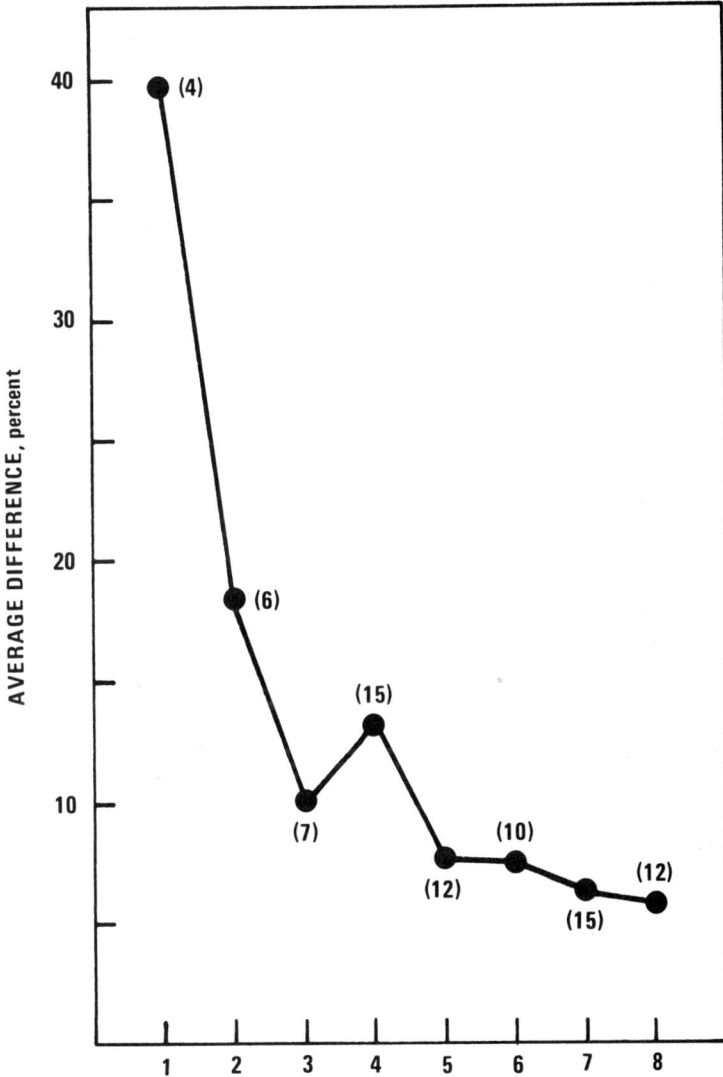

Figure 1. Plot of average difference between slopes and unity for each agency vs the quarterly audit.

contract by the Environmental Monitoring and Services Center of Rockwell International.

Up to this point, the program has been concerned with the air pollution monitoring taking place around energy developments in the western United States. A similar program in the midwest and eastern United States is now under development.

Audits of continuous monitors for NO_2 operated by control agencies of these states is an important part of the energy quality assurance program. Each calendar quarter an audit team visits each monitoring site and performs an audit, as described above. The slope of the regression of analyzer response on audit concentrations is then used to measure performance. Figure 1 shows the impact of the quality assurance program on the quality of the data generated by the participation groups. This figure is a plot of the average difference between the slopes and unity (in terms of percentages) for each agency vs the calendar quarter when the audit was conducted. The figures in parentheses on the plot are the number of agencies taking part in the audit for each quarter. For the first two quarters, the observed errors were approximately 20% or higher. By the time of the fifth quarterly audit, the average error was below 10% and had been stable at this level for the last three audits. This improvement in performance, as demonstrated by these audits, undoubtedly has caused a corresponding improvement in the quality of the NO_2 monitoring data generated in these agencies.

ACKNOWLEDGMENTS

The authors wish to express appreciation to S. M. Bromberg, R. L. Lampe and B. I. Bennett of the Quality Assurance Branch of the Environmental Monitoring and Support Laboratory, EPA, and to Mark Cher and E. P. Parry of Rockwell International, who carried out the work reported here.

REFERENCES

1. Midgett, M. R. "How EPA Validates NSPS Methodology," *Environ. Sci. Technol.* 11(7):655–659 (1977).
2. U.S. Environmental Protection Agency. "Nitrogen Dioxide Measurement Principle and Calibration Procedure," Notice of Final Rulemaking, *Federal Register* 41(232):52686–52695 (December 1, 1976).
3. Bromberg, S. M., B. I. Bennett and R. L. Lampe. "Summary of Audit Performance, Measurement of SO_2, NO_2, CO, Sulfate, Nitrate–1976," EPA-600/4-78-004, Research Triangle Park, NC (1978).
4. Bromberg, S. M., R. L. Lampe and B. I. Bennett. "Summary of Audit Performance, Measurement of SO_2, NO_2, CO, Sulfate, Nitrate, Lead, Hi-Vol Flow Rate–1977," U.S. Environmental Protection Agency, Research Triangle Park, NC (In press).

SECTION II

Effects of Nitrogen Oxides on Experimental Animals

CHAPTER 8

BIOLOGICAL EFFECTS OF NITROGEN DIOXIDE AND NITRIC OXIDE

Taichi Nakajima, Hajime Oda,* Shigeko Kusumoto and Hiroshi Nogami
Department of Environmental Health Research
Osaka Prefectural Institute of Public Health
Osaka, Japan

INTRODUCTION

Nitrogen dioxide (NO_2) and nitric oxide (NO) are the primary hazardous substances among many oxides of nitrogen in the air. Various reports on the effects of NO_2 have been reviewed recently [1,2]. In contrast, the effects of NO have not been well elucidated due to the lack of data at low concentrations, although its concentration is frequently observed to exceed that of NO_2, particularly at roadsides.

The authors have conducted animal exposure experiments to evaluate the health hazard of low concentrations of NO_2 and NO. The studies presented in this chapter will describe long-term effects of NO_2 and NO on mice.

EXPERIMENTAL PROCEDURES

Animal Exposure

The stainless steel animal exposure chambers (1 m³) shown in Figure 1 were used. Concentrated gases supplied in cylinders were diluted with purified air before being introduced into the chamber using an automatic regulation device which controlled the final gas concentration [3]. For NO exposure, NO_2 formed in the mixture of NO and purified air was removed by passing through a soda lime layer before introduction into the chamber [4].

---

*Present address: Division of Basic Medical Sciences, The National Institute for Environmental Studies, Yatabe, Tsukuba, Ibaraki, Japan.

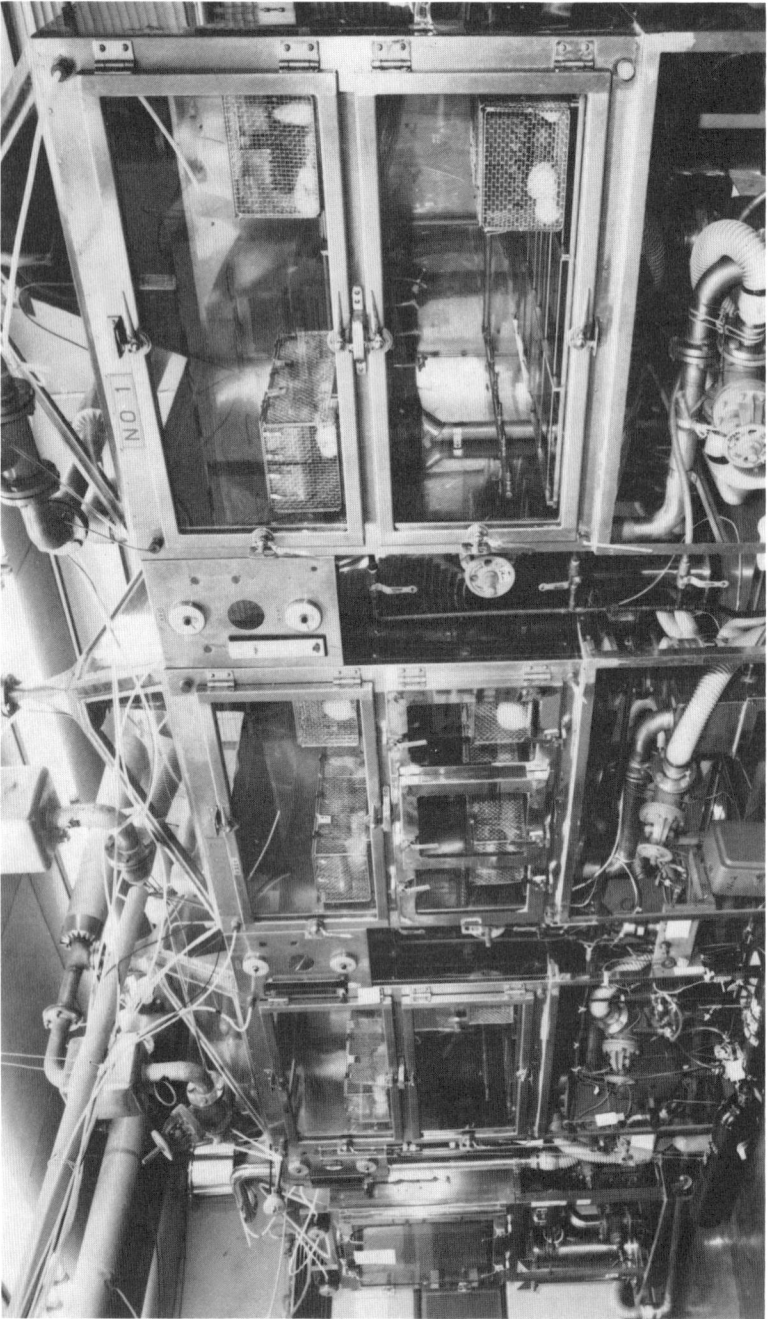

Figure 1. Exposure chambers.

The concentrations of gases were measured with continuous automatic analyzers (Saltzman's method and chemiluminescence method).

The animals (JCL:ICR) were exposed continuously 24 hours per day and 7 days per week. The control groups were exposed to purified air.

NO_2 Exposure

The morphological examination of the resipiratory organs was conducted in 3-week old female mice exposed to 0.7–0.8 ppm NO_2 [5] and in 4-week old female mice exposed to 0.5–0.8 ppm [6] for 1 month and to 0.3–0.5 ppm NO_2 for 6 months [7], respectively.

The recovery process of the morphological changes caused by NO_2 exposure was studied in 3-week old male mice which were maintained for 3 months in a clean air chamber after exposure to 1.0–1.5 ppm NO_2 for 1 month [8].

The susceptibility of exposed mice to respiratory infection with influenza virus was histopathologically studied in 4-week old female mice exposed to 0.5–1.0 ppm NO_2 for 39 days [9] or to 0.3–0.5 ppm NO_2 for 6 months [7] prior to respiratory infection with influenza virus (mouse-adapted A/PR/8 strain).

The alteration of the content of reduced glutathione in the lung was investigated in 7-week old male mice exposed to 0.8 ppm NO_2 for 5 days [10] and in 3-week old.female mice exposed to 0.7–0.8 ppm for 1 month [5] or to 0.3–0.5 ppm NO_2 for 6 months [11]. Methemoglobin in the orbital venous blood of mice exposed to 0.8 ppm NO_2 for 5 days [10], and carboxyhemoglobin in mice exposed to a mixed gas of 0.5–0.8 ppm NO_2 and 50 ppm CO for 1.5 months [12] were determined, respectively.

NO Exposure

For the toxicological study on the long-term effects of NO, 4-week old female mice were exposed to 10 ppm NO containing 1–1.5 ppm NO_2 for 6.5 months [13] or 2.4 ppm NO containing 0.01–0.04 ppm NO_2 for 23–29 months [14].

Histological Examination

After sacrifice of each experimental animal, 10% formalin was infused through the trachea. Then the trachea, bronchi and lungs were removed and fixed in 10% formalin or Zenker's solution. For light microscopy, the tissue specimens were stained with Hematoxylin-Eosin, PAS, Alsian-Blue, Azan-Mallory, van Gieson, Silver and elastic fiber stain. For electron microscopy, the bronchus and lung were cut 1–2 mm in thickness immediately after sacrifice, fixed in 6% glutaraldehyde, pH 7.4 cacodylate buffer and postfixed in 1% osmic acid (the same buffer.) (2.5% glutaraldehyde in pH 7.4 cacodylate has usually been used, with similar results.) After Epon embedding, the sections were stained with uranyl acetate and lead acetate. All experimental animals were subjected to histological examinations.

Hematological and Biochemical Examinations

Using heparinized orbital venous blood, red blood cell counts (RBC), white blood cell counts (WBC), total hemoglobin contents (THb) and hematocrit value (Ht), neutrocyte-lymphocyte ratio (N/L) and Heinz bodies in the red blood cell were examined. Based on the values of RBC, THb and Ht obtained, the mean corpuscular volume (MCV), mean corpuscular hemoglobin level (MCH) and mean corpuscular hemoglobin concentration (MCHC) were calculated. The osmotic resistance of red blood cells was estimated by coil planet centrifuge method [15]. Whole blood was used for determining the levels of 2,3-DPG and cholinesterase activity. The levels of GPT, total bilirubin and urea nitrogen were analyzed using serum. Serum haptoglobin was also determined by the method of Roy et al. [16].

The mice exposed to NO_2 were sacrificed immediately after being taken out of the exposure chamber, and the lungs were removed. The lungs were homogenized in 5% metaphosphoric acid, and then centrifuged after storage at 5°C for 2 hours. GSH content in the supernatant was quantitatively determined using the colorimetric method by 5,5'-dithiobis (2-nitrobenzoic acid) [17].

MetHb and COHb in the orbital venous blood of mice exposed to NO_2 were determined spectrophotometrically by the modified methods of Evelyn-Malloy [10] and Kampen et al. [12], respectively. NOHb and MetHb in mice exposed to NO were measured by ESR (JEOL Model PE-3X) under the modulation width of 6.3 gause and temperature of −140°C [4].

RESULTS

NO$_2$ Exposure

Morphological Changes

By light microscopy, in the bronchioli of mice exposed to 0.7–0.8 ppm for 1 month, hypersecretion, degeneration, desquamation and proliferation of the mucous membrane were focally observed. The bronchiolar cavities were filled with mucus secretions and desquamated epithelia. The alveolar cavities were found to be uneven in size with some enlargement. Proliferation of bronchiolar epithelium (so-called hyperplastic foci) was prominent in the region from the terminal bronchiolus to the alveolus (Figure 2) [6]. These morphological changes were also observed in mice exposed to 0.3–1.0 ppm NO$_2$ for 1–6 months [7,9] and 1.0–1.5 ppm for 1 month]8].

By electron microscopy, edematous swelling and shortening of the cilia were observed on the 10th day of exposure (Figure 3). This swelling was still evident on the 30th day of exposure, but there was no change in the rootlets. Edematous changes of type I alveolar epithelial cells were observed on the 10th day of exposure and persisted to the 45th day (Figure 4). These edematous changes appeared to be prominent in alveolar epithelial cells near the bronchiolus. On the 30th day, bronchial epithelial cells were found

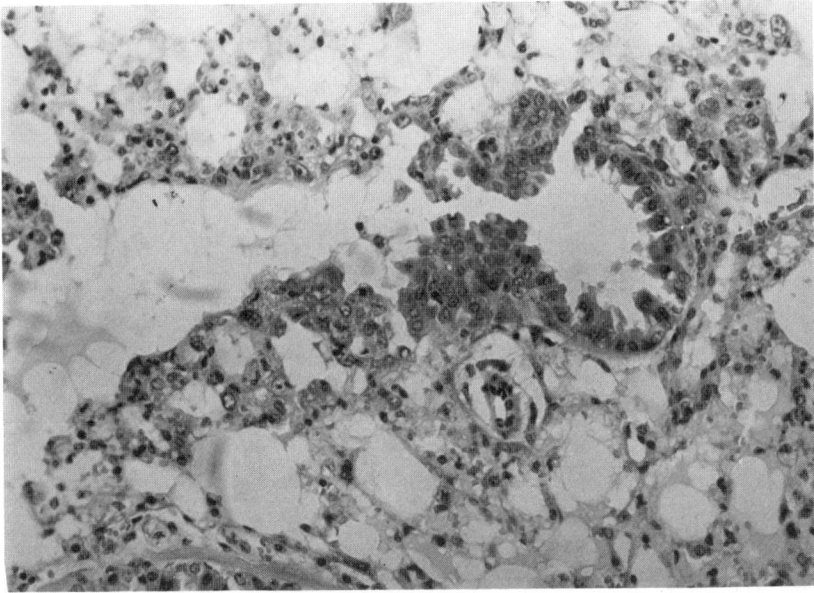

Figure 2. Hyperplasia in the region from the terminal bronchiolus to the alveolus in mouse exposed to 0.5–0.8 ppm NO_2 for 1 month.

to be proliferated toward the alveolar space (Figure 5). Most of the pro-liferating bronchial cells were Clara cells [6].

In the lungs of mice infected with influenza virus after exposure to 0.5–1.0 ppm NO_2 for 39 days, advanced interstitial pneumonia and adeno-matous proliferation of bronchial epithelial cells were noted. Proliferation of epithelial cells extended from the terminal bronchioli toward the alveoli appeared focally as adenomatous in 7 out of 12 mice (Figure 6). The struc-ture of the alveolar wall was fairly well maintained as confirmed through elastic fiber staining (Figure 7). Adenomatous proliferation of the peripheral bronchi as mentioned above was rarely noted in both groups, which were merely infected with virus or exposed to NO_2, whereas a higher incidence of adenomatous proliferation was noted in the group infected with virus after exposure to NO_2 [9]. Similar results were obtained in mice infected with virus after exposure to 0.3–0.5 ppm NO_2 for 6 months [7].

Recovery Processes. Mice killed immediately after exposure to 1.0–1.5 ppm NO_2 for 1 month showed the same morphological changes as described above. However, when the animals were allowed to recover in clean air for 1 month, most of the bronchioles were nearly normal, but desquamation of the mucous membrane, obstruction of the alveolar duct and enlargement of the alveolar lumen still partially persisted, though minimal. Infiltration of inflammatory

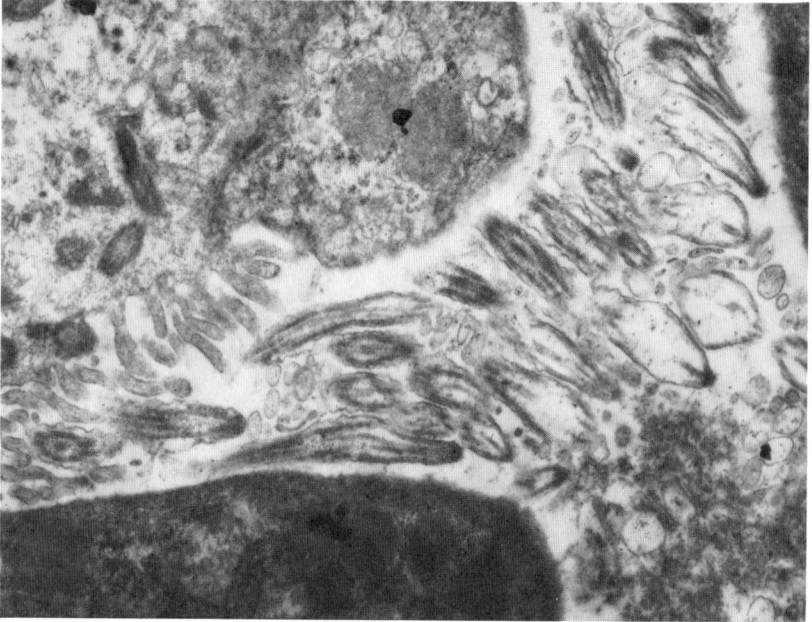

Figure 3. Edematous swelling and shortening of the cilia in mouse exposed to 0.5–0.8 ppm NO$_2$ for 1 month.

cells was rarely observed in the alveolar interstices, but a remarkable cellular infiltration of lymphocytes and monocytes was observed in the trachial and bronchial mucous membrane and submucosa. Focal lymphocytic infiltration occasionally appeared (Figure 8). Lymphocytic infiltration in the group kept for 3 months in clean air was lower in severity than in the group kept for 1 month. These lymphocytic infiltrations were usually observed around the bronchioles but sometimes close to the bronchioles. The histological picture of complete peribronchial infiltration was shown in Figure 9 [8].

Biochemical Changes

GSH content in the lungs of mice exposed to 0.7–0.8 ppm NO$_2$ for 1 month decreased soon after the beginning of exposure, followed by gradual recovery (Figure 10) [5]. Although the animals can recover from the initial reduction of GSH in the lung after a short-term exposure, they tended to lose body weight and showed reduction in GSH level after prolonged exposure to 0.3–0.5 ppm NO$_2$ for 6 months (Table I) [11].

No increase of methemoglobin in the blood of mice exposed to 0.8 ppm NO$_2$ for 5 days [10] and no enhancement of carboxyhemoglobin in the blood of mice exposed to a mixed gas of 0.5–0.8 ppm NO$_2$ and 50 ppm CO for 1.5 months [12] were observed.

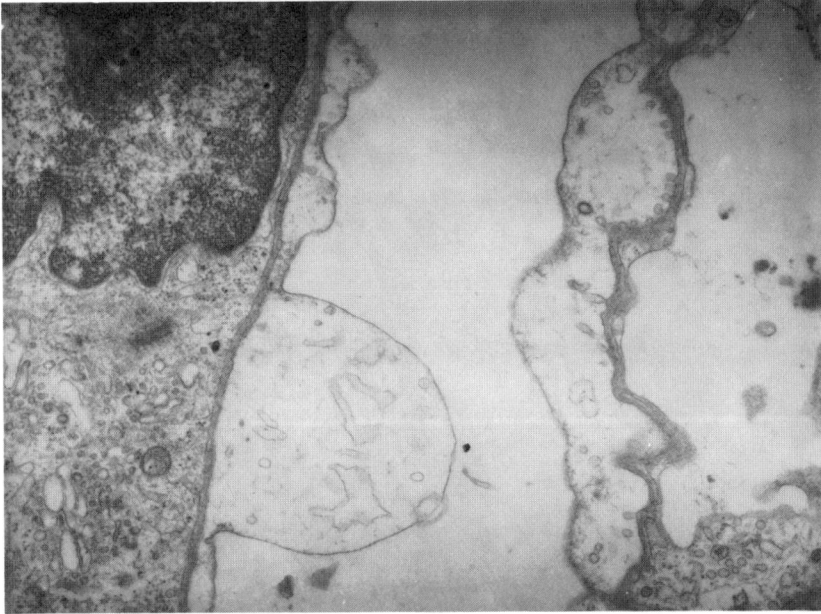

Figure 4. Edematous change of type I alveolar epithelial cell in mouse exposed to 0.5–0.8 ppm NO_2 for 1 month.

Table I. The Content of GSH in the Lung of Mice Exposed to 0.3–0.5 ppm NO_2 for 6 Months (μg/wet weight)

Control group (n = 11)	634 ± 139
Exposed group (n = 9)	875 ± 120

$p < 0.01.$

NO Exposure

10-ppm Exposure

Mice were examined after 2 weeks, 2, 4 and 6.5 months of exposure to 10 ppm NO [13]. No difference was found between the control and exposed groups in body weight and death rate through the 6.5-month period. The exposed group exhibited an increased ratio of organ weight to body weight for lung and spleen after each period with a significant difference observed at 2 weeks and 6.5 months.

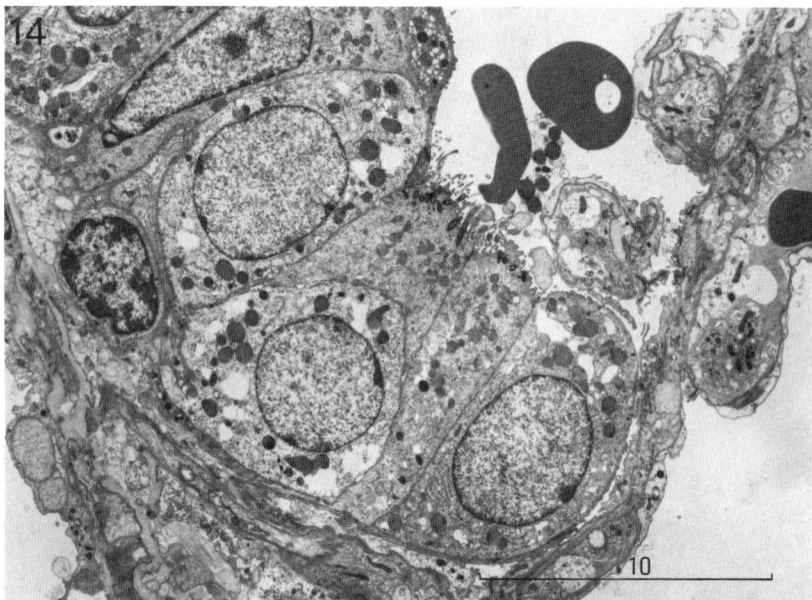

Figure 5. Proliferated bronchial epithelial cells toward the alveolar space in mouse exposed to 0.5–0.8 ppm NO_2 for 1 month.

Hematological tests showed a significant increase in WBC in the exposed group after 2 weeks, and the increase continued. A higher neutrocyte-lymphocyte ratio was seen in the group exposed for 2 months and 6.5 months. No significant differences, however, could be observed between the groups throughout the entire exposure period for RBC, THb, Ht, MCV, MCH and MCHC. Heinz bodies were observed in 11% of RBC of mice exposed for 6.5 months, while none were observed in the control mice. The level of NOHb for both 6.5-month exposure group and the 1-hour exposure group was 0.13%. NOHb was not detected in the control group. MetHb level was 0.2% both in the control and exposed. Measurement of oxygen dissociation curve of the blood suspension showed a slight increase of P_{50} and a decrease of n-value in the 6.5-month exposure group.

Total bilirubin level tended to be higher in the exposed group, and the cholinesterase activity level of the exposed group was lower after 2 months compared with the control group. No significant change was found in GPT, total cholesterol, urea nitrogen and 2,3-DPG.

The histopathological observation of the lungs showed degeneration and necrosis of the mucous membrane along with a subsequent proliferation of the bronchiolar epithelium, and bronchial obstruction by mucous substance.

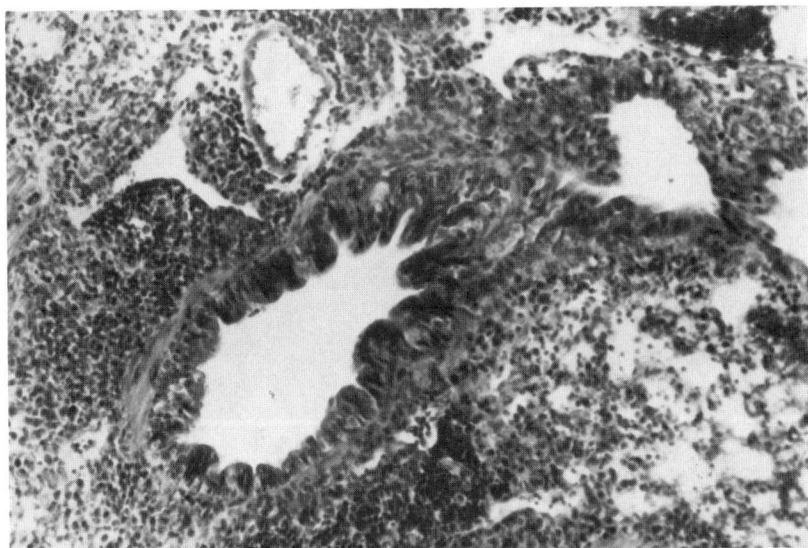

Figure 6. Adenomatous proliferation in mouse infected with influenza virus after exposure to 0.5–1.0 ppm NO_2 for 39 days.

Hypertrophic enlargement of the alveolar septa due to hyperemia and congestion was also observed. These changes were observed even during the acute period and became more prominent particularly as the condition became chronic. No significant difference was seen in histopathology of the heart, liver, kidneys, spleen and brain of the exposed and control groups. Changes indicating the occurrence of anoxia could not be observed.

2.4-ppm Exposure

Mice were exposed for up to 29 months and examined at given intervals [14]. No significant difference was found in body weight and survival rate between the control and exposed groups except for a decreased survival rate after 16 months exposure. Histological examination of the dead mice indicated that almost half of the mice had leukemia, lymphoma, nephritis, lung adenoma or other diseases with no significant difference in their incidences between the control and exposed groups.

In the hematological studies, there was no difference between the control and exposed mice in WBC, neutrocyte-lymphocyte ratio, RBC, THb, Ht, MCV, MCH and MCHC throughout the entire exposure period. Heinz bodies

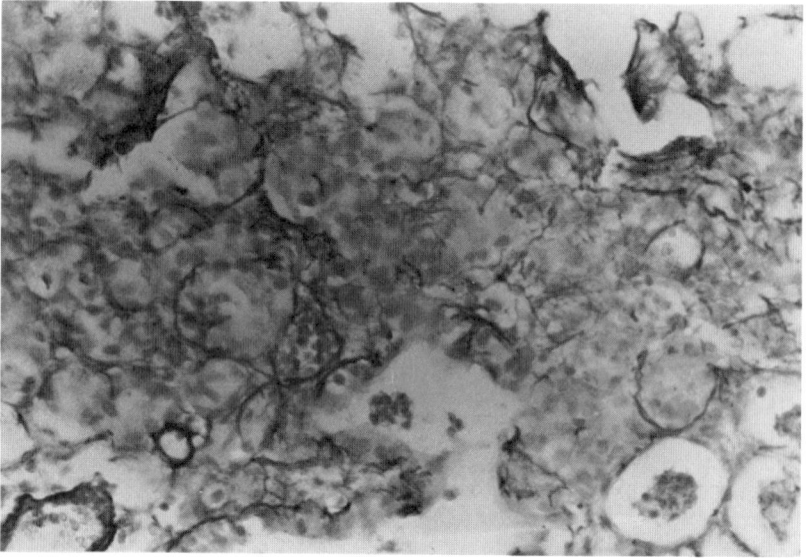

Figure 7. Elastic fiber of alveolar wall in mouse infected with influenza virus after exposure to 0.5–1.0 ppm NO$_2$ for 39 days.

in RBC could not be observed. NOHb was not detected in the control group and it was 0.01% in the exposed group as in the case of mice exposed for 1 hour. The control mice had 0–0.3% MetHb, and no change in MetHb was detected in the exposed group.

In the examination of osmotic resistance of RBC, the 6-month exposure group showed a shift in the start of hemolysis to a lower osmotic pressure, while the 13-month exposure group showed an opposite shift. Haptoglobin of the 13-month exposure group was lower than that of the control mice. No significant change was found in GPT, total cholesterol, urea nitrogen or total bilirubin.

The microscopic observation of the organs indicated that there was no difference between the control and exposed groups in terms of the desquamation and hyperplasia of the bronchiolar epithelium, and obstruction of the lumen. Enlargement of alveolar septa was not seen. The number of mice with adenoma or infiltration of leukemic cells in the lung increased when the exposure was prolonged, but there was also no difference between the control and exposed mice. As for the other organs, changes suggestive of the effects of exposure and in particular, indicative of the occurrence of anoxic changes could not be detected histologically.

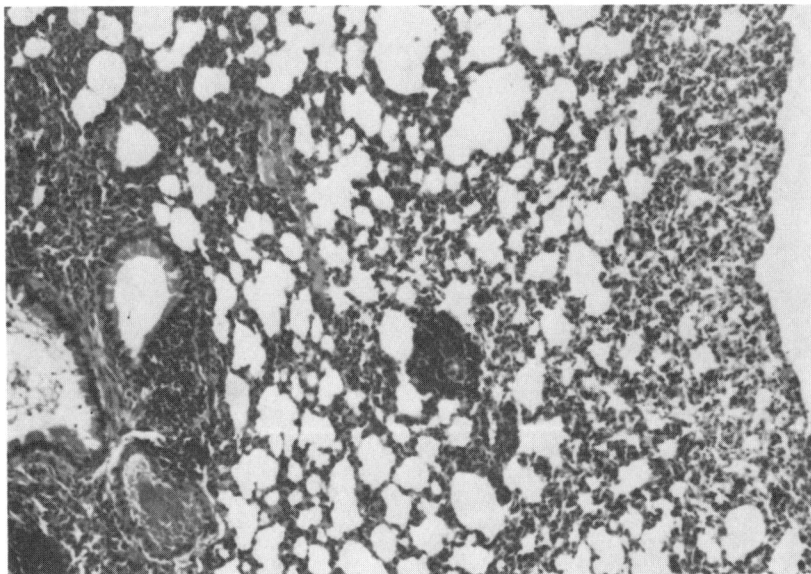

Figure 8. Lymphocytic infiltration appeared islet in form in mouse kept in clean air for 1 month after exposure to 1.0–1.5 ppm NO_2 for 1 month.

DISCUSSION

Long-Term Effects of NO_2

Morphological Changes in the Lung

A variety of histological changes observed in the lung of mice exposed to NO_2 were almost the same as those reported by Blair et al. [18]. Similar structural changes were also found in the rat by Freeman et al. [19]. Edematous swelling of cilia might result in reduction of their ability to transport inhaled particulates, and if edematous changes of alveolar cells should persist, alveolar gas exchange would be impaired. It was interesting that most of the morphological changes were observed in the peripheral air way. These were considered to be the early changes which would possibly lead to chronic obstructive pulmonary disease.

The influence of NO_2 on susceptibility to respiratory infection has been studied in several species of animals [20-22]. Advanced pneumonia observed in mice infected with influenza virus after exposure to NO_2 revealed the morphological evidence supporting the reduction in susceptibility to respiratory infection.

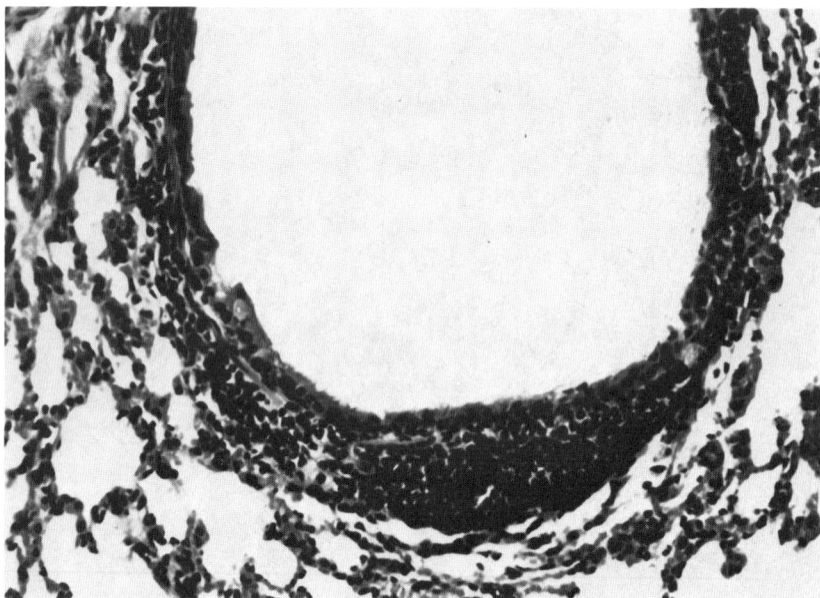

Figure 9. Complete peribronchial lymphocytic infiltration in mouse kept in clean air for 1 month after exposure to 1.0–1.5 ppm NO_2 for 1 month.

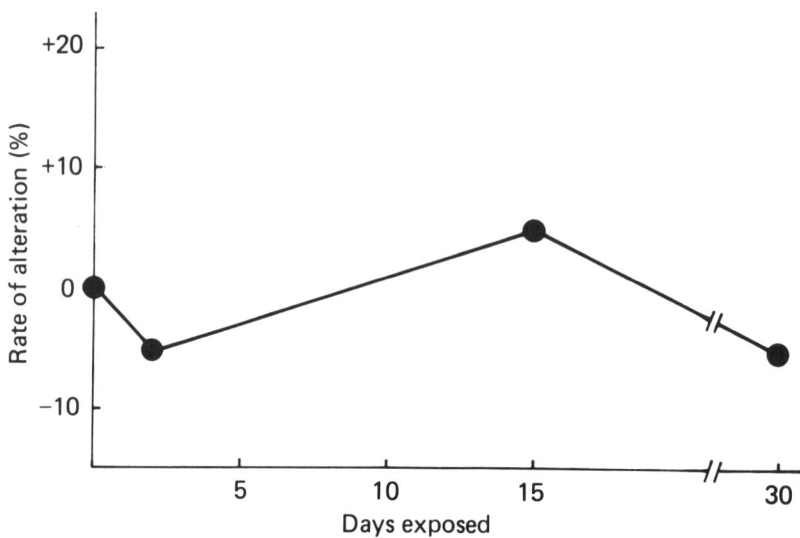

Figure 10. Alteration of GSH content in the lung of mice exposed to 0.5–1.0 ppm NO_2 for 1 month. The number of mice at each period was 12 both in exposed and control groups.

Proliferation of bronchiolar epithelium observed in the mice lungs exposed to 0.5–0.8 ppm NO_2 for 30 days was identical to the early change of experimentally induced pulmonary adenoma in mice as previously reported by Hattori et al. [23]. Furthermore, it was noted that a higher incidence of adenomatous proliferation of the bronchiolar epithelial cells was observed in mice infected with influenza virus after exposure to NO_2. These changes were apparent, but no positive histological evidence was demonstrated to indicate that this adenomatous proliferation would become malignant. In fact, malignant transformation has not been detected even after prolonged exposure to 0.3–0.5 ppm NO_2 for 6 months. Freeman et al. found hypertrophy of bronchiolar and alveolar epithelium in rats [24] and monkeys [25] exposed to low concentrations of NO_2 for long periods. They studied cell renewal in rats and demonstrated that proliferation of type II alveolar cells occurred within 48 hours and returned to normal by the seventh day [26]. However, these findings suggested that long-term inhalation of NO_2 might play some role in the development of lung cancer through synergistic action of NO_2 with carcinogens and tumorigenic viruses.

Recovery Processes of Pulmonary Lesions

A pronounced lymphocytic infiltration around the trachea and bronchus of mice allowed to recover in clean air after exposure to NO_2 resembled that of chronic bronchitis in the rabbit which was autoimmunologically induced by means of injection of tracheobronchial mucosa antigen [27]. Such histologic phenomena have not yet been found in the animals examined either during or immediately following exposure to NO_2. Although the pathogenesis of this response is not yet clear, the following mechanism is suggested. Namely, the degenerated mucous membrane primarily affected as antigen, and consequently, the autoimmune reaction occurred to produce inflammatory reactions characterized by lymphocytic infiltration. Therefore, the effect of NO_2 did not appear to be thoroughly reversible.

Biochemical Changes

The initial reduction of GSH in the lung, soon after the initiation of exposure, might be caused by the primary oxidation of GSH itself by NO_2^-, which was believed to be produced by the infiltration of inhaled NO_2 into the mucus in the air way [10]. Chow et al. [28] and Fukase et al. [29] observed a rise in glutathione reductase activity in rats exposed to NO_2. This compensatory alteration in the enzyme system in exposed mice is attributed to recovery of initially reduced GSH. Mice exposed to NO_2 did not exhibit a rise in MetHb, as previously reported by Wagner et al. [30] in rats and Ehrman et al. [31] in mice.

A combined exposure to NO_2 and CO did not cause an increase in COHb, and this observation agrees with the findings of Lutmer et al. [32].

Long-Term Effects of NO

Since the organs which come into direct contact with air pollutants are generally restricted to the respiratory apparatus, the effects of the pollutants are usually considered in two categories; the effects on the respiratory system organs and the effects to other systems. While the effects of NO_2 are mostly observed in the respiratory organs, various published data shown in Tables II, III and IV, which are relevant to the evaluation of the toxicity of NO, indicate that effects of exposure were observed in the respiratory organs and in the blood.

Formation of Heinz-bodies and increases in spleen weight and total bilirubin after 10 ppm NO exposure suggest that the destruction of red blood cells was facilitated by NO. After 2.4 ppm exposure a slight decrease in osmotic resistance of red blood cells and serum haptoglobin were noted. These changes are known to occur in conjunction with intravascular hemolysis [16]. Since the relationship between a change such as hemolysis and the possible effect on human health is not known at the present time, further investigation should be conducted to ascertain the systemic effect involving the fate of NO in the living body.

We have previously reported that the levels of NOHb in experimental subjects become constant after NO exposure of about 20 minutes [4]. NOHb was observed in long-term exposure and the same level of NOHb was also exhibited in 1-hour exposure groups. These findings have confirmed that the NOHb level does not rise above a certain concentration even after sustained exposure. The small content of NOHb in the blood and the lack of anoxic changes in the organs such as brain and heart have indicated that the ordinary level of NO does not result in hypoxemia in humans, whereas NO was once speculated to cause disturbances in oxygen transport in blood because of its strong affinity to hemoglobin.

Although the increase in lung weight, higher WBC and histological changes of the lung observed in the 10 ppm exposure group indicated some damages to the respiratory organ, these changes could not be attributed to NO alone because of the contamination of 1.0–1.5 ppm NO_2. In the 2.4-ppm exposure group, where NO_2 concentration was much lower, no difference was observed between control and exposed lungs. Therefore, the effect of NO on the lungs is considered to be very slight, and much less than that of NO_2, because the changes in the lung were observed even at 0.5 ppm NO_2 exposure [11]. Table V shows the available data on comparative effects of NO and NO_2 on the lungs. Except for functional changes, effects of NO at low concentrations have not been detected either by histological or biochemical methods. For these reasons, we tentatively conclude that the effects of NO on the respiratory organs are probably less toxic than NO_2.

Table VI shows the summary of the results of both 10 ppm exposure and 2.4 ppm exposure, which suggests that the primary effects of NO are on the blood such as NOHb formation and hemolysis, however, they do not cause anoxia of the other organs. Low concentrations of NO do not affect the respiratory organs.

Table II. Summary of Short-Term Exposure of Animals to NO

Concentration (ppm)	Duration of Exposure	Animal	Observed Effects	Reference No.
6.6	1 hr	Mouse, Rabbit	NOHb was detected by ESR; difference of formation among species	33
10	1 day, 9 days	Rat	NOHb was not detected by ESR	34
11	1 hr	Mouse, Rat	NOHb was detected by ESR; maximum in 20 min at 0.13%; dose-dependent	4
10, 66	1.5 hr, 3 hr	Mouse	No difference in NOHb and COHb between single and combined exposure with CO at 100 ppm and 460 ppm	35
12, 21 11	7 days 30 days	Mouse	No effect on lung peroxidative metabolic pathway	29
16, 50	4 hr	Guinea pig	No change in tidal volume or respiratory frequency	36
20—80	1–2 hr	Mouse	MetHb time-course similar to NOHb; dose-dependent	37
240–4500	3 min–8 hr	Mouse	MetHb level higher; central nervous system symptoms; dead over 320 ppm (76 min); NO four times as toxic as NO$_2$	38
888	1 hr	Rat, Rabbit	^{15}NO distributed mainly in blood just after the exposure; 40% of ^{15}N excreted from urine 1 day after was nitrate	39
5000–20,000	7–50 min	Dog	MetHb level higher; reduced blood oxygen partial pressure; lower blood pH level; death due to pulmonary edema	40
5000–20,000	7–50 min	Dog	Pulmonary edema and hemorrhage; broncho-pneumonia; hyperplasia of bronchial epithelia	41
20,000	4–50 min	Dog	MetHb detected spectroscopically, but not NOHb	42

Table III. Summary of Long-Term Exposure of Animals to NO

Concentration (ppm)	Duration of Exposure	Animal	Observed Effects	Reference No.
1.5–2.0	18 months	Beagle	No change in respiratory function	43
1.5–2.0	61 months	Beagle	No change in cardiovascular function	44
1.5–2.0	61 months	Beagle	Increased residual volume of lung	45
2	6 weeks	Rat	No change in RBC, THb, Ht and oxygen affinity both in single and combined exposure with 2 ppm SO_2	46
2.4	30 months	Mouse	Changes in osmotic resistance of red cell and content of serum haptoglobin; no pathological changes in lung	14
5	30 min/day, 2 days/week, 7 weeks	Guinea pig	Acceleration of experimental asthma sensitized by albumin	47
10	6.5 months	Mouse	Increased lung and spleen weight; increased white cell count; higher total bilirubin value; reduced cholinesterase activity; degeneration, necrosis and proliferation of bronchiolar epithelia; enlargement of alveolar septa	13

Table IV. Summary of Controlled Human Studies on NO

Concentration (ppm)	Duration of Exposure	Subject	Observed Effects	Reference No.
0.33–5	Single breath	Volunteer	80% absorbed in normal respiration; over 90% absorbed in maximum respiration	48
5–30	15 min	Volunteer	Decreased arterial oxygen pressure; increased airway resistance, MetHb level higher	49
10	1 hr	Volunteer	No NOHb in blood	50

Table V. Comparison of the Effects of NO and NO_2 on Respiratory System

Change of Respiratory Organ	NO		NO_2	Reference
Hyperplasia of bronchiolar epithelium (mouse)	−		+	11, 14
Increased airway resistance (human)	+	<	+	49
Decreased arterial oxygen pressure (human)	+	<	+	49
Decreased CO diffusion capacity (human)	−		+	49
Intensified dyspnea by albumin (guinea pig)	+		+	47
Change of peroxidative metabolism (mouse)	−		+	29
Changes of tidal volume and respiratory rate (guinea pig)	−		+	36

Table VI. Summary of 10–ppm and 2.4–ppm Exposure Experiments to NO

Experiment	10 ppm	2.4 ppm
NO (ppm)	9–11	2.3–2.5
NO_2 (ppm)	1–1.5	0.01–0.04
Duration (months)	6.5	29
NOHb (%)	0.13	0.01
Hemolysis	++	+
Anoxia	−	−
Lung damage	+	−

CONCLUSIONS

Long-Term Effects of NO_2

Mice continuously exposed to low concentrations of NO_2 (above 0.3–0.5 ppm) for 1 month or more produced morphological changes in the respiratory organ. These changes included proliferation of the bronchiolar and alveolar epithelium, degeneration and obliteration of mucous membrane, edematous changes in alveolar epithelial cells, loss or shortening of cilia. When the exposed animals were allowed to recover in clean air for 1–3 months, these pulmonary lesions disappeared. However, a pronounced infiltration of lymphocytes, resembling that of autoimmune disease, was observed around the bronchi during the recovery process.

Proliferation of bronchiolar epithelium was identical with the early change of hyperplasia in pulmonary adenoma. Mice infected with influenza virus after exposure to 0.5–1.0 NO_2 for nearly 1 month, showed adenomatous proliferation of bronchial and bronchiolar epithelium.

GSH content in the mice lungs decreased immediately after the initiation of exposure to 0.7–0.8 ppm NO_2, and later returned to normal. However, after prolonged exposure (over 6 months) the GSH level fell again, coincident with the period of body weight loss by the exposed animals. No increase in MetHb was observed in mice exposed to 0.8 ppm NO_2 for 5 days. No enhancement in COHb in animals exposed to a mixed gas of 0.5–0.8 ppm NO_2 and 50 ppm CO for 1.5 months was observed.

Long-Term Effects of NO

Long-term continuous exposures to NO were conducted on mice at 10 ppm and 2.4 ppm. Exposure of 10 ppm for six months resulted in formation of Heinz-bodies in the red blood cells, an increase in spleen weight and total bilirubin, suggesting destruction of red blood cells due to NO exposure. NOHb content (0.13% of total hemoglobin) was constant even after the exposure was continued for 6 months. A slight decrease in the affinity of oxygen to the blood was observed in the oxygen dissociation curve of the blood of the exposed group. However, no anoxic histology of the brain or heart could be observed. Increase in lung weight, degenerative and necrotic changes, and subsequent hyperplasia of the bronchiolar epithelium and enlargement of the alveolar septa were noted, but these changes in the lung cannot be attributed to the effect of NO alone because of NO_2 contamination. When mice were exposed to 2.4 ppm NO, the effect of contaminated NO_2 (0.01–0.04 ppm) was considered to be negligible. There was no difference between the control and exposed with respect to body weight and survival rate except for the decrease in survival rate at the age of 17 months. The causes of death were mostly leukemia, lymphoma, nephritis and lung adenoma, which appeared to occur spontaneously both in control and exposed without significant difference. Decreased osmotic resistance of the red blood cells and serum haptoglobin of the exposed group suggested the acceleration of hemolysis. However, the degree of change was smaller as compared with the 10 ppm exposure. No specific histological changes were found in mice exposed to 2.4 ppm NO and the effect of NO appears to be weaker than that of NO_2 with respect to the effect on the respiratory organs.

ACKNOWLEDGMENTS

The authors express their thanks to Dr. Chentang Chen (Department of Pathology, Kyoto Prefectural University of Medicine), Dr. Shoji Hattori, and Dr. Ryuhei Tateishi (The Center for Adult Disease, Osaka), Dr. Kozo Ito (Department of Public Health, Medical School, Chiba University), Dr. Genshiro Ide, and Dr. Yuji Otsu (Department of Pathology, Medical School, Chiba University), Prof. Hideki Miyaji and Asst. Prof. Akihiko Kurata (Department of Pathology, Hyogo College of Medicine), and Mr. Ryuzo Ohtani (North Osaka Municipal Hospital) for their helpful suggestions and collaboration.

REFERENCES

1. "Environmental Health Criteria for Oxides of Nitrogen," *Environmental Health Criteria, 4,* World Health Organization, Geneva (1977).
2. "Nitrogen Oxides," National Academy of Science, Washington, D.C. (1977).
3. Ando, T., and T. Nakajima. "Trial Construction of an Automatic Gas Concentration Regulation Device for Exposure Experiments," *J. Ind. Hyg. Japan* 12:31–34 (1971).
4. Oda, H., S. Kusumoto and T. Nakajima. "Nitrosyl-hemoglobin Formation in the Blood of Animals Exposed to Nitric Oxide," *Arch. Environ. Health* 30:453–456 (1957).
5. Nakajima, T., C. Chen, S. Kusumoto and K. Okamoto. "Studies on the Contents of Reduced Glutathione, and Histopathological Studies on the Lung of Mice Exposed to Nitrogen Dioxide Gas," *Proc. Osaka Pref. Inst. Publ. Health (Ed. Ind. Health)* 7:35–41 (1969).
6. Hattori, S., R. Tateishi, T. Horai, T. Nakajima and T. Miura. "Morphological Changes in Mouse Bronchial-Alveolar System due to Exposure to Low-Concentration NO_2 and CO Gases," *Jap. J. Thorac. Dis.* 10:16–21 (1972).
7. Motomiya, K., K. Ito, R. Yoshida, G. Ide, Y. Otsu and T. Nakajima. "Effects of NO_2 Gas Exposure on Influenza Virus Infection of Mice," *Rep. Environ. Res. Org. Chiba Univ.* 1:27–33 (1972).
8. Chen, C., S. Kusumoto and T. Nakajima. "The Processes of Histopathological Changes in the Respiratory Organs of Mice after NO_2 Exposure with Special Reference to Chronic Trachititis and Bronchitis," *Proc. Osaka Pref. Inst. Publ. Health (Ed. Ind. Health)* 10:43–49 (1972).
9. Ito, K., K. Motomiya, R. Yoshida, H. Otsu and T. Nakajima. "Effects of NO_2 Inhalation on Influenza Virus Infection in Mice," *Jap. J. Hyg.* 26: 304–314 (1971).
10. Nakajima, T., and S. Kusumoto. "Effects of Nitrogen Dioxide Exposure on the Contents of Reduced Glutathione in Mouse Lung," *Proc. Osaka Pref. Inst. Publ. Health (Ed. Ind. Health)* 6:17–21 (1968).
11. Nakajima, T. "Biological Effects of Nitrogen Oxides–A Summary of Animal Exposure Experiments," *J. Jap. Soc. Air Poll.* 8:223–233 (1973).
12. Nakajima, T., and S. Kusumoto. "Studies on the Carboxyhemoglobin Level of Mice Continuously Exposed to the Mixed Gas of Carbon Monoxide and Nitrogen Dioxide," *Proc. Osaka Pref. Inst. Publ. Health (Ed. Ind. Health)* 8:25–28 (1970).
13. Oda, H., H. Nogami, S. Kusumoto, T. Nakajima, A. Kurata and K. Imai. "Long-term Exposure to Nitric Oxide in Mice," *J. Jap. Soc. Air Poll.* 11:150–160 (1976).
14. Oda, H., H. Nogami, S. Kusumoto, T. Nakajima and A. Kurata. "Lifetime Exposure of 2.4 ppm Nitric Oxide to Mice." In Preparation.
15. Ogita, S., T. Shimamoto, M. Ohnishi, T. Kamei, H. Noma, O. Ishiko, T. Ando and T. Sugawa. "Hemolytic Pattern of Erythrocytes in the Newborn Measured by the Coil Planet Centrifuge System and its Relationship to Neonatal Jaundice," *Eur. J. Pediat.* 127:67–73 (1978).
16. Roy, R. B., R. W. Shaw and G. E. Connel. "A Simple Method for the Quantitative Determination of Serum Haptoglobin," *J. Lab. Clin. Med.* 74:698–704 (1969).

17. Beutler, E., O. Duron and B. M. Kelly. "Improved Method for the Determination of Blood Glutathione," *J. Lab. Clin. Med.* 61:882–888 (1963).
18. Blair, W. H., M. C. Henry and R. Ehrlich. "Chronic Toxicity of Nitrogen Dioxide. II. Effect on Histopathology of Lung Tissue," *Arch. Environ. Health* 18:186–192 (1969).
19. Freeman, G., R. J. Stephens, S. C. Crane and N. J. Furiosi. "Lesion of the Lung in Rats Continuously Exposed to Two Parts per Million of Nitrogen Dioxide," *Arch. Environ. Health* 17:181–192 (1968).
20. Ehrlich, R., and M. C. Henry. "Chronic Toxicity of Nitrogen Dioxide. I. Effect on Resistance to Bacterial Pneumonia," *Arch. Environ. Health* 17:860–865 (1968).
21. Henry, M. C., J. Findlay, J. Spangler and R. Ehrlich. "Chronic Toxicity of NO_2 in Squirrel Monkeys. III. Effect on Resistance to Bacterial and Viral Infection," *Arch. Environ. Health* 20:566–570 (1970).
22. Coffin, D. L., D. E. Gardner and E. J. Blommer. "Time/Dose Response for Nitrogen Dioxide Exposure in an Infectivity Model System," *Environ. Health Perspect.* 13:11–15 (1976).
23. Hattori, S., and M. Matsuda. "Relationship Between Cancer and Tuberculosis, 2) A Retarding Effect of BCG Inoculation for Tumor Induction in Mice with INH or Urethane-4NQO Injections," *Kekkaku (Jap. J. Tubercul.)* 39:449–452 (1964).
24. Freeman, G., S. C. Crane, R. J. Stephens and N. J. Furiosi. "The Subacute Nitrogen Dioxide-induced Lesion of the Rat Lung," *Arch. Environ. Health* 18:609–612 (1969).
25. Furiosi, N. J., S. C. Crane and G. Freeman. "Mixed Sodium Chloride Aerosol and NO_2 in Air. Biological Effects on Monkeys and Rats," *Arch. Environ. Health* 27:405–408 (1973).
26. Evans, M. J., L. J. Cabral, R. J. Stephens and G. Freeman. "Renewal of Alveolar Epithelium in the Rat Following Exposure to NO_2," *Am. J. Pathol.* 70:175–198 (1973).
27. Suhs, R. H., J. L. Lumeng and M. H. Lepper. "An Experimental Immunologic Approach to the Induction and Perpetuation of Chronic Bronchitis," *Arch. Environ. Health* 18:564–573 (1969).
28. Chow, C. K., C. J. Dillard and A. L. Tappel. "Glutathione Peroxidase System and Lysozyme in Rats Exposed to Ozone or Nitrogen Dioxide," *Environ. Res.* 7:311–319 (1974).
29. Fukase, O., K. Isomura and H. Watanabe. "Effects of Nitrogen Oxides on Peroxidative Metabolism of Mouse Lung," *J. Jap. Soc. Air Poll.* 11:65–69 (1976).
30. Wagner, W. D., B. R. Duncan, P. G. Wright and H. E. Stokinger. "Experimental Study of Threshold Limit of NO_2," *Arch. Environ. Health* 10:455–466 (1965).
31. Ehrman, R. A., M. Treshow and I. M. Lytle. "The Hematology of Mice Exposed to Nitrogen Dioxide," *Am. Ind. Hyg. Assoc. J.* 33:751–755 (1972).
32. Lutmer, R. F., K. A. Busch and P. L. Delong. "Effect of Nitric Oxide, Nitrogen Dioxide, or Ozone on Blood Carboxyhemoglobin Concentration during Low Level Carbon Monoxide Exposure," *Atmos. Environ.* 1:45–48 (1967).

33. Oda, H., S. Kusumoto and T. Nakajima. "A Difference in Nitrosyl-hemoglobin Formation among Animal Species," *J. Jap. Soc. Air Poll.* 9:714–716 (1975).
34. Sancier, K. M., G. Freeman and J. S. Mills. "Electron Spin Resonance of Nitric Oxide-Hemoglobin Complexes in Solution," *Science* 137:752–754 (1962).
35. Oda, H., H. Nogami, S. Kusumoto and T. Nakajima. "Nitrosylhemoglobin and Carboxyhemoglobin in the Blood of Mice Simultaneously Exposed to Nitric Oxide and Carbon Monoxide," *Bull. Environ. Contam. Toxicol.* 16:582–587 (1976).
36. Murphy, S. D., C. E. Ulrich, S. H. Frankowitz and C. Xintaras. "Altered Function in Animals Inhaling Low Concentrations of Ozone and Nitrogen Dioxide," *Am. Ind. Hyg. Assoc. J.* 25:246–253 (1964).
37. Oda, H., H. Nogami and T. Nakajima. "Reaction of Hemoglobin with Nitric Oxide and Nitrogen Dioxide in Mice." In Preparation.
38. Pflesser, G. "Die Bedeutung des Stickstoffmonoxyds bei der Verigiftung durch Nitrose Gas," *Arch. Exp. Pathol. Pharmacol.* 179:545–559 (1935).
39. Yoshida, K., M. Imai, K. Kasama, M. Kitabatake and M. Okuda. "Studies on the Metabolic Fate of Nitric Oxide with ^{15}N," *Report Environ. Sci., Mie Univ.* (3):11–16 (1978).
40. Greenraum, R., J. Bay, M. D. Hargreaves, M. L. Kain, G. R. Kelman, J. F. Nunn, C. Prys-Roberts and K. Siebold. "Effects of Higher Oxides of Nitrogen on the Anaesthetized Dog," *Brit. J. Anaesth.* 39:393–404 (1967).
41. Shiel, F. O. "Morbid Anatomical Changes in the Lungs of Dogs After Inhalation of Higher Oxides of Nitrogen During Anaesthesia," *Brit. J. Anaesth.* 39:413–424 (1967).
42. Toothill, C. "The Chemistry of the *in Vivo* Reaction Between Hemoglobin and Various Oxides of Nitrogen," *Brit. J. Anaesth.* 39:405–412 (1967).
43. Vaughan, T. R., L. F. Jennelle and T. R. Lewis. "Long-term Exposure to Low Levels of Air Pollutants," *Arch. Environ. Health* 19:45–50 (1969).
44. Bloch, W. N., Jr., T. R. Lewis, K. A. Busch, J. G. Orthoefer and J. F. Stara. "Cardiovascular Status of Female Beagles Exposed to Air Pollutants," *Arch. Environ. Health* 24:342–353 (1972).
45. Lewis, T. R., W. J. Moorman, Y. Yang and J. F. Stara. "Long-term Exposure to Auto Exhaust and Other Pollutant Mixtures—Effects on Pulmonary Function in the Beagles," *Arch. Environ. Health* 29:102–106 (1974).
46. Azoulay, E., P. Soler, M. C. Blays and F. Basset. "Nitric Oxide Effects on Lung Structure and Blood Oxygen Affinity in Rats," *Bull. Eur. Physiopath. Resp.* 13:629–644 (1977).
47. Kitabatake, M. "The Effect of Air Pollutant on Experimental Provoke of Asthma Attack in Guinea Pigs. III. The Effects of Nitrogen Monoxide and Nitrogen Dioxide," *J. Jap. Soc. Air Poll.* 10:718–721 (1976).
48. Wagner, H. M. "Absorption von NO und NO_2 in MIK- und MAK-Konzentrationen bei der Inhalation," *Staub-Reinhalt. Luft* 30:380–381 (1970).
49. von Nieding, G., und H. M. Wagner. "Vergleich der Wirkung von Stickstoffdioxid und Stickstoffmonoxid auf die Lungenfunktion des Menschen," *Staub-Reinhalt. Luft* 35:175–178 (1975).
50. Nogami, H., H. Oda and T. Nakajima. "Studies on Nitrosylhemoglobin and Methemoglobin in the Blood of Animals and Human," *Proc. 18th Annual Congress of Jap. Soc. Air Poll.* p. 216 (1977).

CHAPTER 9

ABSORPTION AND TRANSPORT OF NITROGEN OXIDES

Elliot Goldstein and Frederic Goldstein
Section of Infectious and Immunologic Diseases
Department of Internal Medicine
School of Medicine
The California Primate Research Center of the College
of Veterinary Medicine

Neal F. Peek
Department of Physics, Letters and Science

Norris J. Parks
Radiobiology Laboratory
University of California at Davis
Davis, California 95616

INTRODUCTION

The health effects of an air pollutant are determined by the absorption, transport and metabolism of the contaminant. Pollutants such as sulfur dioxide (SO_2), which are absorbed within the nasopharynx and upper airways, primarily damage these regions [1-3]. Other pollutants like ozone interact at and damage more distal bronchial and bronchiolar regions [4-6]. In contrast to the aforementioned irritative gases, carbon monoxide penetrates without injuring alveolar–blood barriers and then binds to hemoglobin causing extrapulmonary abnormalities of hypoxemia [1,7]. Investigations of the fate and distribution of inspired nitrogen dioxide (NO_2) have produced similarly valuable toxicological information. Studies conducted in human volunteers [18], rhesus monkeys [9], rabbits [10-12], cats [13] and dogs [12-14], and in model airway systems designed to simulate the

interaction of NO_2 with pulmonary tissues [15], permit reasonable conclusions concerning the effect of NO_2 exposure on pulmonary and extrapulmonary tissues. The fate of inspired nitric oxide (NO), the only other
nitrogen oxide of potential biological importance, has been studied in humans [16], rabbits [16,17], mice [17], rats [17,18] and dogs [19,20]. These
investigations, although less extensive than those of NO_2, have yielded important information concerning the interaction of inspired NO and hemoglobin. This review relates the physical–chemical data of the absorption,
transport and metabolism of NO_2 and NO to toxicity.

NITROGEN DIOXIDE ABSORPTION IN HUMANS

The percentage of inspired NO_2 retained within humans has been measured by von Nieding and his colleagues [8]. These investigators exposed
10 healthy volunteers to 0.55–13.5 mg/m^3 (0.29–7.2 ppm) of NO_2 for brief,
but otherwise unspecified, periods. Nitrogen dioxide concentrations were
measured in inhaled and exhaled air by the Saltzman method. The results
show absorbances of 81–87% during normal respiration and absorbances
of 90% with maximal ventilation. Further studies in humans of the fate
and distribution of the retained NO_2 to our knowledge have not been reported.

NITROGEN DIOXIDE ABSORPTION AND DISTRIBUTION
IN RHESUS MONKEYS

We have studied the absorption and intra- and extrapulmonary distribution of ambient concentrations of NO_2 in rhesus monkeys using NO_2
labeled with tracer quantities of nitrogen-13 ($^{13}NO_2$) [9]. Since these studies
provide much of the quantitative data of the fate of inspired NO_2, the experimental details merit presentation. The ^{13}N (half-time, 10 min) was synthesized by the oxygen-16 (P,α) ^{13}N nuclear reaction in an isochronous
cyclotron [21]. The ^{13}N atoms reacted with oxygen target molecules, radia-
tion-produced chemical species and trace (20 ppm N_2) impurities to form
$^{13}NO_2$, ^{13}N-NO and ^{13}N-N. The $^{13}NO_2$ was separated cryogenically from the
other constitutents (freezing point, $-9.3°C$) and then admixed with the
desired concentrations of carrier NO_2 in a Mylar$^®$ reservoir.

Figure 1 is a schematic diagram of the experimental system used to test
the fate and distribution of continuously inspired NO_2. After anesthesia with
Ketamine$^®$ (10 mg/kg of body weight), a catheter was inserted into the
femoral artery and connected to a heparin source. The animal was then
placed in a restraining chair and allowed to adjust for 15–30 min to a face
mask with a two-way valve. A Wright spirometer measured the volume of
radioactively labeled $^{13}NO_2$ inspired from the Mylar reservoir. Exhaled gas
was collected in a second reservoir. Samples of inspired and expired gas
were removed for radioisotopic and chemical analysis via sampling ports,

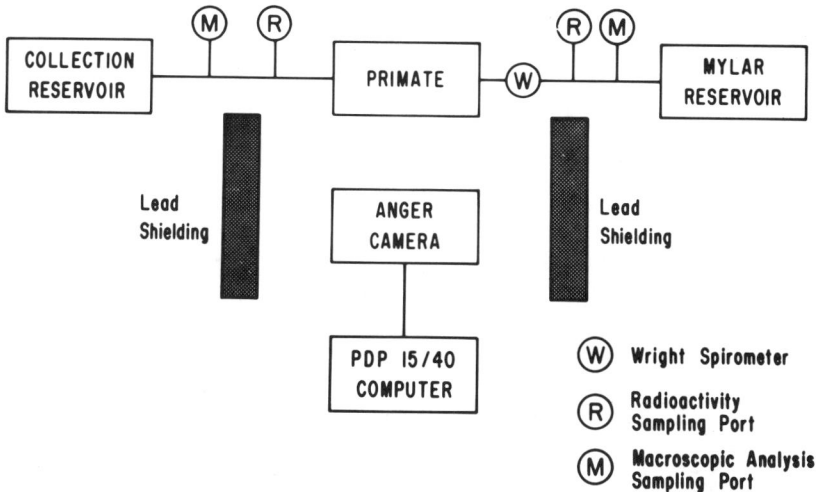

Figure 1. Schematic diagram of experimental system used to determine the fate of inspired $^{13}NO_2$ [9].

which were welded into the interconnecting tubing. The radioactivity concentrations of the 10-ml air samples and the 2.5-ml arterial blood samples were measured with a 2 × 2 in. NaI(Tl) detector. An Anger camera located directly behind the primate continuously measured $^{13}NO_2$ radioactivity in the thorax and surrounding regions. The spatial resolution of the radioisotopic disintegration products was optimized with a high-energy (550 KeV), parallel-hole collimator. The camera was interfaced to a Digital Corporation PDP 15/40 computer so that its information could be accumulated and sequentially stored for 10-sec periods in 64 × 64 arrays on a magnetic tape. Scintiphotographs were also obtained at one-minute intervals. Chemical concentrations of NO_2 were measured by the Saltzman method [22]. Sufficient quantities of radiolabeled $^{13}NO_2$ were produced to permit inspiration of the pollutant for 7–9 min. When the Mylar reservoir was nearly empty, the face mask was disconnected from the Wright spirometer and the animal was allowed to breathe room air. Pulmonary radioactivity was monitored for 10–15 min of this washout period.

A series of control experiments were performed in an identical fashion in the same animals, using Xenon-125 (half-time, 17 hr), a nonreactive gas synthesized by bombarding a sodium iodide (NaI) target with 37 MeV protons to produce the iodine-127 (p,3n) ^{125}Xe nuclear reaction (Figure 2). Cryogenic methods were used to separate the ^{125}Xe from contaminating materials [23].

Table I contains data showing that at concentrations of 0.56–1.71 mg/m³ (0.30–0.91 ppm), 54–63% of inspired NO_2 is retained within the primate.

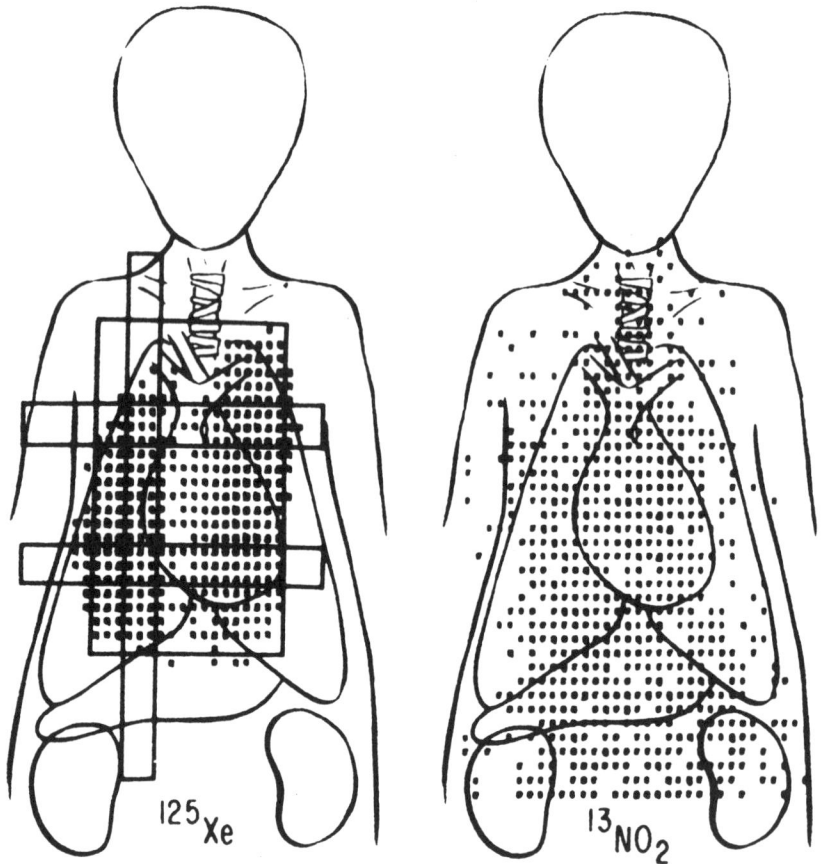

Figure 2. Diagram illustrating the intra- and extrapulmonary radioactivity patterns for ^{125}Xe and ^{13}NO$_2$. The large vertical rectangle encloses the region in which total radioactivity was measured continuously to determine the pulmonary absorption and retention of radioactive gas. The smaller vertical rectangle and two horizontal rectangles enclose regions that were used to study the intrapulmonary distribution of the radioactive gas [9].

Comparison of comparable radioactivity concentrations in samples of inspired and expired air reveals a 31–50% retention of ^{13}N radioactivity. Similar comparisons of inspired and expired ^{125}Xe show retentions of 10% for nonreactive gases over the same time period (Table II).

Figure 3, which compares the curves of pulmonary radioactivity following the inhalation of ^{13}NO$_2$ and ^{125}Xe, shows that the ^{13}NO$_2$ is absorbed at a uniform rate during the 9-min exposure period and that the reactivity is retained throughout the 21-min period following exposure. Since the slopes

Table I. Concentrations of $^{13}NO_2$ in Inspired and Expired Air at Various Times after Onset of Exposure [9]

Primate	Run Number	Time (sec)	Radioactivity (μCi/10 ml) Inspired	Radioactivity (μCi/10 ml) Expired	% $^{13}NO_2$[a] Retention	Chemical Concentrations Inspired	Chemical Concentrations Expired	% [NO_2] Retention
1	1	83	0.36					
		201		0.54				
		272	1.05		31	0.54	0.25	54
		392		0.72				
	2	47	0.28					
		198		0.37				
		283	0.65		34	0.91	---	---
		450	0.65					
		500		0.43				
2	1	90	0.14					
		130		0.07	45	0.48	---	---
		205	0.13					
	2	120	0.28					
		225		0.22	50	0.30	0.11	63
		395	0.60					
		435		0.30				

[a]Based on maximum radioactivity values for inspired and expired air.

Table II. Xenon-125 Radioactivity in Inspired and Expired Air at Various Times after Onset of Exposure [9]

Primate	Time (sec)	Radioactivity Inspired (μCi/10 cm^2)	Radioactivity Expired
1	33	3.78	
	200		3.38
	295	3.86	
	375		3.67
2	45	0.55	
	128		0.65
	210	0.81	

of the initial and terminal segments of the ascending limbs of the radioactivity curves are equal—900 counts/min·min and 900 counts/min·min for run 1 and run 2, saturation levels of NO_2 were not approached; therefore, continued inspiration of NO_2 would result in the absorption of even larger amounts of pollutant. In contrast, the inhalation of ^{125}Xe resulted in a more

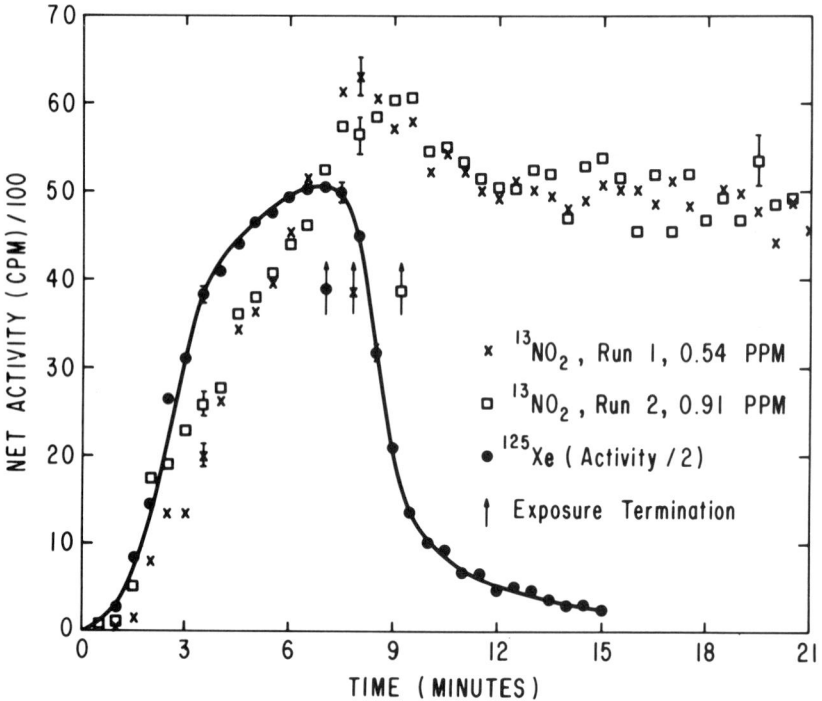

Figure 3. Pulmonary radioactivity following the inhalation of $^{13}NO_2$ and ^{125}Xe in a rhesus monkey [9].

rapid increase in pulmonary radioactivity during the exposure period and a precipitous decrease to negligible levels when the exposure was discontinued. The slope for the initial segment of the ^{125}Xe curve is 2000 count/min·min, and the slope for the final segment approaches zero, proving the attainment of equilibrium concentrations for the nonreactive gas.

The persistence of pulmonary ^{13}N radioactivity at virtually unchanged levels throughout the postexposure period confirms a firm binding of the pollutant or its chemical derivatives to pulmonary tissues. Such binding probably results from the interaction of inspired NO_2 with water in the vapor or liquid phase to form nitric (HNO_3) and nitrous acids (HNO_2), according to the following equations [24]:

$$2NO_2 = N_2O_4 \text{ (fast)} \tag{1}$$

$$N_2O_4 + H_2O = HNO_2 + HNO_3 \text{ (slow)} \tag{2}$$

When water is in the vapor phase, the HNO_2 decomposes to form nitric oxide (NO), NO_2 and water [24].

$$2HNO_2 = NO + NO_2 + H_2O \text{ (fast)} \tag{3}$$

Whether this decomposition occurs in liquid-phase reactions depends on the acidity of the absorbing solution [24,25]. A basic solution favors retention of nitrite ion, whereas in an acidic solution, the nitrous acid will decompose.

The balance overall equation for reactions 1, 2 and 3 is:

$$3NO_2 + H_2O = NO + 2HNO_3 \ . \eqno(4)$$

Although these reactions are unstudied at physiological conditions, the available chemical data suggest that the liquid-phase reactions (1 and 2) are likely to occur. Experiments in which wetted wall tubes were used to simulate pulmonary surfaces have shown that at nitrogen dioxide concentrations and air flowrates similar to those of our experiment [9.4 mg/m^3 (5.0 ppm) NO$_2$; 1.0 liter/min air flowrate], 50% of the NO$_2$ is absorbed as the nitrite [15].

The formation of nitric and nitrous acids by a gas-phase reaction of NO$_2$ and water vapor is less likely; experiments conducted at 35°C with 28–56 mg/m^3 of NO$_2$ (15–30 ppm) and 1.2% by volume of water vapor show that only minute fractions of the equilibrium concentrations of HNO$_3$ and NO are formed in 30 minutes [24]. Because the reaction of water vapor and NO$_2$ follows first-order kinetics in respect to water vapor and second-order kinetics in respect to NO$_2$, even smaller amounts of reaction products will occur at the ambient concentrations of the present experiments.

The chemical and radioactivity measurements of NO$_2$ and ^{13}N concentrations in inspired and expired air further substantiates the probable conversion of NO$_2$ to HNO$_3$ within the lungs. The diminished pulmonary absorption of the ^{13}N labeled species when compared to NO$_2$ can be attributed to transfer of the ^{13}N label from chemically detectable ^{13}NO$_2$ in inspired gas to chemically undetectable H^{13}NO$_3$ or ^{13}NO in expired gas. A liquid-phase reaction of NO$_2$ and water in the lung would result in the formation of H^{13}NO$_3$ and H^{13}NO$_2$; some of these radiolabeled molecules might exit the lung as aerosol droplets. In such a circumstance the H^{13}NO$_2$ would be detected radiochemically and chemically in the expired air, but the H^{13}NO$_3$ would only be detected radiochemically. The same argument holds for the conversion of H^{13}NO$_2$ to ^{13}NO a poorly soluble gas, which would not be detected in these experiments. Since the concentration of NO was unmeasured its quantitative importance as a reaction product is uncertain. A crude estimate of the maximum amount of ^{13}NO radioactivity in expired air can be made if it is assumed that: (1) the relationship of ^{13}N radioactivity to NO$_2$ chemical concentration in inspired air also applies for expired air; and (2) the absorption of three moles of ^{13}NO$_2$ results in the formation of one mole of ^{13}NO and two moles of H^{13}NO$_3$ and only the ^{13}NO exits the lung. Such assumptions result in the following equation:

$$^{13}NO = \frac{A\left[[NO_{2I}] - [NO_{2E}]\right]}{3}$$

where   A = ^{13}N radioactivity in inspired air divided by the chemical concentration of NO_2 in inspired air ($\mu Ci/10$ ml/mg/m^3)

$[NO_{2_I}]$ = chemical concentration of NO_2 in inspired air (mg/m^3)

$[NO_{2_E}]$ = chemical concentration of NO_2 in expired air (mg/m^3)

Substitution of the appropriate values in the above formula reveals ^{13}NO concentrations of 0.19 $\mu Ci/10$ ml and 0.13 $\mu Ci/10$ ml in run 1—primate 1 and run 2—primate 2. When these concentrations are subtracted from the ^{13}N concentrations in the expired air, ^{13}NO$_2$ retention becomes 50% in run 1—primate 1, and 74% in run 2—primate 2, values that are similar to the chemically measured absorptions of 54 and 63%.

The amount of HNO_3 formed within the lung from the continuous inhalation of NO_2 can also be calculated from Equation 4 if the tidal volume, dead space and respiratory rate are known or assumed.

$$[HNO_3]min = \frac{2}{3} \frac{(TV-DS) \times RR \times \left[[NO_{2_I}] - [NO_{2_E}] \right] \times HNO_3 \text{ mol wt}}{NO_2 \text{ mol wt}}$$

where $[HNO_3]min$ = concentration of nitric acid formed (mg/min)

TV = tidal volume (m^3)

DS = dead space (m^3)

RR = respiratory rate (breaths/min)

$[NO_{2_I}]$ = concentration of NO_2 in inspired air (mg/m^3)

$[NO_{2_E}]$ = concentration of NO_2 in expired air (mg/m^3)

HNO$_3$ mol wt = 63

NO$_2$ mol wt = 46

If the tidal volume in the primate is assumed to be 43.5 ml, the deadspace 12.6 ml, and the respiratory rate 42 breaths/min [26], then the inhalation of 0.56 mg/m^3 (0.30 ppm) of NO$_2$ for one minute results in the intrapulmonary formation of 0.0004 mg of HNO$_3$. Similar computations for 7- and 60-minute exposures to 0.56 mg/m^3 (0.30 ppm) of NO$_2$ show the formation of 0.0028 and 0.024 mg of HNO$_3$.

The reactivity of NO$_2$ with pulmonary tissues is also apparent in the differences in distribution patterns of the pollutant and the chemically inert ^{125}Xe control. Figures 4 and 5 show that more ^{13}N radioactivity is in the upper lobe portions of vertical sections from apex to the base of the lung than in lower lobe sections, a result opposite to that observed with ^{125}Xe, in which radioactivity concentrations are higher in lower lobes because of their better ventilation [27,28]. The differences in ^{13}N and ^{125}Xe radioactivity are

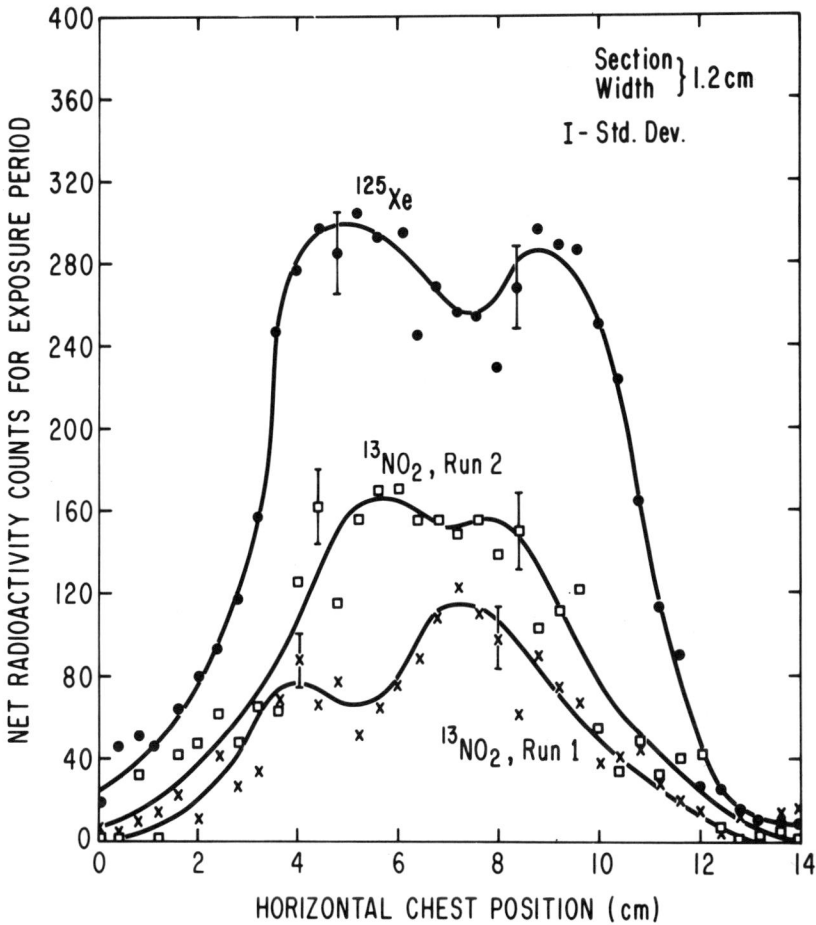

Figure 4. Comparison of ^{13}N and ^{125}Xe radioactivity in horizontal sections through the upper (A) and lower (B) pulmonary lobes of a rhesus monkey [9].

attributable to conversion of NO_2 to HNO_3 and HNO_2 and the circulatory removal of these compounds. Blood flow in the upright lung increases in a linear fashion from the apex to the base [28,29]. Hence, even though more NO_2 entered the lower lobes, this increase was offset by the enhanced rate at which the pollutant and its chemical derivatives were removed via the blood. The absorption of $^{13}NO_2$ or its chemical derivatives accounts for the rapid dissemination of these compounds throughout the body and the presence of blood radioactivity long after the exposure has been discontinued (Figure 6).

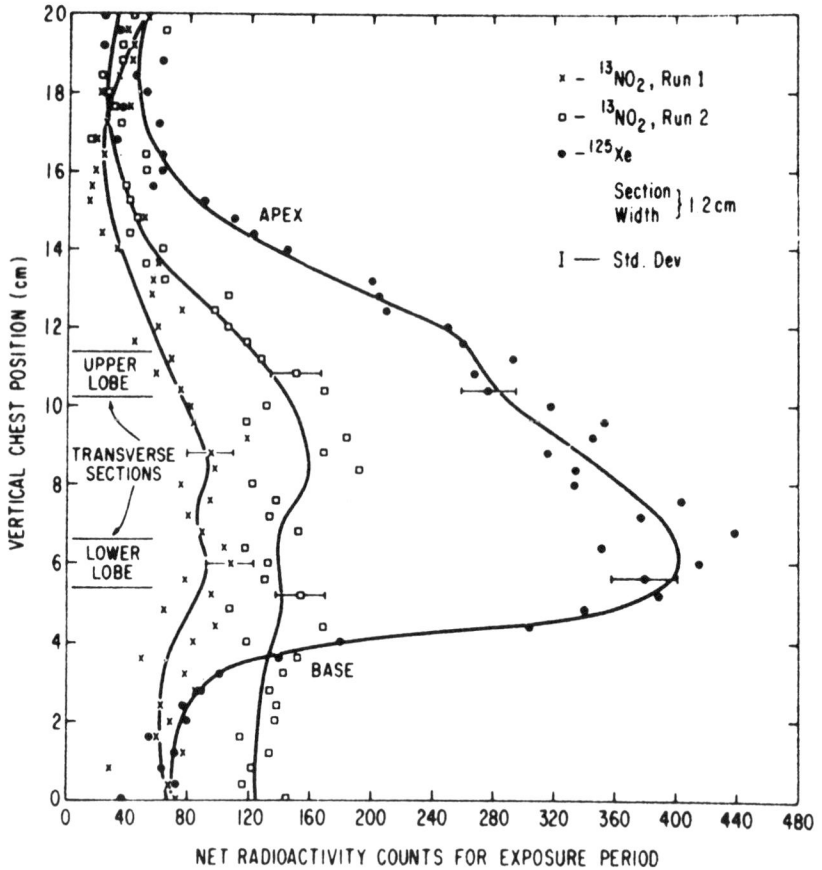

Figure 5. Comparison of ^{13}N and ^{125}Xe radioactivity in a vertical section from the apex to the base of the lung of a rhesus monkey [9].

NITROGEN DIOXIDE ABSORPTION IN OTHER MAMMALS

That high percentages of inspired NO_2 are absorbed at uniform rates throughout the primate lung and that once absorbed, NO_2 or its chemical derivatives (HNO_3 and HNO_2) remain within the lung or disseminate to extrapulmonary sites, accords with experiments in other animal models. The uptake of NO_2 within the upper respiratory tract of tracheotomized rabbits and dogs has been measured by Dalhamn and Sjoholm [10], Yokoyama [11] and Vaughn and co-workers [14]. Dalhamn and Sjoholm [10] showed that the continuous inhalation of 240–412 mg/m^3 (130–220 ppm) of NO_2 for

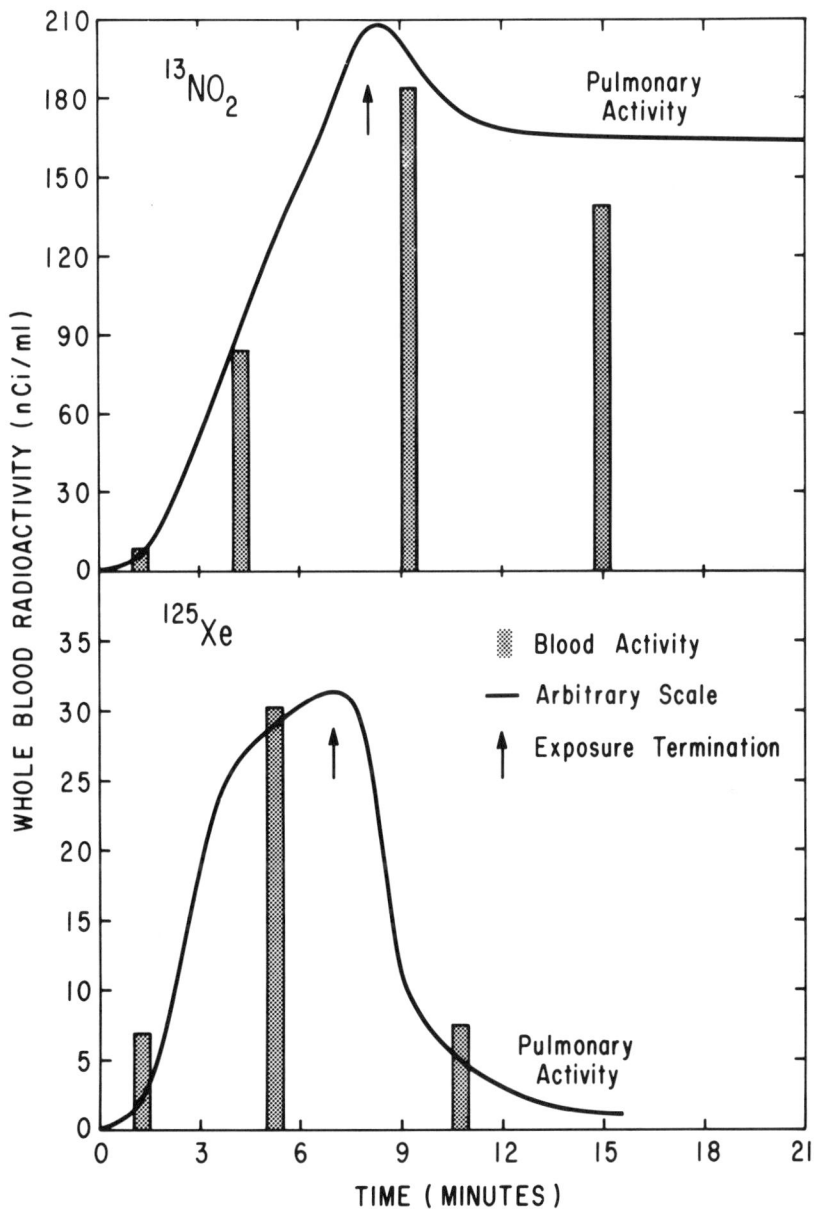

Figure 6. Relationship of pulmonary and blood radioactivity in a rhesus monkey at various times after the inhalation of $^{13}NO_2$ or ^{125}Xe [9].

periods of 15–45 minutes resulted in the retention of 31–78% of the pollutant in the rabbit. Similar retentions of approximately 42% were reported by Yokoyama [11] in rabbits and dogs for lower concentrations of 7.5–75 mg/m^3 (5–41 ppm) of NO_2. At very low concentrations, 0.19–1.8 mg/m^3 (0.1–1.0 ppm), NO_2 absorption is even higher with retention rates of 90% in studies testing the absorption of NO_2 from automobile exhausts [14]. Whether the quantitative differences in NO_2 absorption between the first two and the third studies are due to the concentrations tested or reflects technical differences in these experiments is uncertain. Regardless of the cause for the variations in absorption, the three studies confirm the retention of high percentages of inspired NO_2.

The finding of a uniform rate of absorption of NO_2 in nonhuman primates correlates with results obtained in studies of NO_2 absorption within the upper respiratory tract of tracheotomized cats [13]. Corn and co-workers [13] have shown that the mass transfer coefficient for NO_2 (a sophisticated measurement of absorption) is relatively constant for 30-minute exposures to NO_2 concentrations of 4.1–10.3 mg/m^3 (2.2–5.5 ppm). These studies also demonstrated that the absorption of NO_2 is independent of the frequency and depth of breathing.

The conclusion that NO_2 is converted to nitric and nitrous acids within the lungs and that these chemical derivatives enter the blood stream has been confirmed by Scorcova and Kaut [12] who measured nitrite and nitrate concentrations in arterial and venous blood and urine of rabbits exposed to 45 mg/m^3 (24 ppm) of NO_2 for 4 hours. These investigators reported increased concentrations of nitrites and nitrates in blood and urine and also the formation of methemoglobin.

NITROGEN DIOXIDE ABSORPTION IN MODEL AIRWAY SYSTEMS

Studies by Ichioka [15] of model airway systems designed to simulate the dynamic behavior of NO_2 within the respiratory system confirm the propensity of the pollutant to react at distal sites. This investigator constructed a system of glass tubes, filter papers and terminal bubblers filled with NO_2 absorbing reagents, in which the different parts corresponded to bronchial airways and pulmonary alveoli. Nitrogen dioxide concentrations of 9.4 mg/m^3 (5.0 ppm) were passed through the system, and the amount of NO_2 absorbed in the individual glass tubes and in the bubblers was measured by the Saltzman method. The results clearly showed that most of the NO_2 reached the terminal bubblers mimicking passage through the upper airways into the alveoli [15].

TOXICOLOGICAL CORRELATIONS

The formation of increasing amounts of highly injurious nitric and nitrous acids throughout the lung following the inhalation of NO_2 probably accounts for the irritative properties of the pollutant. As a result of this irritation, increases in airway resistance are common in humans exposed for brief periods to concentrations of NO_2 of greater than 2.0 ppm [9,30−32]. The irritative properties of NO_2 are even more marked in vulnerable individuals such as asthma patients [33]. The propensity of NO_2 to penetrate to alveolar areas may explain, in part, impairments in bacterial defense mechanisms noted in animals and humans exposed to above ambient concentrations of the pollutant [34−36]. In this situation the injurious acids probably damage the alveolar macrophage [35,37], the cell most involved in the initial defense of the lung against bacterial infection [38,39]. The acids also cause extravasation of fluids from the vascular space producing pulmonary edema which further compromises bacterial defense systems [40].

Since the chemical nature of all the nitrogen containing compounds that enter the blood is uncertain, only inferential statements can be made regarding hematological and extrapulmonary toxicity. If nitrates are the principal compound, hematological and extrapulmonary toxicity is probably minimal because the rapid excretion of this anion renders it nontoxic to mammals [41]. Nitrites, on the other hand, can affect hemoglobin to form methemoglobin [41,43]. Absorption of nitrites into the bloodstream would account for the increased concentrations of methemoglobin that have been reported in persons exposed to high ambient concentrations of the pollutant [42] and laboratory animals exposed to artificaly high concentrations (188−282 mg/m^3, 100−150 ppm) [43]. Because maximal methemoglobin concentrations in humans exposed to high ambient concentrations of NO_2 are much below the 10% threshold for methemoglobin toxicity [41], significant oxygen deprivation is unlikely as a consequence of nitrite ion formation.

Nitrates and nitrites formed from the inhalation of NO_2 can theoretically be involved in the formation of N-nitroso compounds (nitrosamines and nitrosamides), which are potent carcinogens [44]. Recent studies have shown that N-nitroso compounds are formed in vivo in the gastrointestinal tract by the interaction of nitrite with secondary and tertiary amino compounds [44]. Since ingested nitrates can be reduced to nitrites by intestinal bacteria, some investigators have attempted to link environmental nitrates as well as nitrites with human cancer [45,46]. At present, this hypothesized relationship remains highly speculative. In view of the miniscule amounts of nitrite and nitrate derived from inspired NO_2 when compared with amounts ingested, NO_2 is unlikely to be a significant source for these anions even if they are linked to human cancer via the formation of N-nitroso compounds.

NITRIC OXIDE ABSORPTION AND METABOLISM

Nitric oxide (NO), the precursor of NO_2 [47], is a much less soluble gas (4.7 ml/100 ml of water at 20°C and 1 atm pressure) [48] than is NO_2. As such, its absorption following inspiration is probably much less than the

50–90% values reported for NO_2. Because, to the authors' knowledge, NO absorption has not been measured in humans or animal models, the actual absorption is unknown. The intrapulmonary distribution of the pollutant is also unknown.

Physiological studies of the reaction of inspired NO with mammalian tissues have concentrated on the reaction of the gas with hemoglobin to form methemoglobin (met-Hb) and nitrosyl-hemoglobin (NO-Hb). The older studies which were performed with very high concentrations of NO {24,600 mg/m³ (20,000 ppm) [19,20]; 12.3 mg/m³ (10 ppm) [18]} showed that met-Hb was formed under these conditions. Nitrosyl hemoglobin was not detected even by electron spin resonance spectrometry [18].

Recent studies in rabbits and humans confirmed the formation of met-Hb at lower, ambient concentrations of pollutant mixtures [1.2–3.7 mg/m³ (1.0–3.0 ppm) of NO; 0.56 mg/m³ (0.30 ppm) of NO_2; 55 mg/m³ (50 ppm) of CO] [16]. Although specifically sought, NO-Hb was not detected in these tests using electron paramagnetic resonance spectroscopy. In one experiment that merits emphasis, a volunteer stationed near the entrance of an automobile tunnel for 3 hours manifested increases in met-Hb concentration from 0.5–1.5% of his total hemoglobin. Further, the slope of the rising met-Hb concentration indicated that continued exposure would result in even higher concentrations of met-Hb. Because the met-Hb was in the high spin state, these investigators postulated the presence of NO-Hb as an undetected chemical intermediate in the formation of met-Hb. They attributed their inability to detect NO-Hb to its existence at levels below 0.01%.

Nitrosyl-hemoglobin has been detected using electron spin resonance spectrometry in the blood of mice and rats following 1-hour exposures to NO concentrations of 2.4–13.0 mg/m³ (2–10.6 ppm) [17]. In these experiments, corresponding increases were observed for NO and NO-Hb concentrations; at exposure levels of 13.0 mg/m³ (10.6 ppm) of NO, NO-Hb represented 0.13% of total hemoglobin. Since only 0.01% of the total hemoglobin was NO-Hb in the low-level NO exposures [2.4 mg/m³ (2 ppm)], the inability to detect NO-Hb in the previously cited human experiments may indeed reflect technical difficulties in measuring small concentrations. That such is the case is suggested by recent evidence in which NO-Hb was detected in human blood by an isotopic dilution technique using [15]NO followed by field-ionization mass spectrometry [49].

The finding that continued exposure to low levels of NO results in increased met-Hb and NO-Hb has important implications for health. Methemoglobin is an ineffective oxygen carrier; therefore, in vulnerable individuals such as those with hypoxia, secondary to pulmonary or cardiac disease, the replacement of hemoglobin by met-Hb will reduce already critically low oxygen-carrying capacities. Although biological studies confirming the deleterious nature of met-Hb formation secondary to exposure to ambient concentrations of NO are not available, such reactions are theoretically capable of augmenting tissue damage in hypoxia individuals.

ACKNOWLEDGMENTS

This study was supported by U.S. Public Health Service Research Grants No. ES 01327 and ES 00628 from the National Institute of Environmental Health; Grant No. RP-680 from the Electric Power Research Institute; and Grant No. 1116 from the Air Resources Board of California.

The authors gratefully acknowledge the editor and publisher of the journal *American Review of Respiratory Diseases* for permission to reprint the tables and figures used in this chapter.

REFERENCES

1. Coffin, D. L., and H. E. Stokinger. *Biological Effects of Air Pollutants in Air Pollution*, 3rd ed., A. C. Stern, Ed. (New York: Academic Press, Inc., 1977), pp. 231–260.
2. Speizer, F. E., and N. R. Frank. "The Uptake and Release of SO_2 by the Human Nose," *Arch. Environ. Health* 12:725–728 (1966).
3. Frank, N. R., R. E. Yoder, E. Yokoyama and F. E. Speizer. "The Diffusion of $^{35}SO_2$ from Tissue Fluids Into the Lungs Following Exposure of Dogs to $^{35}SO_2$," *Health Phys.* 13:31–38 (1967).
4. Fisher, B. R. "Models of Soluble Gas Retention in the Nasal-Pharyngeal Airways," *Health Phys.* 23:430 (1972).
5. Mellick, P. W., D. L. Dungworth, L. W. Schwartz and W. S. Tyler. "Short-Term Morphologic Effects of High Ambient Levels of Ozone on Lungs of Rhesus Monkeys," *Lab Invest.* 36:82–90 (1977).
6. Goldstein, E., H. C. Bartlema, M. van der Ploeg, P. van Duijn, J. G. M. M. van der Stap and W. Lippert. "Effect of Ozone on Lysosomal Enzymes of Alveolar Macrophages Engaged in Phagocylosis and Killing of Inhaled *Staphylococcus aureus*," *J. Infect. Dis.* 138:299–311 (1978).
7. Goldsmith, J. R., and E. P. Redford. "Medical Aspects of Air Pollution," in *Harrison's Principles of Internal Medicine*, 6th ed., M. M. Wintrobe, Ed. (New York: McGraw-Hill Book Company, 1970), pp. 1329–1332.
8. von Nieding, G., M. Wagner, H. Krekeler, U. Smidt and K. Muysers. "Adsorption of NO_2 in Low Concentrations in the Respiratory Tract and its Acute Effects on Lung Function and Circulation," Second International Clean Air Congress at the Union of Air Pollution Prevention Association, Washington, D.C., December 6–11, 1970.
9. Goldstein, E., N. F. Peek, N. J. Parks, H. H. Hines, E. P. Steffey and B. Tarkington. "Fate and Distribution of Inhaled Nitrogen Dioxide in Rhesus Monkeys," *Am. Rev. Resp. Dis.* 115:403–412 (1977).
10. Dalhamn, T., and J. Sjoholm. "Studies on SO_2, NO_2, and NH_3: Effect on Ciliary Activity in Rabbit Trachea of Single *In Vitro* Exposure and Reabsorption in Rabbit Nasal Cavity," *Acta Physiol. Scand.* 58:287–291 (1963).
11. Yokoyama, E. "Uptake of SO_2 and NO_2 by the Isolated Upper Airways," *Bull. Inst. Publ. Health* 17:302–306 (1968).
12. Svorcova, S., and V. Kaut. "The Arterio-Venous Differences in the Nitrite and Nitrate Ions Concentrations After the Nitrogen Oxides Inhalation," *Cs. Hyg.* 16(2/3):71–76 (1971) (in Czechoslovakian).

13. Corn, M., N. Kotsko and D. Stanton. "Mass-Transfer Coefficient for Sulfur Dioxide and Nitrogen Dioxide Removal in Cat Upper Respiratory Tract," *Ann. Occup. Hyg.* 19:1–12 (1976).
14. Vaughan, R. R., L. E. Jennelle and T. R. Lewis. "Long Term Exposure to Low Levels of Air Pollutants," *Arch. Environ. Health* 19:45–50 (1969).
15. Ichioka, M. "Model Experiments on Absorbability of the Airway Mucous Membrane of SO_2 and NO_2 Gasses," *Bull. Tokyo Med. Dent. Univ.* 19: 361–375 (1972).
16. Case, G. D., J. C. Schooley and J. S. Dixon. "Uptake and Metabolism of Nitrogen Oxides in Blood," paper presented at the 20th Annual Meeting of the Biophysical Society. Seattle, Washington, February 24–27, 1976.
17. Oda, H., S. Kusumoto and T. Nakajima. "Notrosyl-Hemoglobin Formation in the Blood of Animals Exposed to Nitric Oxide," *Arch. Environ. Health* 30:453–456 (1975).
18. Sancier, K. M., G. Freeman and J. S. Mills. "Electron Spin Resonance of Nitric Oxide-hemoglobin Complexes in Solution," *Science* 137:752–754 (1963).
19. Greenbaum, R., J. Bay, M. D. Hargraeves, M. L. Kain, G. R. Kelman, J. F. Nunn, C. Prys-Roberts and K. Siebold. "Effects of Higher Oxides of Nitrogen on the Anesthetized Dog," *Brit. J. Anesthetol.* 39:393–404 (1967).
20. Toothill, C. "The Chemistry of the *In Vivo* Reaction Between Hemoglobin and Various Oxides of Nitrogen," *Brit. J. Anesthetol.* 39:405–412 (1967).
21. Parks, N. J., N. F. Peek and E. Goldstein. "The Synthesis of [13]N Labeled Atmospheric Gasses via Proton Irradiation of a High Pressure Oxygen Target," *Int. J. Appl. Radiat. Isot.* 26:683–687 (1975).
22. Saltzman, B. E. "Colorimetric Microdetermination of Nitrogen Dioxide in the Atmosphere," *Anal. Chem.* 26:1949–1955 (1954).
23. Hines, H. H., N. F. Peek, G. L. DeNardo and A. L. Jansholt. "Production and Characteristics of [125]Xe: A New Noble Gas for *In Vivo* Studies," *J. Nucl. Med.* 16:143–147 (1975).
24. England, C., and W. H. Corcoran. "Kinetics and Mechanisms of the Gas-Phase Reaction of Water Vapor and Nitrogen Dioxide," *Ind. Eng. Chem. Fund.* 13:373–384 (1974).
25. Wendel, M. M., and R. L. Pigford. "Kinetics of Nitrogen Tetroxide Adsorption in Water," *Am. Ind. Chem. Eng. J.* 4:249–256 (1958).
26. Lees, M. H., M. R. Malinow and J. T. Parer. "Cardiorespiratory Function of the Rhesus Monkey During Phencyclidine Anesthesia," in *DeCourt, LU Temas de Medicina* (Sao Paulo, Brazil: Livro Homenagem, 1965), pp. 61–69.
27. Ball, W. C., Jr., P. B. Stewart, L. G. S. Newsham and D. V. Bates. "Regional Pulmonary Function Studied with Xenon[133]," *J. Clin. Invest.* 41:519–531 (1962).
28. Newhouse, M. T., F. J. Wright, G. K. Ingham, N. P. Archer, L. B. Hughes and O. L. Hopkins. "Use of Scintillation Camera and [135]Xenon for Study of Topographic Pulmonary Function," *Resp. Physiol.* 4:141–253 (1963).
29. Dollery, C. T., and P. M. S. Gillam. "The Distribution of Blood and Gas Within the Lungs Measured by Scanning After Administration of ^{133}Xe," *Thorax* 18:316–325 (1963).

30. Abe, M. "Effects of Mixed NO_2-SO_2 Gas on Human Pulmonary Functions. Effects of Air Pollution on the Human Body," *Bull. Tokyo Med. Dent. Univ.* 14:415-433 (1967).

31. Yokoyama, E. "Effects of Acute Controlled Exposure to NO_2 on Mechanics of Breathing in Healthy Subjects," *Bull. Inst. Public Health* 17: 337-346 (1968) (in Japanese).

32. von Nieding, G., H. Krekeler, R. Fuchs, M. Wagner and K. Koppenhagen. "Studies of the Effects of NO_2 on Lung Function: Influence on Diffusion, Perfusion and Ventilation in the Lungs," *Int. Arch. Arbeits Med.* 31:61-72 (1971).

33. Orehek, J., P. Massari, P. Gayrard, C. Grimard and J. Charpin. "Effect of Short-Term, Low-Level Nitrogen Dioxide Exposure on Bronchial Sensitivity of Asthmatic Patients," *J. Clin. Invest.* 57:301-307 (1976).

34. Purvis, M. R., and R. Ehrlich. "Effect of Atmospheric Pollutants on Susceptibility to Respiratory Infection. Effect of NO_2," *J. Infect. Dis.* 113: 72-76 (1963).

35. Goldstein, E., M. C. Eagle and P. D. Hoeprich. "Effect of Nitrogen Dioxide on Pulmonary Bacterial Defense Mechanisms," *Arch. Environ. Health* 26:202-204 (1973).

36. Ramirez, R. J., and A. R. Dowell. "Silo-Filler's Disease: Nitrogen Dioxide-Induced Lung Injury. Long-Term Follow-Up and Review of the Literature," *Ann. Int. Med.* 74:569-576 (1971).

37. Gardner, A. E., R. S. Holzman and D. L. Coffin. "Effects of Nitrogen Dioxide on Pulmonary Cell Population," *J. Bacteriol.* 98:1041-1043 (1969).

38. Goldstein, E., and H. C. Bartlema. "Role of the Alveolar Macrophage in Pulmonary Bacterial Defense," *Bull. Eur. Physiopathol. Resp.* 13:57-67 (1977).

39. Green, G. M., G. J. Jakab, R. B. Low and G. S. Davis. "Defense Mechanisms of the Respiratory Membrane," *Am. Rev. Resp. Dis.* 115:479-514 (1977).

40. Guidotti, T. L. "The Higher Oxides of Nitrogen: Inhalation Toxicology," *Environ. Res.* 15:443-472 (1978).

41. National Academy of Science Committee on Nitrate Accumulation. "Hazards of Nitrate, Nitrite and Nitrosamines to Man and Livestock," pp. 46-75 (1972).

42. MacQuiddy, E. L., L. W. Latowsky, J. P. Tollman and A. I. Finlayson. "Toxicology of Oxides of Nitrogen. Physiological Effects and Symptomatology," *J. Ind. Hyg. Toxicol.* 23:134-147 (1941).

43. Szponar, L., and L. Bilczuk. "Methemoglobin Level Among the Population of a Village in an Area Exposed to Pollutants Emitted by a Hydrogen Plant," *Pol. Tyg. Lek.* 28:9-10 (1973) (in Polish).

44. Lijinsky, W. "Nitrosamines and Nitrosamides in the Etiology of Gastrointestinal Cancer," *Cancer* 40:2446-2449 (1977).

45. Zaldivar, R., and W. H. Wetterstand. "Further Evidence of a Positive Correlation Between Exposure to Nitrate Fertilizers ($NaNO_3$ and KNO_3) and Gastric Cancer Death Rates: Nitrites and Nitrosamines," *Experientia* 31:1354-1355 (1975).

46. Hill, M. J., G. Hawksworth and G. Tettersall. "Bacteria, Nitrosamines and Cancer of the Stomach," *Brit. J. Cancer* 28:562-567 (1973).

47. Pitts, J. N., and A. C. Lloyd. "Sources of Nitrogenous Compounds and Methods of Control: Discharges into the Atmosphere," in *Nitrogenous Compounds in the Environment*, Report of the U.S. Environmental Protection Agency Hazardous Materials Advisory Committee (Washington, D.C.: U.S. Government Printing Office, 1973), pp. 43–65.
48. *The Merck Index*, 8th ed., Merck and Co., Inc., Rahway, NJ (1968), p. 735.

CHAPTER 10

BIOCHEMICAL EFFECTS OF NITROGEN DIOXIDE ON ANIMAL LUNGS

Mohammad G. Mustafa

Division of Environmental and Nutritional Sciences
School of Public Health, and
Division of Pulmonary Diseases, Department of Medicine
University of California
Los Angeles, California 90024

Edward J. Faeder

Research and Development Department
Southern California Edison Company
Rosemead, California 91770

Si Duk Lee

Environmental Criteria and Assessment Office
U.S. Environmental Protection Agency
Cincinnati, Ohio 45268

INTRODUCTION

Oxides of nitrogen (NO_x), most notably nitric oxide (NO) and nitrogen dioxide (NO_2), are produced during fuel combustion. NO_x may arise as a direct combustion product from fuel nitrogen or may be generated from atmospheric nitrogen as a function of burning conditions, especially temperature. Although as many as eight oxides of nitrogen may be present in ambient air, NO and NO_2 are the major ones. These two gases coexist, but NO_2 is the usual end product since NO is oxidized to NO_2 either during the combustion process or subsequently in the air. Elevated levels of NO_x have occurred in ambient air of many American cities because of the increased consumption of fossil fuels for energy during the last three to four decades.

Motor vehicles and fossil fuel power generation are the major sources of NO_x emissions, but contribution from other sources may become significant locally.

The presence of NO_2 in the air has been of concern primarily because in the presence of hydrocarbons and solar radiation it undergoes a complex series of reactions giving rise to photochemical smog [1-5]. The potential adverse human reactions to photochemical smog, particularly to the powerful oxidant ozone (O_3) [6-11], and eye irritants, peroxyacyl nitrates (PAN) [12,13], have remained an important issue in the field of air pollution. Although NO_2 as an oxidant is much less powerful than ozone [9,11,14,15], its ambient concentration in ppm is often as high as, or higher than, that of ozone [16,17]. The elevated and somewhat ubiquitous presence of NO_2 therefore raises the question whether it would have any direct health effects, in addition to being an ingredient for the formation of photochemical smog.

This chapter reviews the relative short- and long-term effects of NO_2 as an air pollutant. It assesses the biological effects of NO_2 in terms of metabolic alterations in the lung. It should be noted that only limited information exists on the low-level NO_2 effects that have direct relevance to air pollution regulation. Many of the experimental studies with animals have dealt with relatively high levels of this pollutant [14,18-24], and the data generated are mostly appropriate for NO_2 effects in industrial or occupational environments, where accidental exposures leading to lung injury have been reported [14,22-26]. Nonetheless, data from dose-dependence and time-course studies involving various levels of NO_2 are important for understanding the mechanism of its action. Other considerations include: (1) the degree of NO_2 effects relative to ozone and PAN, (2) the fate and/or transformation of inhaled NO_2, and (3) extrapulmonary effects.

MODELS FOR STUDY

Biological effects of NO_2 have been studied in numerous animal species. Acute toxicity, including mortality, varies among species. Guinea pigs, mice and rats appear to be more sensitive than rabbits, dogs and others [27]. Typically, animal mortality may not result from a 1-hr exposure to 50 ppm (94 mg/m^3) NO_2, but an increase of either time or concentration will lead to mortality [27]. We have observed that a 24-hr exposure to 30 ppm NO_2 or a 12-hr exposure to 50 ppm (94 mg/m^3) NO_2 caused 50% mortality in 2-month old rats [28]. Therefore, in terms of mortality, ozone is at least 10 times more toxic than NO_2; a 3-4 ppm (60-80 mg/m^3) ozone exposure will cause more than 50% mortality in rats within 8 hr [29]. Although guinea pigs are found to be more sensitive to NO_2 than others [27], rats are commonly used in biochemical and morphological studies involving the lung because specific pathogen free (SPF) animals are preferred. SPF rats are routinely available from various established colonies.

Studies of NO_2 effects on the lung include biochemical, morphologic and physiologic approaches. They have been designed from three different views:

(1) those that elucidate biochemical or molecular mechanisms of NO_2 injury [8,9,11,14,15,23,30–35]; (2) those that delineate the sequence of tissue or cellular injury, including methods for detection of early damage [8,9,11,14, 15,19,23,36–48]; and (3) those that examine factors or conditions that modulate the toxic effects of NO_2 exposure [8,11,14,49–56]. Most studies with NO_2 have examined pulmonary effects [27,30,36,37,41,42,47,48, 53,57–65], although extrapulmonary effects have also been examined and reported [39,43,45,66]. In early classical toxicology studies, direct effects of NO_2 were demonstrated, e.g., acute lung injury and animal mortality. More subtle studies, which may be relevant when considering air pollution effects, have included criteria that may show rather indirect effects of NO_2 exposure, e.g., some of the extrapulmonary changes [67–70] and postexposure infection in the lung [71–77]. Several studies used relatively low concentrations of NO_2 (less than 1 ppm or 1.9 mg/m^3) and noted significant alterations in postexposure mortality.

MODE OF ACTION

The mechanisms by which NO_2 or O_3 causes lung tissue damage appear to be similar and are based on oxidation properties and free radical potential of each of these gases. Oxidation of functional groups, i.e., -SH, -NH$_2$, -CHO and phenolic-OH, and peroxidation of unsaturated lipids either directly or via free radical-mediated reactions, have been studied or postulated [6–11,14, 30,31,46,49,52,53,56,78–85]. Free radicals are particularly able to abstract hydrogen from biomolecules and initiate a chain of damaging reactions in tissue, including disruption of nucleic acids and membrane lipids and proteins [8,10,11,86,87]. Nitrogen dioxide can react with olefinic compounds and induce autoxidation of biological compounds at concentrations as low as 0.1 ppm (0.2 mg/m^3) [88]. Although these reactions of NO_2 or ozone with plant and animal tissues (including lung tissue) under in vitro conditions have been amply documented [6–11,14,30,31,46,49,52,53,56,78–85], their occurrence under in vivo conditions has not been demonstrated with certainty. To date, only a few laboratories have reported evidence for the occurrence of lipid peroxidation in vivo [30,53,84,89]. Since enzymatic or nonenzymatic mechanisms are known to exist in lung and other tissues that destroy lipid peroxides [84,90,91], it is possible that lipid peroxidation products do not accumulate in lung or other tissues during oxidant exposure. The occurrence of in vivo lipid peroxidation due to NO_2 or ozone exposure is thus difficult to demonstrate, but the concept persists of lipid peroxidation as the mechanism for oxidant effects.

Other mechanisms exist that may be linked to dissolved and/or transformation products of NO_2. Being a reactive gas, inhaled NO_2 will react predominantly with the surface cells and membranes of the lung. However, NO_2, being soluble to a certain degree in tissue fluid, may cause pulmonary endothelial as well as extrapulmonary damages. As much as 60% of inhaled NO_2 may be retained in the lung [70,92]. Our preliminary study with isolated perfused lung suggests that approximately 50% of inhaled NO_2 appears as NO_2^- ion in the perfusate [92]. Under in vivo conditions, the acidic products of dissolved NO_2, i.e., nitrous and nitric acids and the corresponding NO_2^- and NO_3^- ions formed, are likely to reach the extrapulmonary sites. These products generated in tissue fluid in situ may therefore additionally contribute to the overall NO_2 effects.

BIOCHEMICAL AND METABOLIC EFFECTS

Biochemical and metabolic changes in the lung have been studied as a means of delineating lung injury from a variety of conditions. Some of the more elaborate studies to date have involved exposure to ozone [9-11,93] and high concentrations of oxygen [94-98]. There are selected studies conducted with both inhaled and ingested lung toxins [99-104]. NO_2 perturbation studies with lung metabolism are limited. In general, these studies are composed of substrate utilization, enzyme activities representative of metabolic pathways, and analysis of products or metabolites of biological importance.

Oxidative and Energy Metabolism

Oxygen consumption is among the fundamental metabolic processes of a cell and is a relatively sensitive index for metabolic integrity. Measurement of oxygen consumption in lung tissue can be routinely carried out using tissue slices, tissue homogenate or isolated mitochondria. Mitochondria are the major sites for oxygen utilization in cells or tissues. Mitochondrial functions are critical to cellular terminal oxidative processes and energy production. Inhaled oxidant, e.g., NO_2, causing lung mitochondrial injury would therefore have detrimental consequences on the metabolism of lung cells.

Mitochondrial oxidative metabolism can be determined in lung homogenate or isolated mitochondria. Most commonly, the oxygen consumption is assayed by a polarographic or manometric method in the presence of a substrate, e.g., 2-oxoglutarate or succinate [85,105]. Effects of NO_2 and ozone on mitochondrial oxygen consumption has been documented using an in vitro system, i.e., directly exposing tissue homogenate or isolated mitochondria to one of these gases [33,85]. In general, these oxidants depress oxygen utilization, possibly by altering the mitochondrial membrane structure or inhibiting various dehydrogenases [83].

Ozone and NO_2 effects on oxygen consumption in lung tissue or mitochondria can be divided into two categories: high-level, short-term effects and low-level, long-term effects. Short-term exposure of rats to ozone (2-4 ppm or 40-80 mg/m³ for several hours) caused a significant depression of lung mitochondrial oxygen consumption [85,105,106]. However, exposure of rats to 30-50 ppm (56-94 mg/m³) NO_2 for up to 12 hr did not seem to alter the rate of oxygen consumption in lung homogenate from exposed animals relative to control [92]. Exposure of rats to lower levels of NO_2

(3–15 ppm or 6–28 mg/m^3) for 4–7 days resulted in an increase of oxygen consumption in lung tissue (Table I) [92], similar to that observed for low-level ozone exposure [15,107].

Increased oxidative metabolism in lung tissue has been shown for a variety of pathological conditions, i.e., succinate dehydrogenase and succinate oxidase activities in mouse lungs infected with tubercle bacilli [108]; oxidase and dehydrogenase activities pertaining to succinate, malate, 2-oxo-glutarate and isocitrate in guinea pig lungs after exposure to silica dust [109,110]; and oxygen consumption in human lung tissue with a proliferative disease [111,112]. Increased oxidative metabolism after low-level oxidant exposures seems to offer another example of proliferative condition in lung tissue.

Glucose Metabolism

Glucose metabolism in the lung is of particular interest to investigators because even though the lung exists in a relatively more aerobic environment than other organs, a substantial amount of glucose breakdown occurs with the formation of lactate, commonly a product of anaerobic metabolism. Glucose metabolism may provide various chemical intermediates, including energy source (ATP) and reducing equivalents (NADPH), for biosynthesis that may be particularly important for the repair of injured lung tissue.

Glucose metabolism in the lung is conveniently studied using isolated perfused lung and tissue slices [113]. Any injury to the lung seems to stimulate glucose consumption and lactate production in the tissue. Acute lung injury due to high-level NO$_2$ or ozone exposure causes at least 50% of the glucose consumed to be converted into lactate [92]. Glucose consumption is also stimulated after low-level NO$_2$ or ozone exposure (Table II) [92]. Rats receiving 5, 7 or 15 ppm (9, 13 or 28 mg/m^3) NO$_2$ for up to 4 days show a significant increase of glucose consumption [41,92]. In conjunction with glucose consumption there is an increased production of both pyruvate and lactate (Table II). Although under these conditions as much as 50% of the

Table I. Effect of NO$_2$ Exposure on O$_2$ Consumption in Lung Tissue [92]

Tissue Preparation and Parameter Used	% Increase Over Control[a] After NO$_2$ Exposure			
	3 ppm	7 ppm	10 ppm	15 ppm
Tissue Slices	–	11	38	57[b]
Homogenate				
Succinate Oxidase	2	12	30	71[b]
Cytochrome Oxidase	7	15	6	35[b]

[a]Exposure to NO$_2$ was for 7 days at 3 ppm and 4 days at other concentrations. Values were computed on a per lung basis.
[b]Significantly different from control (p < 0.05).

glucose consumed is converted to lactate, the rest is oxidized and/or incorporated into protein and lipid fractions of lung tissue, as determined from the $^{14}CO_2$ produced and ^{14}C fixed from labeled glucose [41,92].

Increased glucose utilization in the lung after NO_2 exposure seems to correlate with cellular changes in the alveolar epithelium. The histochemical study of Sherwin et al. [62] has shown that alveolar type 2 cells stain for lactate dehydrogenase (LDH), and that after a 2-ppm (3.8 mg/m^3) NO_2 exposure for 3 weeks there is a significant increase in LDH staining, indicating an increased number of type 2 cells.

Glucose consumption is also stimulated through the hexose monophosphate (HMP) shunt pathway in the lung after NO_2 exposures (Table III) [46,92]. This stimulation is usually determined by measuring increased activities of glucose-6-phosphate dehydrogenase (G6PD) and 6-phosphogluconate dehydrogenase (6PGD), two important enzymes in this pathway.

Table II. Effect of NO_2 Exposure on Glucose Metabolism [92]

	% Increase Over Control[a] After NO_2 Exposure at		
Metabolite	7 ppm	10 ppm	15 ppm
Glucose	22[b]	59[b]	88[b]
Pyruvate	16[b]	53[b]	60[b]
Lactate	18	51	76[b]

[a]Exposure to NO_2 was at 4 days. Values were computed on a per lung basis.
[b]Significantly different from control (p < 0.05).

Table III. Effect of NO_2 Exposure on Enzyme Activities [92]

	% Increase Over Control[a] After NO_2 Exposure at			
Enzyme Activity	3 ppm	7 ppm	10 ppm	15 ppm
Glucose-6-P dehydrogenase	9	42[b]	83[b]	162[b]
6-P-gluconate dehydrogenase	14	30	70[b]	110[b]
Glutathione Reductase	15	25[b]	50[b]	100[b]
Disulfide Reductase	20	–	–	126[b]
Glutathione Peroxidase	4	62[b]	–	45[b]

[a]Exposure to NO_2 was for 7 days at 3 ppm and 4 days at other concentrations. Values were computed on a per lung basis.
[b]Significantly different from control (p < 0.05).

The ratio of $^{14}CO_2$ from glucose-1-^{14}C to that from glucose-6-^{14}C gives another measure of the operation of this pathway [114,115]. Augmentation of G6PD and 6PGD activities, i.e., increased formation of NADPH form $NADP^+$ via the HMP shunt, has been observed in a variety of tissue injury, inflammation and repair processes, including lung tissue injury and repair [116-118]. The significance of agumented G6PD and 6PGD activities probably relates to an increased production of NADPH for biosynthetic processes, including lipids and proteins, and increased production of pentoses for nucleic acid synthesis.

Lipid Metabolism

Lipid metabolism in the lung has been studied extensively because of various unique features of lung lipids and pulmonary surfactant [119-122]. Pulmonary surfactant, composed of proteins and phospholipids rich in dipalmitoyl lecithin, lines the alveolar surface and lowers the surface tension of the individual alveoli during the inflation–deflation cycle of the breathing process. Synthesis and turnover of surfactant are important aspects of the overall lipid metabolism in the lung. NO_2 is of concern because of its oxidative and peroxidative reactions with lung lipids and surfactant. Alterations in fatty acid composition of lung lipids and surfactant (including changes in the rate of lipid synthesis) after NO_2 or O_3 exposure have been reported by various investigators [30,31,34,48,51,52,123].

Nucleic Acid and Protein Metabolism

Metabolic studies of nucleic acids and proteins in normal lung are limited, but changes in the metabolism of these two parameters in lung injury have been reported [124-136]. DNA content of adult animal lungs is found to increase during the repair phase of lung injury, e.g., from oxygen [136] or NO_2 exposure [127]. Conventionally, radioactive precursors, thymidine for DNA and uridine for RNA, are injected into animals, and incorporation into nucleic acids of lung tissue is then determined [137,138]. By using radioactive precursors and autoradiographic procedures, it has been shown that after NO_2 or ozone exposure an increased labeling of DNA occurs predominantly in the type 2 cells of the alveoli adjacent to the terminal bronchioles [127,139].

Protein synthesis is particularly active in the alveolar type 2 cells, as shown by the amino acid incorporation and autoradiographic procedures [140,141]. Low-level ozone exposures have been shown to cause an increase of radiolabeled amino acid incorporation into lung proteins [131], including collagen [134,135]. Similar studies with NO_2 exposure have not been reported, except that ^{14}C-proline incorporation into lung collagen has been found in rats after a low-level NO_2 exposure [42].

The increase in nucleic acid and protein synthesis in the lung after low-level NO_2 exposures may be related to hyperplasia of certain types of lung cells or recruitment of inflammatory cells. This, in turn, may be related to increased protein and nucleic acid content of the lung, including its weight, observed after the exposure (Table IV) [92,142].

Metabolism of Sulfhydryls

Soluble sulfhydryl compounds or nonprotein sulfhydryls (NPSH), particularly reduced glutatione (GSH), constitute a major pool of cellular reducing substances. A rat lung weighing 1 g contains approximately 1.6 μmoles of NPSH, up to 90% of which may be GSH [143]. The levels of GSH and NPSH in lung tissue are altered in response to oxidant exposure [9,38,83,84,91,94, 143]. It has been suggested [15,28] that low-level NO_2 exposure causes an increase of NPSH and GSH levels in the lung. In conjunction with the increased level of these reducing substances, the activities of several enzymes, which involve sulfhydryl metabolism, are found to increase after NO_2 exposure (Table III) [92]. These are glutathione reductase, disulfide reductase and glutathione-disulfide transhydrogenase, although they have been studied more elaborately for ozone exposures [9,83,84,92,143]. The significance of sulfhydryl metabolism in relation to oxidant stress has been discussed [11].

Metabolism Related to Cellular Detoxification

A variety of toxic intermediates (free radicals, superoxide anion, hydroxyl radical, hydrogen peroxide and lipid peroxides) are possibly generated in lung tissue during NO_2 or ozone exposure and eliminated via enzymatic or nonenzymatic detoxification mechanisms. Activities of such enzymes as glutathione peroxidase, superoxide dismutase, peroxidase and catalase may be associated with the enzymatic mechanisms and found to respond to oxidant exposure [9,15,38,53,84,91,95].

Influence of Nutrition

Exposure of animals to NO_2 results in metabolic changes in the lung. The magnitude of metabolic changes, which are essentially a response to injury,

Table IV. Effect of NO_2 Exposure on Lung Weight, Protein and DNA [92]

Parameter	% Increase Over Control[a] After NO_2 Exposure at			
	3 ppm	7 ppm	10 ppm	15 ppm
Lung Weight	–	7[b]	35[b]	50[b]
DNA Content	0	8	36	41[b]
Protein Content	9	0	10	34

[a]Exposure to NO_2 was for 7 days at 3 ppm and 4 days at other concentrations. Values were computed on a per lung basis.
[b]Significantly different from control ($p < 0.05$).

depends on the concentration of oxidant gas and/or length of exposure; however, the nutritional status of the animals has been found to influence the metabolic changes during a given oxidant exposure. Animals poorly supplied with vitamin E exhibit much greater sensitivity to ozone or NO_2 exposure compared with animals receiving a nutritionally sound or supplemented dosage of vitamin E [11,38,52–55,84,91,107,145–147]. While a high concentration of NO_2 or ozone causes a greater mortality in vitamin E-deficient animals [53], low-level oxidant exposures cause an increased lung injury, which can be monitored in terms of metabolic alterations [38,53,84,107, 145–147]. Vitamin C, glutathione and other antioxidant compounds have also been shown to be protective against oxidant toxicity [11,49,144,148]. These findings are consistent with the thesis that dietary intake of antioxidant compounds may lessen the sensitivity of the lung to inhaled oxidants [8,11,52,56,86,93,107,145–147].

Extrapulmonary Effects

Any extrapulmonary effects implies that NO_2 or its transformed products can leave the lung compartment and reach extrapulmonary sites where they produce cellular or metabolic changes. Most commonly, serum or plasma is examined for such effects of NO_2, although liver, kidney and brain have been referred to at times. Activities of a number of marker enzymes, e.g., lactate dehydrogenase (LDH), creatine phosphokinase (CPK), glutamate oxalate transaminase (GOT), glutamate pyruvate transaminase (GPT), lysosyme and aldolase, in serum or plasma have been measured after NO_2 exposures [38, 56,149]. Variable increases in the activities of these enzymes have been observed in the blood and other tissues of exposed animals relative to control [56,149–151]. Other parameters examined include hematocrit, hemoglobin, globulins, electrolytes, blood urea, etc., but without any significant changes attributable to NO_2 exposure [150,151]. Increased urinary excretion of proteins and hydroxyproline, a breakdown product of collagen, have been reported as a result of NO_2 exposures [67,152–154]. Although many of these extrapulmonary parameters may serve as the indices of NO_2 effects, the actual mechanisms by which NO_2 produces such effects are still a matter of speculation.

DISCUSSION AND CONCLUSIONS

This chapter has considered various biochemical changes in animal lungs caused by NO_2 exposures. In general, lung metabolism increases after NO_2 exposure. A time-course study indicates that the metabolic parameters increase within two days of exposure, reaching the peak in four days [28,41, 92]. Thereafter, the increases level off during a continued exposure. The metabolic increases, as studied for 3, 5, 7, 10 and 15 ppm NO_2 exposures, are fairly dose-dependent [28,41,92]. These findings seem to correlate well with many of the morphological and histological changes in the lung [61–64].

In general, NO_2 is less toxic than ozone, but the basic mechanisms of its action are very similar as both are oxidants and have free radical potential. The lung is the target organ for inhaled NO_2 reaction, but any portion of the gas that goes into solution and reaches the blood stream will be responsible for extrapulmonary effects observed.

There are many questions on the effects of NO_2 exposure that are yet to be answered and for which further research must be carried out. The threshold limit for NO_2 effects, i.e., the concentration of NO_2 in the breathing air that will not produce measurable health effects, is an important question to consider. A series of biological parameters must be tested [148], and some of the appropriate ones may be selected to indicate pollutant damage. In our laboratory, a measurable alteration in lung metabolism has been observed for 3 ppm NO_2 using male adult rats [92], but further studies are required to find the lowest limit of NO_2 effects. The time-course of biochemical effects involving short- and long-term exposures, dose-response involving continuous and intermittent exposures, reversibility vs irreversibility of effects, and development of tolerance or adaptation, are also important areas that require careful study.

Finally, NO_2 occurs in the atmosphere in combination with many other pollutants with which it constantly interacts. Determination of NO_2 effects in the lung or in extrapulmonary tissue must consider the influence of other pollutants that may be present simultaneously.

ACKNOWLEDGMENTS

This study was supported in part by Southern California Edison Company, Contract No. U2277914; U.S. Environmental Protection Agency, Contract No. 68-03-2221; and the U.S. Public Health Service–National Institutes of Health, Grant No. HL-17719 and RCD Award No. HL-00301 to Mohammad G. Mustafa.

The collaborative efforts of Mr. Nabil Elsayed, Mr. Edward Postlethwait and Dr. Joyce S. T. Lim are gratefully acknowledged.

REFERENCES

1. Dimitriades, B. "Effects of Hydrocarbons and Nitrogen Oxides on Photochemical Smog Formation," *Environ. Sci. Technol.* 6:253–260 (1972).
2. Kerr, J. A., J. G. Calvert and K. L. Demerjian. In: *Free Radicals in Biology*, Vol. II, W. A. Pryor, Ed. (New York: Academic Press, Inc., 1976), p. 159.
3. Seinfeld, J. H. *Air Pollution: Physical and Chemical Fundamentals* (New York: McGraw-Hill Book Co., 1975).
4. Heiklen, J. *Atmospheric Chemistry* (New York: Academic Press, Inc., 1976).

5. Stern, A. C., Ed. *Air Pollution*, Vol. III (New York: Academic Press, Inc., 1977).

6. Stokinger, H. E. "Ozone Toxicology: A Review of Research and Industrial Experience: 1954–1964," *Arch. Environ. Health* 10:719–731 (1965).

7. Nasr, A. N. M. "Biochemical Aspects of Ozone Intoxication: A Review," *J. Occup. Med.* 9:589–597 (1967).

8. Menzel, D. B. In: *Free Radicals in Biology*, Vol. II, W. A. Pryor, Ed. (New York: Academic Press, Inc., 1976), p. 181.

9. Mustafa, M. G., A. D. Hacker, J. J. Ospital, M. Z. Hussain and S. D. Lee. In: *Biochemical Effects of Environmental Pollutants*, S. D. Lee, Ed. (Ann Arbor, MI: Ann Arbor Science Publishers, Inc., 1977), p. 59.

10. Mudd, J. B., and B. A. Freeman. In: *Biochemical Effects of Environmental Pollutants*, S. D. Lee, Ed. (Ann Arbor, MI: Ann Arbor Science Publishers, Inc., 1977), p. 97.

11. Mustafa, M. G., and D. F. Tierney. "State of the Art: Biochemical and Metabolic Changes in the Lung with Oxygen, Ozone, and Nitrogen Dioxide Toxicity," *Am. Rev. Resp. Dis.* 118:1061–1090 (1978).

12. Mudd, J. B. In: *Free Radicals in Biology*, Vol. II, W. A. Pryor, Ed. (New York: Academic Press, Inc., 1976), p. 203.

13. Dungworth, D. L., G. L. Clarke and R. L. Plata. "Pulmonary Lesions Produced in A-strain Mice by Long-Term Exposure to Peroxyacetyl Nitrate," *Am. Rev. Resp. Dis.* 99:565–574 (1969).

14. Stokinger, H. E., and D. L. Coffin. In: *Air Pollution*, Vol. I, 2nd ed., A. C. Stern, Ed. (New York: Academic Press, Inc., 1968), p. 445.

15. Mustafa, M. G., N. Elsayed, J. S. T. Lim, E. Postlethwait and S. D. Lee. In: *Nitrogenous Air Pollutants: Chemical and Biological Implications*, D. Grosjean, Ed. (Ann Arbor, MI: Ann Arbor Science Publishers, Inc., 1979), p. 165.

16. State of California Air Resources Board. "Ten-Year Summary of California Air Quality Data, 1963–1972,"Sacramento, CA (1974).

17. Trijonis, J. "Empirical Relationships Between Atmospheric Nitrogen Dioxide and its Precursors," EPA-600/3-78-018, Environmental Sciences Research Laboratory, Office of Research and Development, U.S. Environmental Protection Agency, Research Triangle Park, NC (1978).

18. U.S. Environmental Protection Agency. "Air Quality Criteria for Nitrogen Oxides," Air Pollution Control Office, U.S. Environmental Protection Agency, Washington, D.C. (1971).

19. Morrow, P. E. "An Evaluation of Recent NO_x Toxicity Data and an Attempt to Derive an Ambient Air Standard for NO_x by Established Toxicological Procedures," *Environ. Res.* 10:92–112 (1975).

20. National Academy of Sciences. "Medical and Biological Effects of Environmental Pollutants: Nitrogen Oxides," Washington, D.C. (1977).

21. World Health Organization and U.N. Environment Program. "Oxides of Nitrogen," Geneva (1977).

22. Patty, F. A., Ed. *Industrial Hygiene and Toxicology*, Vol. 2 (New York: Wiley-Interscience, 1962).

23. Gray, E. L. "Oxides of Nitrogen: their Occurrence, Toxicity, Hazard–A Brief Review," *Arch. Ind. Health* 19:479–486 (1959).

24. Darke, C. S., and A. J. N. Warrack. "Bronchiolitis from Nitrous Fumes," *Thorax* 13:327–333 (1958).

25. Lowry, T., and L. M. Schuman. "Silofiller's Disease–a Syndrome Caused by Nitrogen Dioxide," *J. Am. Med. Assoc.* 162:153–160 (1956).
26. Adley, F. E. "Exposure to Oxides of Nitrogen Accompanying Shrinking Operations," *J. Ind. Hyg. Toxicol.* 28:17–20 (1946).
27. Hine, C. H., F. H. Meyers and R. W. Right. "Pulmonary Changes in Animals Exposed to Nitrogen Dioxide: Effects of Acute Exposure," *Toxicol. Appl. Pharmacol.* 16:201–213 (1970).
28. Ospital, J. J., A. D. Hacker, N. Elsayed and M. G. Mustafa. "Effects of Nitrogen Dioxide on Rat Lung Metabolism," Unpublished results (1979).
29. Mustafa, M. G., N. Elsayed, A. D. Hacker and J. J. Ospital. "Influence of Animal Weights on Ozone Effects in Rat Lungs," Unpublished results (1979).
30. Thomas, H. V., P. K. Mueller and R. L. Lyman. "Lipoperoxidation of Lung Lipids in Rats Exposed to Nitrogen Dioxide," *Science* 159:532–534 (1968).
31. Roehm, J. N., J. C. Hadley and D. B. Menzel. "Oxidation of Unsaturated Fatty Acids by O_3 and NO_2: a Common Mechanism of Action," *Arch. Environ. Health* 23:142–148 (1971).
32. Ramazzotto, L. T., and L. J. Rappaport. "The Effect of Nitrogen Dioxide on Aldolase Activity," *Arch. Environ. Health* 22:379–380 (1971).
33. Ramazzotto, L., C. R. Jones and F. Cornell. "Effect of Nitrogen Dioxide on the Activities of Cytochrome Oxidase and Succinic Dehydrogenase on Homogenates of Some Organs of the Rat," *Life Sci.* 10:601–604 (1971).
34. Arner, E. C., and R. O. Rhoades. "Long-Term Nitrogen Dioxide Exposure. Effects on Lung Lipids and Mechanical Properties," *Arch. Environ. Health* 26:156–160 (1973).
35. Law, F. P. P., J. C. Drach and J. E. Sinsheimer. "Effect of NO_2 and 3-Methylcholanthrene on Pulmonary Enzymes," *J. Pharm. Sci.* 64:1421–1422 (1975).
36. Sherwin, R. P., and V. Richters. "Lung Capillary Permeability: Nitrogen Dioxide Exposure and Leakage Tritiated Serum," *Arch. Intern. Med.* 128:61–68 (1971).
37. Sherwin, R. P., and D. A. Carlson. "Protein Content of Lung Lavage Fluid of Guinea-Pigs Exposed to 0.4 ppm Nitrogen Dioxide," *Arch. Environ. Health* 27:90–93 (1973).
38. Chow, C. K., C. J. Dillard and A. L. Tappel. "Glutathione Peroxidase System and Lysozyme in Rats Exposed to Ozone or Nitrogen Dioxide," *Environ. Res.* 7:311–319 (1974).
39. Tusl, M., A. Vyskocil and V. Sesinova. "Changes in Plasma Corticosterones of Rats After Inhalation of Nitrogen Oxides," *Staub-Reinhalt Luft* 35:210–211 (1975).
40. Kosimder, S. "Electrolytes and Lipid Disturbances in Chronic Intoxication with Nitrogen Oxides," *Int. Arch. Occup. Environ.* 35:217–232 (1975).
41. Ospital, J. J., N. Elsayed, A. D. Hacker, M. G. Mustafa and D. F. Tierney. "Altered Glucose Metabolism in Lungs of Rats Exposed to Nitrogen Dioxide," *Am. Rev. Resp. Dis.* 113:108 (1976).

42. Hacker, A. D., N. Elsayed, M. G. Mustafa, J. J. Ospital and S. D. Lee. "Effects of Short-Term Nitrogen Dioxide Exposure on Lung Collagen Synthesis," *Am. Rev. Resp. Dis.* 113:107 (1976).
43. Kleinerman, J., and D. Rynbrandt. "Lung Proteolytic Activity and Serum Protease Inhibition After NO_2 Exposure," *Arch. Environ. Health* 31:37–41 (1976).
44. Mustafa, M. G., J. J. Ospital and A. D. Hacker. "Effect of Ozone and Nitrogen Dioxide Exposure on Lung Metabolism," *Environ. Health Persp.* 16:184 (1976).
45. Gooch, R. C., A. E. Luippold, D. A. Creasia and J. G. Brewen. "Observation on Mouse Chromosomes Following Nitrogen Dioxide Inhalation," *Mut. Res.* 48:117–119 (1977).
46. Menzel, D. B., M. B. Abou-Donia, C. R. Roe, R. Ehrlich, D. E. Gardner and D. L. Coffin. In: *Proc. Int. Conf. on Photochemical Oxidant Pollution and its Control*, Vol. II, B. Dimitriades, Ed. (Research Triangle Park, NC: U.S. Environmental Protection Agency, 1977), p. 577.
47. Sherwin, R. P., D. Okimoto and D. Mundy. "Sequestration of Exogenous Peroxidase in the Lungs of Animals Exposed to Continuous 0.5 ppm Nitrogen Dioxide," *Fed. Proc.* 36:1079 (1977).
48. Blank, M. L., W. Dalbey, P. Nettesheim, J. Price, D. Creasia and F. Snyder. "Sequential Changes in Phospholipid Composition and Synthesis in Lungs Exposed to Nitrogen Dioxide," *Am. Rev. Resp. Dis.* 117:273–280 (1978).
49. Fairchild, E. J., S. D. Murphy and H. E. Stokinger. "Protection by Sulfur Compounds Against the Air Pollutants O_3 and NO_2," *Science* 130:861–862 (1959).
50. Fairchild, E. J., and S. L. Graham. "Thyroid Influence on the Toxicity of Respiratory Irritant Gases, Ozone and Nitrogen Dioxide," *J. Pharm. Exp. Ther.* 139:177–184 (1963).
51. Roehm, J. N., J. C. Hadley and D. B. Menzel. "Antioxidant vs. Lung Disease," *Arch. Intern. Med.* 128:88–93 (1971).
52. Menzel, D. B., J. N. Roehm and S. D. Lee. "Vitamin E: the Biological and Environmental Antioxidant," *J. Agric. Food Chem.* 20:481–486 (1972).
53. Fletcher, B. L., and A. L. Tappel. "Protective Effects of Dietary alpha-tocopherol in Rats Exposed to Toxic Levels of Ozone and Nitrogen Dioxide," *Environ. Res.* 6:165–175 (1973).
54. Csallany, A. S. "The Effect of Nitrogen Dioxide on the Growth of Vitamin E Deficient, Vitamin E Supplemented and DPPD Supplemented Mice," *Fed. Proc.* 34:913 (1975).
55. Ayaz, K. L., and A. S. Csallany. "The Effect of Continuous Low Level NO_2 Exposure and Dietary Vitamin E upon Lipofuscin Pigment Concentrations and Glutathione Peroxidase Activity in Mice," *Fed. Proc.* 36:1079 (1977).
56. Menzel, D. B., D. H. Donovan and M. Sabransky. In: *Nitrogenous Air Pollutants: Chemical and Biological Implications*, D. Grosjean, Ed. (Ann Arbor, MI: Ann Arbor Science Publishers, Inc., 1979), p. 149.
57. Freeman, G., and G. B. Haydon. "Emphysema After Low-Level Exposure to Nitrogen Dioxide," *Arch. Environ. Health* 8:125–128 (1964).

58. Freeman, G., R. J. Stephens, S. C. Crane and N. J. Furiosi. "Lesion of the Lung in Rats Continuously Exposed to Two Parts per Million of Nitrogen Dioxide," *Arch. Environ. Health* 17:181–192 (1968).

59. Freeman, G., R. J. Stephens and N. J. Furiosi. "The Subacute Nitrogen Dioxide-Induced Lesion of the Rat Lung," *Arch. Environ. Health* 18:609–612 (1969).

60. Blair, W. H., M. C. Henry and R. Ehrlich, "Chronic Toxicity of Nitrogen Dioxide: II. Effect on Histopathology of Lung Tissue," *Arch. Environ. Health* 19:186–192 (1969).

61. Yuen, T. G. H., and R. P. Sherwin. "Hyperplasia of Type 2 Pneumocytes and Nitrogen Dioxide (10 ppm) Exposure," *Arch. Environ. Health* 22:178–188 (1971).

62. Shwerwin, R. P., J. Dibble and J. Weiner. "Alveolar Wall Cells of the Guinea-Pig: Increase in Response to 2 ppm Nitrogen Dioxide," *Arch. Environ. Health* 24:43–47 (1972).

63. Evans, M. J., L. J. Cabral, R. J. Stephens and G. Freeman. "Renewal of Alveolar Epithelium in the Rat Following Exposure to NO_2," *Am. J. Pathol.* 70:175–190 (1973).

64. Evans, M. J., L. J. Cabral, R. J. Stephens and G. Freeman. "Transformation of Alveolar Type 2 Cells to Type 1 Cells Following Exposure to NO_2," *Exp. Mol. Pathol.* 22:142–150 (1975).

65. Kleinerman, J. "Some Effects of Nitrogen Dioxide on the Lung," *Fed. Proc.* 36:1714–1718 (1977).

66. Mersch, J., B. J. Dice, B. J. Haverback and R. P. Sherwin. "Diphosphoglycerate Content of Red Blood Cells: Measurements in Guinea-Pigs Exposed to 0.4 ppm Nitrogen Dioxide," *Arch. Environ. Health* 27:94–95 (1973).

67. Sherwin, R. P., and L. J. Layfield. "Proteinuria in Guinea-Pigs Exposed to 0.5 ppm Nitrogen Dioxide," *Arch. Environ. Health* 28:336–341 (1974).

68. Oda, H., S. Kusomoto and T. Nakajima. "Nitrosyl-Hemoglobin Formation in the Blood of Animals Exposed to Nitric Oxide," *Arch. Environ. Health* 30:453–455 (1975).

69. Drozdz, M., E. Kucharz, K. Rudyga and T. Molska-Drozdz. "Studies on the Effect of Long-Term Exposure to Nitrogen Dioxide on Serum and Liver Proteins Level and Enzyme Activity in Guinea-Pigs," *Eur. J. Toxicol.* 9:287–293 (1976).

70. Goldstein, E., N. F. Peele, N. J. Parks, H. H. Hines, E. P. Steffey and B. Tarkington. "Fate and Distribution of Inhaled Nitrogen Dioxide in Rhesus Monkeys," *Am. Rev. Resp. Dis.* 115:403–412 (1977).

71. Ehrlich, R. "Effect of Nitrogen Dioxide on Resistance to Respiratory Infection," *Bact. Rev.* 30:604–614 (1966).

72. Ehrlich, R., and M. C. Henry. "Chronic Toxicity of Nitrogen Dioxide. I. Effect on Resistance to Bacterial Pneumonia," *Arch. Environ. Health* 17:860–865 (1968).

73. Giordano, A. M., and P. E. Morrow. "Chronic Low-Level Nitrogen Dioxide Exposure and Mucociliary Clearance," *Arch. Environ. Health* 25:443–449 (1972).

74. Goldstein, E., M. C. Eagle and P. D. Hoeprich. "Effect of Nitrogen Dioxide on Pulmonary Bacterial Defense Mechanisms," *Arch. Environ. Health* 26:202–204 (1973).

75. Coffin, D. L., D. F. Gardner and E. J. Blommer. "Time-Dose Response for Nitrogen Dioxide Exposure in an Infectivity Model System," *Environ. Health Persp.* 13:11–15 (1976).

76. Gardner, D. E., and J. A. Graham. In: *Pulmonary Macrophage and Epithelial Cells, Proceedings of the Sixteenth Annual Hanford Biology Symposium*, R. P. Schneider, G. E. Doyle and H. A. Ragan, Eds., Richland, Washington (1976), p. 1.

77. Ehrlich, R., J. C. Findlay, J. D. Fenters and D. E. Gardner. "Health Effects of Short-Term Exposure to Inhalation of NO_2-O_3 Mixtures," *Environ. Res.* 14:223–231 (1977).

78. Thompson, W. W., W. M. Dugger, Jr. and R. L. Palmer, "Effects of Ozone on the Fine Structure of the Palisade Parenchyma Cells of Bean Leaves," *Can. J. Bot.* 44:1677–1682 (1966).

79. Little, C., and P. J. O'Brien. "The Effectiveness of a Lipid Peroxide in Oxidizing Protein and Nonprotein Thiols," *Biochem. J.* 106:419–423 (1968).

80. Mudd, J. B., R. Leavitt, A. Ongun and T. T. McManus. "Reaction of Ozone with Amino Acids and Proteins," *Atmos. Environ.* 3:669–682 (1969).

81. Mudd, J. B., T. T. McManus, A. Ongrin and T. E. McCullough. "Inhibition of Glycolipid Biosynthesis in Chloroplasts by Ozone and Sulfhydryl Reagents," *Plant Physiol.* 48:335–339 (1971).

82. Chang, C. W. "Effect of Ozone on Sylfhydryl Groups of Ribosomes in Pinto Bean Leaves: Relationship with Ribosome Dissociation," *Biochem. Biophys. Res. Commun.* 44:1429–1435 (1971).

83. DeLucia, A. J., P. M. Hoque, M. G. Mustafa and C. E. Cross. "Ozone Interaction with Rodent Lung. I. Effect on Sulfhydryls and Sulfhydryl-Containing Enzyme Activities," *J. Lab. Clin. Med.* 80:559–566 (1972).

84. Chow, C. K., and A. L. Tappel. "Activities of Pentose Shunt and Glycolytic Enzymes in Lungs of Ozone-Exposed Rats," *Arch. Environ. Health* 26:205–208 (1973).

85. Mustafa, M. G., and C. E. Cross. "Effects of Short-Term Ozone Exposure on Lung Mitochondrial Oxidative and Energy Metabolism," *Arch. Biochem. Biophys.* 162:585–594 (1974).

86. Menzel, D. B. "Toxicity of Ozone, Oxygen and Radiation," *Ann. Rev. Pharmacol.* 10:379–394 (1970).

87. Mead, J. F. In: *Free Radicals in Biology*, Vol. I, W. A. Pryor, Ed. (New York: Academic Press, Inc., 1976), p. 51.

88. Chao, S. C., and G. Jaffe. "Gas Phase Reaction of Nitrogen Dioxide and Ethylene at 25°C," *J. Chem. Phys.* 56:1987–1988 (1972).

89. Sevanian, A., R. A. Stein and J. F. Mead. In: *The Molecular Bases of Environmental Toxicity*, R. S. Bhatnagar, Ed. (Ann Arbor, MI: Ann Arbor Science Publishers, Inc., 1979).

90. Little, C., and P. J. O'Brien. "An Intracellular CSH Peroxidase with a Lipid Peroxide Substrate," *Biochem. Biophys. Res. Commun.* 31:145–150 (1968).

91. Chow, C. K., and A. L. Tappel. "An Enzymatic Protective Mechanism Against Lipid Peroxidation Damage to Lungs of Ozone-Exposed Rats," *Lipids* 7:518–524 (1972).

92. Postlethwait, E., N. Elsayed, J. S. T. Lim and M. G. Mustafa. "Biochemical Effects of Nitrogen Dioxide Exposure in Rat Lungs," Unpublished results (1979).

93. Cross, C. E., A. J. DeLucia, A. K. Reddy, M. Z. Hussain, C. K. Chow and M. G. Mustafa. "Ozone Interaction with Lung Tissue: Biochemical Approaches," *Am. J. Med.* 60:929–935 (1976).

94. Clark, J. M., and C. J. Lambertsen. "Pulmonary Oxygen Toxicity: A Review," *Pharmacol. Rev.* 23:37–133 (1971).

95. Kimball, R. E., K. Reddy, T. H. Peirce, L. W. Schwartz, M. G. Mustafa and C. E. Cross. "Oxygen Toxicity: Augmentation of Antioxidant Defense Mechanisms in Rat Lung," *Am. J. Physiol.* 230:1425–1431 (1976).

96. Tierney, D. F., L. Ayers, S. Herzog and J. Yang. "Pentose Pathway and Production of Reduced Nicotinamide Adenine Dinucleotide Phosphate," *Am. Rev. Resp. Dis.* 108:1348–1351 (1973).

97. Crapo, J. D., and D. F. Tierney. "Superoxide Dismutase and Pulmonary Oxygen Toxicity," *Am. J. Physiol.* 226:1401–1407 (1974).

98. Tierney, D. F., L. Ayers and R. S. Kasuyama. "Altered Sensitivity to Oxygen Toxicity," *Am. Rev. Resp. Dis.* 115 (Suppl. 2):59–65 (1977).

99. Witschi, H. P. In: *Essays in Toxicology*, Vol. 6, W. J. Hayes, Ed. (New York: Academic Press, Inc., 1975), p. 125.

100. Witschi, H. P., and M. G. Cote. "Biochemical Pathology of Lung Damage Produced by Chemicals," *Fed. Proc.* 35:89–94 (1976).

101. Witschi, H. P. "The Biochemical Pathology of Rat Lung After Acute Paraquat Poisoning," *Toxicol. Appl. Pharmacol.* 25:485–486 (1973).

102. Witschi, H. P., and W. Saheb. "Stimulation of DNA Synthesis in Mouse Lung Following Intraperitoneal Injection of Butylated Hydroxytoluene," *Proc. Soc. Exp. Biol. Med.* 147:690–693 (1974).

103. Aso, Y., K. Yoneda and Y. Kikkawa. "Morphologic and Biochemical Study of Pulmonary Changes Induced by Bleomycin in Mice," *Lab. Invest.* 35:558–568 (1976).

104. Kimbrough, R. D., and T. B. Gaines. "Toxicity of Paraquat to Rats and its Effects on Rat Lungs," *Toxicol. Appl. Pharmacol.* 17:679–690 (1970).

105. Mustafa, M. G. "Augmentation of Mitochondrial Oxidative Metabolism in Lung Tissue During Recovery of Animals from Acute Ozone Exposure," *Arch. Biochem. Biophys.* 165:531–538 (1974).

106. Mustafa, M. G., A. J. DeLucia, G. K. York, C. Arth and C. E. Cross. "Ozone Interaction with Rodent Lung. II. Effects on Oxygen Consumption of Mitochondria," *J. Lab. Clin. Med.* 82:357–365 (1973).

107. Mustafa, M. G., and S. D. Lee. "Pulmonary Biochemical Alterations Resulting from Ozone Exposure," *Ann. Occup. Hyg.* 19:17–26 (1976).

108. Segal, W. "Enhancement of Succinate Oxidation in Lung and Liver Mitochondria of Tuberculous Mice," *Arch. Biochem. Biophys.* 113:750–757 (1966).

109. Kilroe-Smith, T. A., and M. G. Breyer. "Changes in Activities of Respiratory Enzymes in Lungs of Guinea-Pigs Exposed to Silica Dust," *Brit. J. Ind. Med.* 20:243–247 (1963).

110. Breyer, M. G., T. A. Kilroe-Smith and H. Prinsloo. "Changes in Activities of Respiratory Enzymes in Lungs of Guinea-Pigs Exposed to Silica Dust. II. Comparison of the Effects of Quartz Dust and Lampblack on the Succinate Oxidase System," *Brit. J. Ind. Med.* 21:32–34 (1964).

111. Fritts, H. W., Jr., B. Strauss, W. Wichern, Jr. and A. Courand. "Utilization of Oxygen in Lung of Patients with Diffuse, Non-obstructive Pulmonary Disease," *Trans. Assoc. Am. Phys.* 76:302–311 (1963).

112. Strauss, B. "*In vitro* Respiration of Normal and Pathologic Human Lung," *J. Appl. Physiol.* 19:503–509 (1964).

113. Tierney, D. F., and S. E. Levy. In: *The Biochemical Basis of Pulmonary Function*, Vol. 2, R. G. Crystal, Ed. (New York: Marcel Dekker, Inc., 1976), p. 105.

114. Katz. J., and H. G. Wood. "The Use of $^{14}CO_2$ Yields from Glucose-1- and -6-^{14}C for the Evaluation of the Pathways of Glucose Metabolism," *J. Biol. Chem.* 238:517–523 (1963).

115. Tierney, D. F. "Intermediary Metabolism of the Lung," *Fed. Proc.* 33: 2232–2237 (1974).

116. Braasch, W., S. Gudbjarnason, P. S. Puri, K. G. Ravens and R. J. Bing. "Early Changes in Energy Metabolism in the Myocardium Following Acute Coronary Artery Occlusion in Anesthetized Dogs," *Circ. Res.* 23:429–438 (1968).

117. Vorne, M., and P. Arvela. "Effect of Carbon Tetrachloride-Induced Progressive Liver Damage on Drug Metabolizing Enzymes and Cytochrome P-450 in Rat Liver," *Acta Pharmacol. Toxicol.* 29:417–427 (1971).

118. Beaconsfield, P., and A. Capri. "Localization of an Infectious Lesion and Glucose Metabolism via the Pentose Phosphate Pathway," *Nature* 201: 825–827 (1964).

119. Naimark, A. "Cellular Dynamics and Lipid Metabolism in the Lung," *Fed. Proc.* 32:1967–1971 (1973).

120. Mason, R. J. In: *The Biochemical Basis of Pulmonary Function*, Vol. 2, R. G. Crystal, Ed. (New York: Marcel Dekker, Inc., 1976), p. 127.

121. Scarpelli, E. M. *The Surfactant System of the Lung* (Philadelphia, PA: Lea and Febiger, 1968), p. 269.

122. Clements, J. A., and R. J. King. In: *The Biochemical Basis of Pulmonary Function*, Vol. 2, R. G. Crystal, Ed. (New York: Marcel Dekker, Inc., 1976), p. 363.

123. Trzeciak, H. I., S. Kosimder, K. Kryk and A. Kryk. "The Effects of Nitrogen Oxides and Their Neutralization Products with Ammonia on the Lung Phospholipids of Guinea-Pigs," *Environ. Res.* 14:87–91 (1977).

124. Scheel, L. D., O. J. Dobrogorski, J. T. Mountain, J. L. Svirbely and H. E. Stokinger. "Physiologic Biochemical Immunologic and Pathologic Changes Following Ozone Exposure," *J. Appl. Physiol.* 14:67–80 (1959).

125. Hacker, A. D. "Oxygen Toxicity and the Formation of Collagen in the Lung," *Am. Rev. Resp. Dis.* 117:345 (1968).

126. Chvapil, M., and J. Hurych. "Control of Collagen Biosynthesis," *Int. Rev. Conn. Tissue Res.* 4:67–196 (1968).

127. Kleinerman, J. "Effects of NO_2 in Hamsters: Autoradiographic and Electron Microscopic Aspects," *Atomic Energy Commission Symp. Ser.* 18: 271–279 (1970).

128. Witschi, H. P. "The Effects of Diethylnitrosamine on Ribonuclei Acid and Protein Synthesis in the Liver and Lung of the Syrian Golden Hamster," *Biochem. J.* 136:789–794 (1973).

129. Massaro, D. "Protein Synthesis in Lung: Recovery from Exposure to Hyperoxia," *J. Appl. Physiol.* 35:32–34 (1973).
130. Werthamer, S., P. D. Penha and L. Amaral. "Pulmonary Lesions Induced by Chronic Exposure to Ozone," *Arch. Environ. Health* 29:164–166 (1974).
131. Hussain, M. Z., M. G. Mustafa, C. E. Cross and W. S. Tyler. "Increased Protein Synthesis in Lung After Subacute Exposure of Rats and Monkeys to Ozone," *Fed. Proc.* 33:1468 (1974).
132. Chvapil, M., and Y. M. Peng. "Oxygen and Lung Fibrosis," *Arch. Environ. Health* 30:528–532 (1975).
133. Välmäki, M., K. Juva, J. Rantanen, T. Ekfors and J. Nünikoski. "Collagen Metabolism in Rat Lungs During Chronic Intermittent Exposure to Oxygen," *Aviation Space Environ. Med.* 46:684–690 (1975).
134. Hussain, M. Z., M. G. Mustafa, C. K. Chow and C. E. Cross. "Ozone-Induced Increase in Lung Proline Hydroxylase Activity and Hydroxyproline Content," *Chest* 69(Suppl.):273–275 (1976).
135. Hussain, M. Z., C. E. Cross, M. G. Mustafa and R. S. Bhatnagar. "Collagen Synthesis in Lungs of Rats Exposed to Low Levels of Ozone," *Life Sci.* 18:897–904 (1976).
136. Crapo, J. D., K. Sjostrom and R. T. Drew. "Tolerance and Cross-Tolerance Using NO_2 and O_2. I. Toxicology and Biochemistry," *J. Appl. Physiol.* 44:364–369 (1978).
137. Witschi, H. P. "A Comparative Study of *In Vivo* RNA and Protein Synthesis in Rat Liver and Lung," *Cancer Res.* 32:1686–1694 (1972).
138. Witschi, H. P. "Qualitative and Quantitative Aspects of the Biosynthesis of Ribonucleic Acid and Protein in the Liver and the Lung of the Syrian Golden Hamster," *Biochem. J.* 136:781–788 (1973).
139. Evans, M. J., W. Mayr, R. F. Bils and C. G. Loosli. "Effects of Ozone on Cell Renewal in Pulmonary Alveoli of Aging Mice," *Arch. Environ. Health* 22:450–453 (1971).
140. Massaro, G. D., and D. Massaro. "Granular Pneumocytes: Electron Microscopic Radioautographic Evidence of Intracellular Protein Transport," *Am. Rev. Resp. Dis.* 105:927–931 (1972).
141. Chevalier, G., and A. J. Collet. "*In Vivo* Incorporation of Choline–3H, Leucine–3H and Galactoce–3H in Alveolar Type II Pneumocytes in Relation to Surfactant Synthesis: A Quantitative Radioautographic Study in Mouse by Electron Microscopy," *Anat. Rec.* 174:289–310 (1972).
142. Mustafa, M. G., and S. D. Lee. In: *Assessing Toxic Effects of Environmental Pollutants,* S. D. Lee and J. B. Mudd, Eds. (Ann Arbor, MI: Ann Arbor Science Publishers, Inc., 1979), p. 105.
143. DeLucia, A. J., M. G. Mustafa, M. Z. Hussain and C. E. Cross. "Ozone Interaction with Rodent Lung. III. Oxidation of Reduced Glutathione and Formation of Mixed Disulfides Between Protein and Nonprotein Sulfhydryls," *J. Clin. Invest.* 55:794–802 (1975).
144. Matzen, R. N. "Effect of Vitamin C and Hydrocortisone on the Pulmonary Edema Produced by Ozone in Mice," *J. Appl. Physiol.* 11:105–109 (1957).
145. Mustafa, M. G. "Influence of Dietary Vitamin E on Lung Cellular Sensitivity to Ozone," *Nutr. Reports Int.* 11:473–476 (1975).

146. Mustafa, M. G., A. D. Hacker, J. J. Ospital, N. Elsayed and S. D. Lee. "Prophylactic Effect of Dietary Vitamin E on the Metabolic Response of Lung Tissue to Low-Level Ozone Exposure," *Am. Rev. Resp. Dis.* 113:98 (1976).

147. Mustafa, M. G., N. Elsayed, R. Kass, J. J. Ospital, A. D. Hacker and S. D. Lee. "Influence of Dietary Vitamin E on Ozone Effects in Rodent Lung," *Fed. Proc.* 35:1708 (1976).

148. Wagner, W. D., R. B. Duncan, P. G. Wright and H. E. Stokinger. "Experimental Study of Threshold Limit of NO_2," *Arch. Environ. Health* 10: 455–466 (1965).

149. Kosimder, S., M. Luciak and M. Drozdz. "The Influence of Ammonia on Some Disturbances in Protein, Carbohydrate and Lipid Metabolism Caused by Chronic Intoxication with Combustion Gases," *Intern. Arch. Occup. Health* 35:37–59 (1975).

150. Cole, W. B., and D. W. Badger. "Physiological Effects of Nitrogen Dioxide Exposure and Heat Stress in Cynomologous Monkeys," *Toxicol. Appl. Pharmacol.* 29:130 (1974).

151. Fenters, J. D., R. Ehrlich, J. Spencer and V. Tolkacz. "Serologic Response in Squirrel Monkeys Exposed to Nitrogen Dioxide and Influenza Virus," *Am. Rev. Resp. Dis.* 104:448–451 (1971).

152. Block, W. N., S. Lassiter, J. F. Stara and T. R. Lewis. "Blood Rheology of Dogs Chronically Exposed to Air Pollutants," *Toxicol. Appl. Pharmacol.* 25:576–581 (1973).

153. Kosimder, S., M. Luciak, K. Zajusz, A. Misiewicz and J. Szygula. "Studies on Emphysogenic Action of Nitrogen Oxides," *Patal. Polska* 24:107–125 (1973).

154. Hatton, D. J., C. S. Leach, A. E. Nicogossian and M. DiFerrante. "Collagen Breakdown and Nitrogen Dioxide Inhalation," *Arch. Environ. Health* 32:33–36 (1977).

CHAPTER 11

COMBINED EFFECTS OF NITROGEN OXIDES AND OZONE ON MICE

Hiromu Watanabe

Environmental Science Institute of Hyogo Prefecture
Kobe 654, Japan

Osamu Fukase and Kimio Isomura

Public Health Institute of Hyogo Prefecture
Kobe 652, Japan

INTRODUCTION

Studies of biochemical alterations in lung tissues resulting from short-term and prolonged exposure to oxidant air pollutants have revealed that these changes precede many of the structural indications of tissue degradation and offer a sensitive method for detection prior to actual irreversible damage [1]. Chow and Tappel [2] and Chow et al. [3] demonstrated that exposure to nitrogen dioxide (NO_2) or ozone (O_3) resulted in increased activities of enzymes relating to the glutathione peroxidase system that is a protective mechanism against lipid peroxidation. They revealed that the alteration in the activities of these enzymes is a more sensitive and specific indicator of tissue damage than other known biochemical measurements. Studies [4,5] have shown also that dose-related increases in the activities of the enzymes of the glutathione peroxidase system and a dose-related elevation of the levels of reduced glutathione (GSH) occurred in the lungs of mice exposed to nitrogen dioxide or ozone.

This chapter investigates the combined effects of repeated exposure to nitrogen oxides (NO_x) and ozone by using the elevation of GSH levels in lungs as an indicator of the effect of pollutants. Nitrogen oxides and ozone are the major constituents of photochemical air pollution that have presented widespread potential public health problems. To interpret the biological importance of this pollution, it is necessary to study the biological effects of each pollutant independently, as well as in combination.

MATERIALS AND METHODS

Male, 4-week old ICR-JCL mice obtained from a breeding laboratory (CLEA Japan, Inc., Tokyo) were used in the experiments. The animals were maintained on commercial chow and water ad libitum, except during exposure periods. The temperature was controlled at $22 \pm 2°C$.

Groups of animals (8–10 animals in each) were exposed repeatedly to nitrogen dioxide, nitric oxide (NO) or ozone 3 for 4 hr a day for 7 consecutive days. For combined exposure to NO_2 and O_3, 40 animals were divided equally into 4 groups. The first group was exposed to NO_2 for 3 hr and O_3 for 3 hr on 7 consecutive days. Two other groups were exposed to the same concentration of NO_2 or O_3 3 hr a day for 7 consecutive days. The fourth groups served as a control. The combined exposure to NO and O_3 was performed in the same manner as that exposure to NO_2 and O_3. Exposures of all animals were begun when they were 5 weeks of age. For the exposure, the animals were housed in wire mesh cages in a stainless steel chamber with a volume of about 0.3 m^3. NO_2 or NO was introduced from a cylinder that contained an appropriate concentration of the gas in nitrogen to the airflow of the chamber and diluted to the desired concentrations with the filtered room air. Ozone was generated by passing oxygen through a neon tube silent arc generator, and the resulting effluent was mixed with the filtered room air before introduction into the chamber. NO_2 and NO concentrations within the chamber were measured by the method of Saltzman [6]. Acidic permanganate solution was used to convert NO to NO_2. Ozone concentrations were measured by the neutral-buffered potassium iodide method [7].

Approximately 20 hr after each exposure the animals were killed by cervical dislocation. The lungs were quickly removed, blotted, weighed and then homogenized in 8.0 ml per lung of ice cold 0.02 M disodium ethylenediaminetetraacetate (EDTA) solution. The homogenate was filtered through a filter paper and GSH levels in the filtrate were determined as nonprotein sulfhydryls according to the method of Sedlak and Lindsay [8]. Independent determinations of GSH gave values that accounted for 95% of the nonprotein sulfhydryls present in mouse lung homogenates [9]. GSH levels were expressed as per whole lung.

RESULTS

The effect of a 7-day repeated exposure to 3–48 ppm NO_2 on GSH levels in lungs is shown in Figure 1. GSH levels increased linearly as a function of the degree of exposure (concentration in ppm X exposure time in hours), although there was a lack of observable response until a critical level was

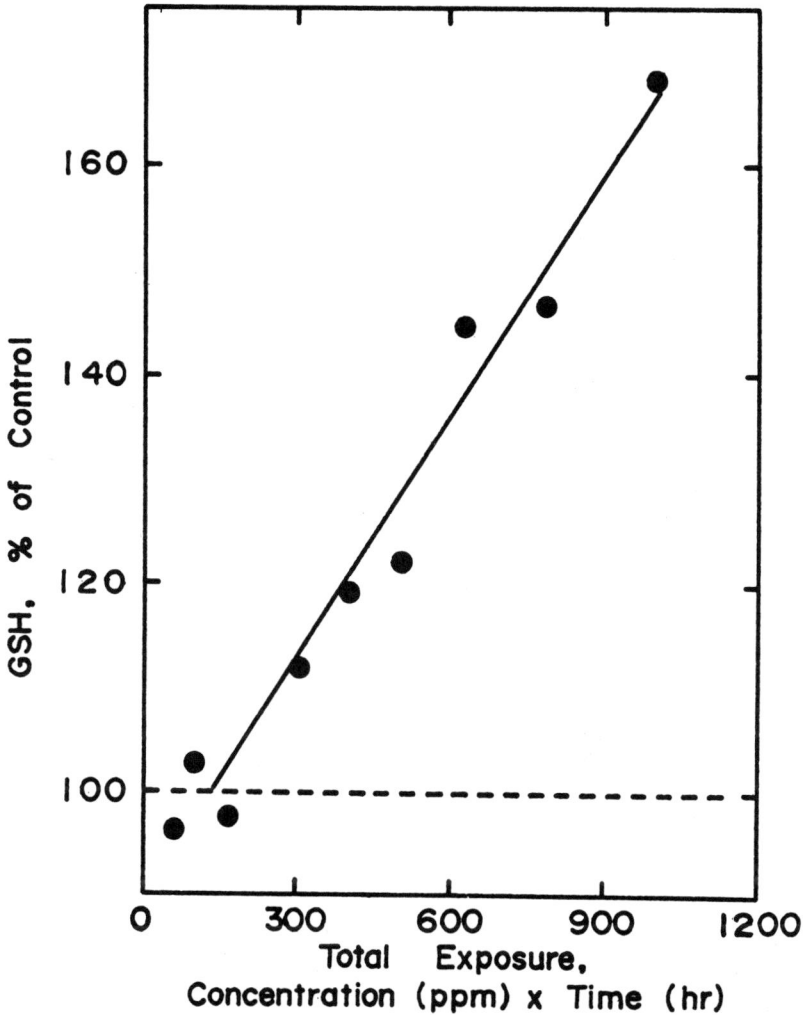

Figure 1. Effect of 7-day repeated exposure to NO_2 on GSH levels in lungs of mice. Each point represents the mean for 10 experiments. Control values are 389 ± 4 (mean ± SE) nmoles per whole lung.

reached. The effect of 7-day repeated exposure to 0.2–3.6 ppm O_3 on GSH levels in lungs is shown in Figure 2. A linear, dose-related increase in GSH levels was observed; however, the increase attained the plateau level at about 45% over control. There was no increase in GSH levels in lungs in 7-day repeated exposure to 10–21 ppm NO. The results indicate that the elevation of GSH levels in lungs can be used as a quantitative indicator of the effect of

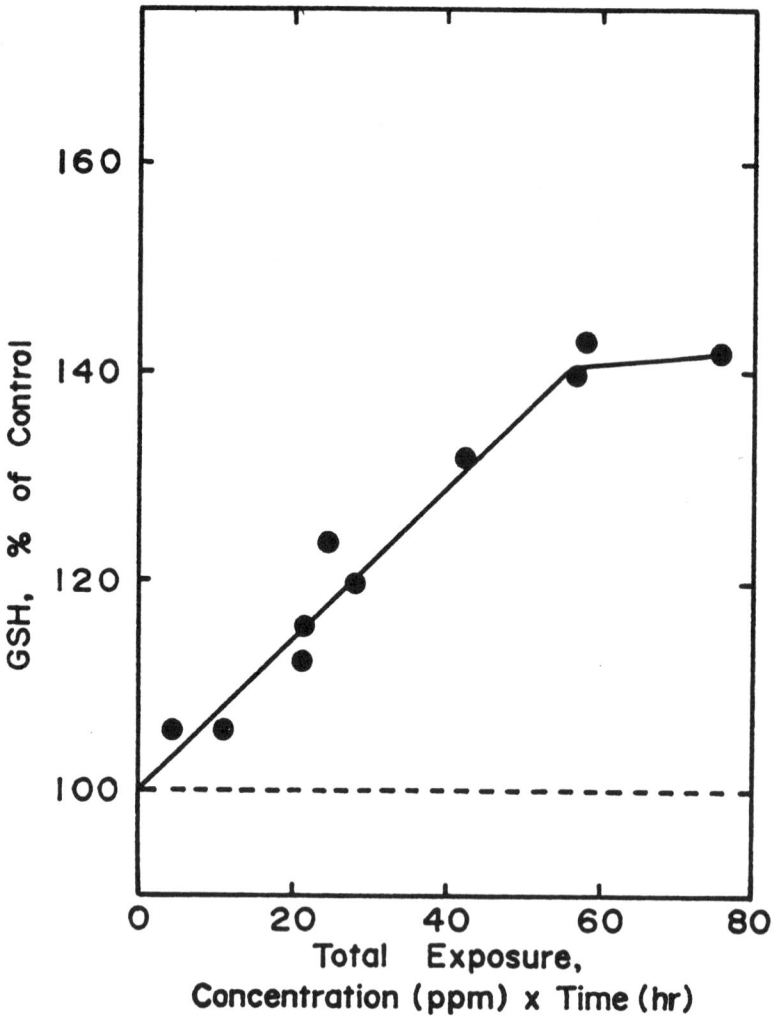

Figure 2. Effect of 7-day repeated exposure to O_3 on GSH levels in lungs of mice. Each point represents the mean for 8–10 experiments. Control values are 394 ± 4 (mean ± SE) nmoles per whole lung.

the repeated exposure (3 hr/day for 7 consecutive days) to 5–50 ppm NO_2 or below 3 ppm O_3.

Figure 3 shows the effect of 7-day repeated exposure to NO_2 and O_3 in combination on GSH levels in lungs. Exposure to 7 or 15 ppm NO_2 in combination with 0.5 or 1.0 ppm O_3 resulted in significant increases of 8.3–26.6% in GSH levels. The effect of combining the two pollutants was equal to

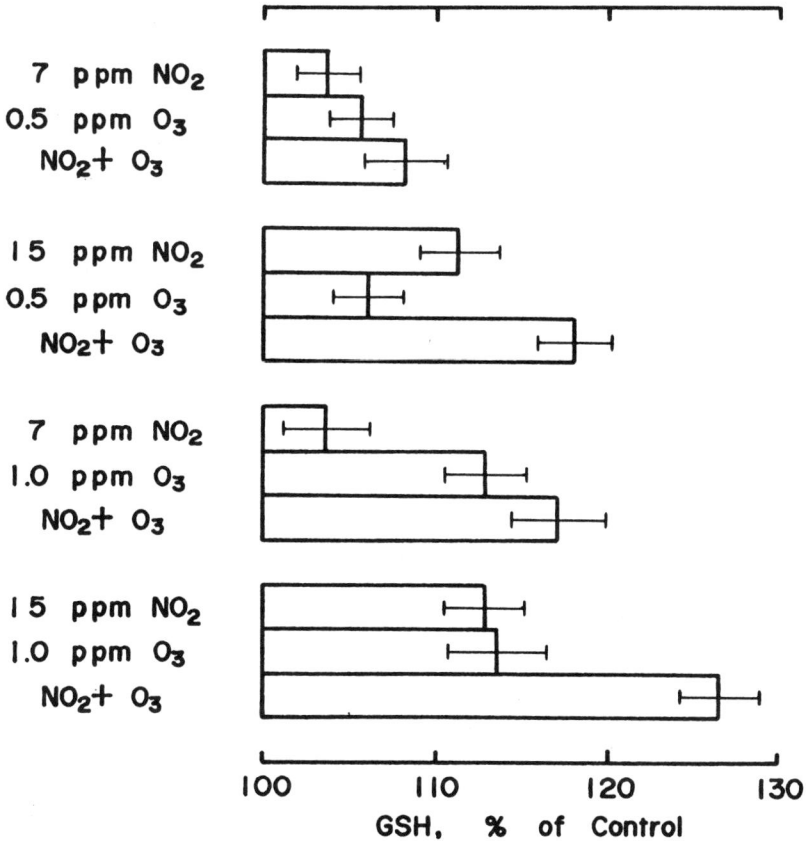

Figure 3. Effect of 7-day repeated exposure to NO_2 and O_3 in combination on GSH levels in lungs of mice. Each column with bars represents the mean and SE for 10 experiments. Control values are 378 ± 5 (mean $\pm$ SE) nmoles per whole lung.

the arithmetic sum of the effects of the two individual pollutants. The same result was obtained when the daily combined exposure was performed with O_3 followed by NO_2. In the experiment in which the animals were exposed to 3 ppm NO_2 and 1.0 ppm O_3, the effect of combined exposure was only equal to the effect of O_3; GSH levels of the 3-ppm NO_2-exposed animals were equal to those of the control. In the experiment in which the animals were exposed to 30 ppm NO_2 and 1.0 ppm O_3, the effect of combined exposure was less than the effect obtained by the sum of that of the two individual pollutants. As a 45% increase in GSH levels occurred in the 30-ppm NO_2-exposed animals, it seemed that the increase in GSH levels attained the plateau level in the NO_2- and O_3-exposed animals.

In the combined exposure to 10 ppm NO and 1.0 ppm O_3, NO made no difference in the increased GSH levels in lungs caused by the exposure to O_3 alone.

DISCUSSION

The data presented here demonstrate that repeated exposure to nitrogen dioxide or ozone results in a dose-related elevation of GSH levels in lungs of mice. The changes in GSH levels were found to be accompanied by increased activities of enzymes relating to the glutathione peroxidase system [4,5]. It has been suggested that this system serve as the protective mechanism against oxidizing damage by oxidants [2,4,5,10]. Both nitrogen dioxide and ozone are similar in that they have oxidizing abilities, including lipid peroxidation, which have been proposed as mechanisms for their toxicities [11–13].

The observations made in this study indicate that ozone is more biologically active than nitrogen dioxide. According to the data on acute toxicity [14], effects on pulmonary antibacterial activity [15], cellular responses at lower levels [16] or biochemical responses [3], ozone has been 10–20 times more toxic than nitrogen dioxide. The present data also show ozone to be 10–15 times more active than nitrogen dioxide. Higher toxicity of ozone may be partly related to its oxidative nature [13].

The mode of the elevation of GSH levels in lungs also shows two dissimilarities between the effect of nitrogen dioxide and ozone. One is that the response to ozone can be extrapolated to as low as ambient concentrations, whereas no response to nitrogen dioxide is observed at ambient concentrations. The other is that the ozone-exposed animals attain a plateau level at about 45% over control. It is suggested that biochemical responses to protect against oxidants are inhibited at higher concentrations of ozone. These findings indicate there are mechanistic differences between the action of nitrogen dioxide and ozone, as previously pointed out [3,5,13]. Stokinger and Coffin [14] have outlined the comparison of toxicological actions of nitrogen dioxide and ozone, in which they have indicated that one characteristic, distinguishing the former from the latter, is the lack of observable acute tissue response until a critical concentration is reached. Ozone has been shown to have measurable biological effects at levels converging on a no-threshold dose-response, like ionizing radiation [17].

Of the various oxides of nitrogen found in the air, both nitrogen dioxide and nitric oxide are important as air pollutants. Nitric oxide has been recognized to be less acutely toxic [14]. That GSH levels in lungs were unaffected by the exposure to relatively higher concentrations of nitric oxide indicates that it may be less active than nitrogen dioxide on the pulmonary metabolic responses. The effects of extremely high concentrations of nitric oxide on the GSH levels could not be investigated because of its rapid conversion to nitrogen dioxide.

Considerable experimental attention has been given to the biological effects of mixtures of air pollutants. With regard to the effects of mixtures of nitrogen dioxide and ozone, Goldstein et al. [18] showed that pulmonary consequence of exposure to both pollutants was equivalent to the injury that would be expected from each individual pollutant. Ehrlich et al. [19] showed that the effect of a single exposure to the mixture on the resistance to bacterial pneumonia was additive. They also suggested that a synergistic effect might be present on repeated inhalation of the mixtures. Freeman et al. [16] showed that pulmonary lesions induced with the mixtures reflected mainly the effects of ozone and suggested that potential injury from ambient smog would result mainly from it, whereas injury from tobacco smoke would be due largely to its relatively high concentration of nitrogen dioxide. They indicated that smoking and residing in smoggy areas would be additively injurious. Although nitrogen dioxide and ozone were not used simultaneously and relatively higher concentrations of nitrogen dioxide were used in this experiment, the data obtained in our study also show that nitrogen dioxide and ozone act additively and that the effects of their mixtures would depend mainly on the effects of ozone if the concentrations of both pollutants are the same level. These findings seem to be important in the evaluation of health hazards from photochemical smog.

This study confirms that the elevation of GSH levels in lungs of mice can be used as a quantitative indicator for monitoring the pulmonary effects of oxidants. In addition [20], the above indicator is available for evaluating various factors that may accentuate or reduce the pulmonary effects of oxidants.

CONCLUSIONS

The combined effects of repeated exposure to nitrogen dioxide or nitric oxide and ozone were studied by using the elevation of GSH levels in lungs of mice as an indicator for monitoring the pulmonary effects of oxidants. The 7-day repeated exposure to nitrogen dioxide or ozone resulted in a dose-related increase in GSH levels in lungs. Ozone was more biologically active than nitrogen dioxide, and certain differences were observed between the action of nitrogen dioxide and ozone. In the 7-day repeated exposure to 3, 7 or 15 ppm nitrogen dioxide in combination with 0.5 or 1.0 ppm ozone, the effect of combining the 2 pollutants was equal to the sum of the effects of the individual pollutants. Repeated exposure to nitric oxide had no affect on the GSH levels or on the alteration in GSH levels caused by the exposure to ozone. The observations suggest that the combined effects of nitrogen oxides and ozone are additive.

ACKNOWLEDGMENTS

This study was supported in part by a grant from the Environmental Agency C-8 entitled "Studies of Biological Effects of Air Pollution" (1975–1977).

REFERENCES

1. Hueter, F. G., and M. Fritzhand. "Oxidants and Lung Biochemistry. A Brief Review," *Arch. Intern. Med.* 128:48–53 (1971).
2. Chow, C. K., and A. L. Tappel. "An Enzymatic Protective Mechanism against Lipid Peroxidation Damage to Lungs of Ozone-Exposed Rats," *Lipids* 7:518–524 (1972).
3. Chow, C. K., C. J. Dillard and A. L. Tappel. "Glutathione Peroxidase System and Lysozyme in Rats Exposed to Ozone or Nitrogen Dioxide," *Environ. Res.* 7:311–319 (1974).
4. Fukase, O., K. Isomura and H. Watanabe. "Effect of Ozone on Glutathione *in Vivo*," *Taiki Osen Kenkyu (J. Japan Soc. Air Poll.)* 10:58–62 (1975).
5. Fukase, O., K. Isomura and H. Watanabe. "Effects of Nitrogen Oxides on Peroxidatve Metabolism of Mouse Lung," *Taiki Osen Kenkyu (J. Japan Soc. Air Poll.)* 11:65–69 (1976).
6. Saltzman, B. E. "Colorimetric Microdetermination of Nitrogen Dioxide in the Atmosphere," *Anal. Chem.* 26:1949–1955 (1954).
7. Saltzman, B. E., and N. Gilbert. "Iodometric Microdetermination of Organic Oxidants and Ozone. Resolution of Mixtures by Kinetic Colorimetry," *Anal. Chem.* 31:1914–1920 (1959).
8. Sedlak, J., and R. H. Lindsay. "Estimation of Total, Protein-Bound, and Nonprotein Sulfhydryl Groups in Tissue with Ellman's Reagent," *Anal. Biochem.* 25:192–205 (1968).
9. Fukase, O., and K. Isomura. "Effect of Metallic Cadmium Fume on Peroxidative Metabolism of Mouse Lung," *Hyogo-ken Eisei Kenkyu-sho Kenkyu Hokoku (Bull. Public Health Inst. Hyogo Prefecture)* 11:1–3 (1976).
10. Chow, C. K. "Biochemical Responses in Lungs of Ozone-Tolerant Rats," *Nature* 260:721–722 (1976).
11. Thomas, H. V., P. K. Mueller and R. L. Lyman. "Lipoperoxidation of Lung Lipids in Rats Exposed to Nitrogen Dioxide," *Science* 159:532–534 (1968).
12. Goldstein, B. D., C. Lodi, C. Collinson and O. J. Balchum. "Ozone and Lipid Peroxidation," *Arch. Environ. Health* 18:631–635 (1969).
13. Roehm, J. N., J. G. Hadley and D. B. Menzel. "Antioxidants vs Lung Disease," *Arch. Intern. Med.* 128:88–93 (1971).
14. Stokinger, H. E., and D. L. Coffin. "Biologic Effects of Air Pollutants," In: *Air Pollution*, Vol. 1, A. C. Stern, Ed. (New York: Academic Press, Inc., 1968), pp. 445–546.
15. Goldstein, E., M. C. Eagle and P. D. Hoeprich. "Effect of Nitrogen Dioxide on Pulmonary Bacterial Defense Mechanisms," *Arch. Environ. Health* 26:202–204 (1973).
16. Freeman, G., L. T. Juhos, N. J. Furiosi, R. Mussenden, R. J. Stephens and M. J. Evans. "Pathology of Pulmonary Disease From Exposure to Interdependent Ambient Gases (Nitrogen Dioxide and Ozone)," *Arch. Environ. Health* 29:203–210 (1974).
17. Brinkman, R., H. B. Lamberts and T. S. Veninga. "Radiomimetic Toxicity of Ozonized Air," *Lancet* 1:133–136 (1964).

18. Goldstein, E., D. Warshauer, W. Lippert and B. Tarkington. "Ozone and Nitrogen Dioxide Exposure. Murine Pulmonary Defense Mechanisms," *Arch. Environ. Health* 28:85–90 (1974).
19. Ehrlich, R., J. C. Findlay, J. D. Fenters and D. E. Gardner. "Health Effects of Short-Term Inhalation of Nitrogen Dioxide and Ozone Mixtures," *Environ. Res.* 14:223–231 (1977).
20. Fukase, O., K. Isomura and H. Watanabe. "Effects of Exercise on Mice Exposed to Ozone," *Arch. Environ. Health* 33:198–201 (1978).

CHAPTER 12

INITIATION OF PEROXIDATION BY NITROGEN DIOXIDE IN NATURAL AND MODEL MEMBRANE SYSTEMS

J. F. Mead and M. Gan-Elepano

Laboratory of Nuclear Medicine and Radiation Biology
University of California
Los Angeles, California 90024
and
Department of Biological Chemistry
School of Medicine
University of California at Los Angeles
Los Angeles, California 90024

F. Hirahara

The National Institute of Nutrition
Tokyo, Japan

INTRODUCTION

Although the lipid bilayer of most biomembranes would seem to be an ideal site for a radical-propagated peroxidation reaction, the proof of its occurrence or the course it might take are difficult to find in living organisms.

For this reason, it is often convenient to use simple model systems for the initial studies, with the hope that information derived from them might aid in formulating intelligent questions about the more complex systems and in interpreting the answers. With this aim in mind, our laboratory is using a series of model systems of increasing complexity to aid in the interpretation of in vivo results [1].

Among these are the erythrocyte and liver microsomal membrane systems, which are, in effect, complete living membranes; a phospholipid vesicle system, which can be formulated to contain all the lipid components of membranes without the complicating proteins; and some simpler systems. Results obtained with these models have helped to devise experiments with which to examine the membrane systems of lung and other tissues.

The simplest of these models is a monolayer of linoleic acid adsorbed on silica gel, and this has been used by our group [2,3] and others [4,5] with good effect.

Using this system, we have found that:

1. the reaction is initiated by atmospheric oxygen (or its complex with the iron contained in the silica gel) or by other agents, as NO_2 [2];
2. the disappearance of linoleic acid follows first-order kinetics, unlike bulk-phase peroxidation [2];
3. the rate of peroxidation is decreased by adding saturated fatty acids to the monolayer [6];
4. tocopherols and other antioxidants introduce a lag period during which the oxidation rate is exceedingly slow, followed by the normal rate, when the antioxidant is almost exhausted [7];
5. the products of the reaction are not the usual hydroperoxides of bulk-phase autoxidation but are epoxides (in monolayers with linoleate to silica ratios of greater than 0.02) and, increasingly, hydroxy epoxides at ratios smaller than this [2]; and
6. olefins other than lineoleate, such as oleate or cholesterol, are epoxidized during the peroxidation even though, by themselves, they would yield quite different products at a lower rate [6].

Progressing directly from the simple model system to the living animal, it was found that in the lungs of rats breathing air or, particularly, air containing 5–10 ppm of nitrogen dioxide (NO_2), the same products were produced [1]. Although the evidence indicated that they were formed by a mechanism similar to that operating in the model system, proof could not readily be obtained. A model system of intermediate complexity was clearly needed and the liver microsomal membrane was initially chosen for this. Even this system is somewhat difficult to understand clearly and in this case recourse to a simpler model system was also necessary.

MATERIALS AND METHODS

Materials

Linoleic acid, 99% pure, and methyl palmitate were obtained from Applied Science Labs., Inc. Nitrogen dioxide, 100 ppm in nitrogen, was supplied by the Matheson Gas Co. D,α-Tocopherol, found to be 99% pure by thin-layer chromatography and ultraviolet spectrophotometry, was purchased from Eastman Organic Chemicals Co. L-Ascorbic acid was from Sigma Chemical Co.

Treatment of Animals

Rats were fed a fat-free diet [8] containing 5% stripped lard from weaning. Vitamin E (D,α-tocopherol, 280 ppm) was either added or omitted in the control and experimental diets. Rats maintained on a lab chow diet were also used.

Incubation Experiments

Livers were used from rats that had been on the control and tocopherol-deficient diets for 6–9 months. The microsomal fractions were prepared by the method of Brenner and Peluffo [9] and were suspended in 0.05 M *tris* buffer solution, pH 7.4. Incubations were carried out in 25-ml flasks under air at 37°C in an Aquatherm water bath shaker, according to Bidlack and Tappel [10]. Aliquots containing 25 mg of microsomal protein were incubated with 10 mM sodium ascorbate and 10 mM FeCl$_3$ in a total volume of 1 ml with 10 mM *tris*-HCl buffer, pH 7.4.

When NO$_2$ was used as the peroxidation initiator, the microsomes in buffer solution alone were incubated under a constant flow of 9–14 ppm NO$_2$ in air. In both cases, flasks were removed from the shaker for analysis at stated intervals.

Extraction and Treatment of Lipids

The total lipids from the incubation mixtures were extracted with chloroform–methanol (2:1, v/v) according to the method of Folch et al. [11]. They were subjected to methanolysis by heating with 5% methanolic HCl at 80°C under nitrogen for 2.5 hr. The fatty acid methyl esters were extracted with pentane, washed free of methanol, dried and analyzed on a Beckman Gas–Liquid Chromatographic Apparatus.

Incubations of Monolayers Adsorbed on Silica Gel

The preparation, incubation and extraction procedures for the linoleic acid monolayer systems were carried out according to the methods of Wu et al. [3,7]. Solutions of linoleic acid plus α-tocopherol were stirred with silica gel G, (washed with HCl and treated with EDTA) to make monolayers 0.04 mol% in tocopherol. The dry-coated silica was incubated at 60°C for the desired length of time under air or air containing 10–15 ppm NO$_2$, and then extracted with methanol. The unchanged acid was methylated and quantitated by GLC as previously described [3]. The α-tocopherol in the control and incubated samples was extracted twice with 3 ml each of ethanol. The fatty acids were removed by passing the mixture through small Florisil columns with ether:pentane (25:85) as solvent. The tocopherol was then analyzed either on a Varian Aerograph Gas–Liquid Chromatographic Apparatus, using QF1 as the stationary phase, or on a Chromatronix High-Pressure Liquid Chromatogram H466, using LiChroprep[TM] Si 60 (Cole Scientific, Inc., Calabasas, CA) columns.

RESULTS

Figure 1 shows that incubation of the liver microsomal preparation under air in the presence of ferrous ion and ascorbate results in the rapid disappearance of polyunsaturated fatty acid (in this case 22:6) with preparations from animals on tocopherol-deficient or chow diets. In this respect,

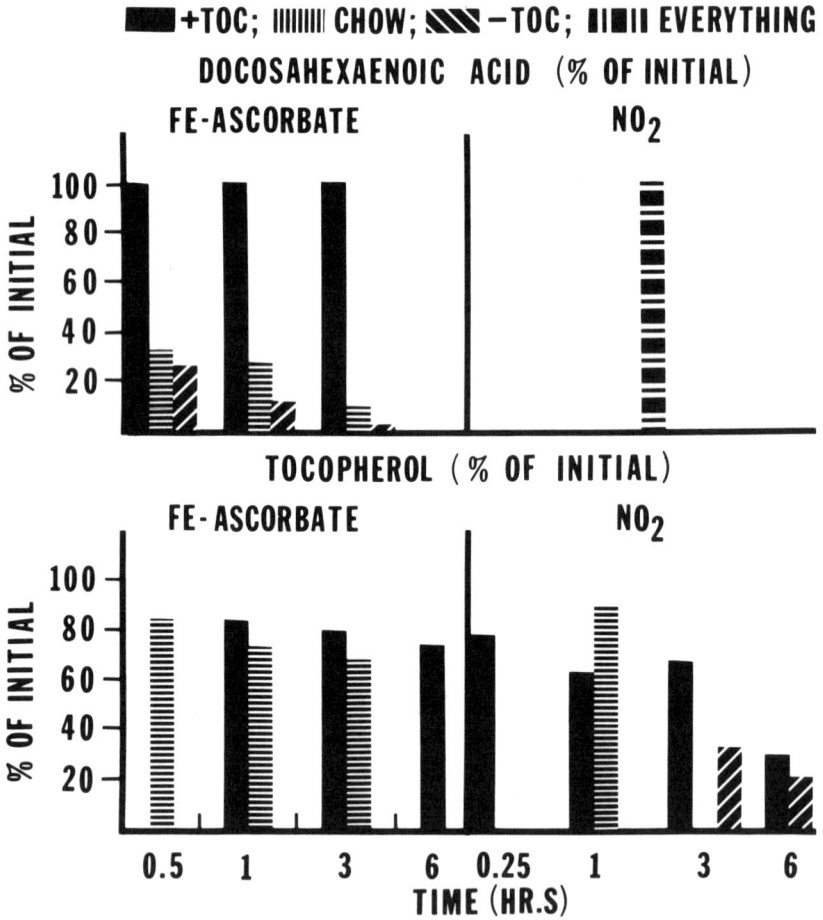

Figure 1. Changes in polyunsaturated fatty acids and tocopherol on incubation of liver microsomal fractions in the presence of Fe-ascorbate or NO_2.

it appears that the chow diet is tocopherol-deficient. In the preparations from tocopherol-supplemented animals, however, the fatty acid was completely protected during the incubation period. When air containing NO_2 was used in the absence of iron and ascorbate, no oxidation of fatty acid occurred.

On the other hand, tocopherol was slowly oxidized at about the same rate in both systems, slightly more rapidly in the presence of NO_2. Again it is evident that at higher concentrations of tocopherol its oxidation rate was lower. These data are incomplete because in the livers of the tocopherol-deficient and some of the chow-fed animals tocopherol was initially present in amounts too small for accurate measurement.

To aid in interpreting these results, the simple linoleate monolayer was studied under comparable conditions. Figure 2 shows that, as has been shown many times [5,7,12], inclusion of tocopherol in an unsaturated fatty acid preparation results in a lag period during which little oxidation takes place. In the absence of antioxidant, however, monolayer autoxidation proceeds rapidly with first-order kinetics [3]. When exposed to NO_2-containing air, the monolayer linoleic acid is oxidized without any noticeable delay. At the same time, tocopherol oxidation proceeds rapidly, as it does in the presence of air alone.

DISCUSSION

Interpretation of the results of the microsomal incubation would at first appear rather simple. Radicals generated by the Fe^{2+}-ascorbate system ($\cdot OH$ or equivalent species) are capable of attacking both polyunsaturated fatty acids and tocopherol with equal avidity. Thus, in a membrane containing low concentrations of tocopherol, the apparent result will be a rapid oxidation

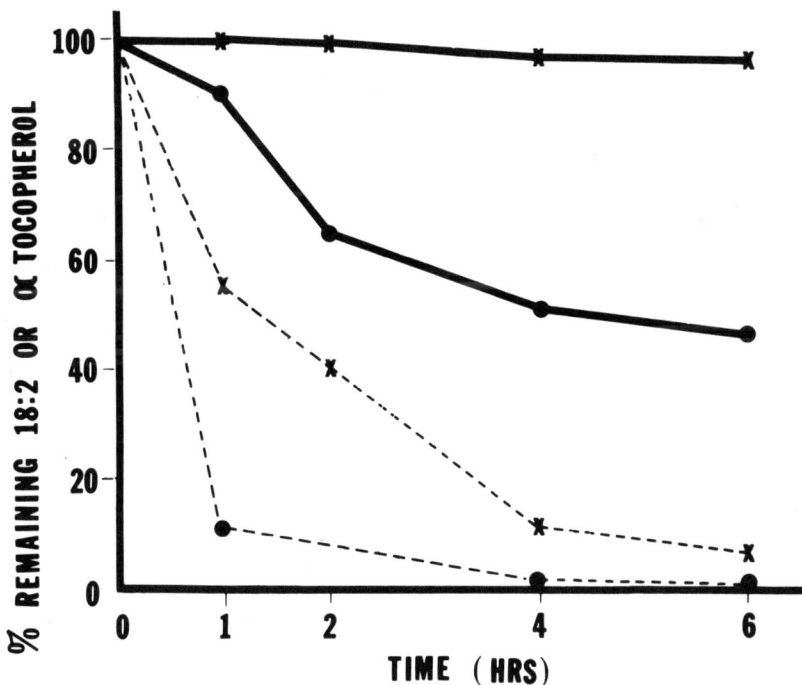

Figure 2. Oxidation of linoleic acid (——) and α-tocopherol (– – –) from the monolayer on DTPA-treated silica gel G under air (×) or 10 ppm NO_2 (•) at 60°C.

of the fatty acids of the membrane bilayer. NO_2, however, reacts slowly with the unsaturated fatty acids but rapidly with tocopherol so that any tocopheral present in the bilayer will be attacked first and afford protection to the fatty acids.

The results with the linoleic acid monolayer are not so easy to understand. The two systems should be somewhat comparable since the iron contained in the silica gel matrix probably gives rise to initiating species similar to those operating in the bilayer. This is true even in acid-washed, DTPA-treated silica gel, since the iron is not totally removed or blocked by this treatment. It thus remains in fixed sites for initiation of peroxidation. The NO_2, on the other hand, is a very mobile initiator, giving rise to tocopherol oxidation and more slowly to fatty acid oxidation, even in the presence of tocopherol bound in the monolayer. This mobility of the initiator, compared with the mobilities of fatty acid and antioxidant, permits the oxidation of the fatty acid to proceed. It is also possible that the more rapid oxidation of the tocopherol in the presence of NO_2 reduces its effective antioxidant capacity to a low level at the early stages of the reaction.

CONCLUSIONS

NO_2, although a less efficient initiator of peroxidation than the oxygen or metal-oxygen radicals, is effective by virtue of its greater mobility and its facile destruction of the membrane antioxidant tocopherol.

The implications of these findings are that the action of NO_2 on tissues may well be largely in the destruction of antioxidant, thus leaving the tissue unprotected against the action of oxidants, possibly including NO_2 itself. In the case of tissues high in tocopherol, both the rate of destruction of tocopherol and the consequent oxidation of other tissue lipids is decreased. A practical means of protection against this type of pollutant would thus seem to be to increase the tissue content of tocopherol by dietary means. Certainly, even a mild deficiency in tissue tocopherol content should not be allowed to occur in areas subjected to higher than normal concentrations of NO_2.

REFERENCES

1. Sevanian, A., J. F. Mead and R. A. Stein. "Epoxides as Products of Lipid Autoxidation in Rat Lungs," *Lipids* 14:634–643 (1979).
2. Wu, G.-S., and J. F. Mead. "Autoxidation of Fatty Acid Monolayers Adsorbed on Silica Gel: I. Nature of Adsorption Sites," *Lipids* 12:965–970 (1977).
3. Wu, G.-S., R. A. Stein and J. F. Mead. "Autoxidation of Fatty Acid Monolayers Adsorbed on Silica Gel: II. Rates and Products," *Lipids* 12:971–978 (1977).
4. Honn, F. J., I. I. Bezman and B. F. Dauberg. "Autoxidation of Drying Oils Adsorbed on Silica," *J. Am. Oil Chem. Soc.* 28:129–133 (1951).
5. Porter, W. L., L. A. Levasseur and A. S. Herick. "An Addition Compound of Oxidized Tocopherol and Linoleic Acid," *Lipids* 6:1–8 (1971).

6. Wu, G.-S., R. A. Stein and J. F. Mead. "Autoxidation of Fatty Acid Monolayers Adsorbed on Silica Gel: III. Effects of Saturated Fatty Acids and Cholesterol," *Lipids* 13:517–524 (1978).
7. Wu, G.-S., R. A. Stein and J. F. Mead. "Autoxidation of Fatty Acid Monolayers Adsorbed on Silica Gel: IV. Effects of Antioxidants," *Lipids* 14:644–650 (1979).
8. Mohrhauer, H., and R. I. Holman. "The Effect of Dose Level of Essential Fatty Acids Upon Fatty Acid Composition of Rat Liver," *J. Lipid Res.* 4:151–159 (1963).
9. Brenner, R. R., and R. O. Peluffo. "Effect of Saturated and Unsaturated Fatty Acids on the Desaturation *in vitro* of Palmitic, Stearic, Oleic, Linoleic and Linolenic Acids," *J. Biol. Chem.* 241:5213–5219 (1966).
10. Bidlack, W. R., and A. L. Tappel. "Damage to Microsomal Membrane by Lipid Peroxidation," *Lipids* 8:177–182 (1973).
11. Folch, P. J., M. Lees and G. H. Sloane-Stanley. "A Simple Method for the Isolation and Purification of Total Lipids from Animal Tissues," *J. Biol. Chem.* 226:497–509 (1957).
12. Porter, W. L., L. A. Levasseur and A. S. Henick. "Evaluation of Some Natural and Synthetic Phenolic Antioxidants in Linoleic Acid Monolayers on Silica," *J. Food Sci.* 42:1533–1535 (1977).

CHAPTER 13

PHARMACOLOGICAL MECHANISMS IN THE TOXICITY OF NITROGEN DIOXIDE AND ITS RELATION TO OBSTRUCTIVE RESPIRATORY DISEASE

Daniel B. Menzel

Director
Laboratory of Environmental Pharmacology and Toxicology
Departments of Pharmacology and Medicine
Duke University Medical Center
Durham, North Carolina 27710

INTRODUCTION

The work of Freeman and co-workers [1,2] allows little doubt that continuous exposure of rats to nitrogen dioxide (NO_2) leads to a pathological condition similar, if not identical, to human emphysema. The connection between human emphysema, obstructive respiratory disease and exposure to nitrogen oxides (NO_x) is then a tantalizing one, and this chapter presents some speculative ideas about the potential connection between them. In view of the high levels of NO_x in cigarette smoke and the impact of cigarette smoking on the incidence of obstructive respiratory disease, the potential role of NO_x in the pathogenesis of this disease is important. The largest impact in terms of total number of people may be the subtle effect of air pollution on the lung, undoubtedly exacerbated by cigarette smoking. To be sure, NO_x is but one part of urban air pollution, and the response of man to this complex mixture of toxicants may be quite different from the response of animals to a single component. The toxicity of the complex mixture is likely to be greater, as indicated by epidemiological studies. Thus, extrapolation from animals to man on the basis of the toxicity of NO_2 alone is likely to be a conservative estimate of human toxicity.

ANATOMICAL LOCALIZATION OF PATHOLOGICAL
EFFECTS OF NO$_2$

In animals, the striking effect of NO$_2$ exposure is the localization of pathological effects occurring within the lung. The most affected area is that portion of the respiratory bronchiole joining the alveolus. Miller et al. [3] have suggested that a similar effect for ozone is due, in part, to the dose delivered to that region of the respiratory tract through the combined effects of pulmonary anatomy, radial and axial diffusion, and the reaction of ozone (O$_3$) with tracheal mucous. Similar considerations may apply to NO$_2$ as it has only slightly greater solubility in water than ozone and is moderately less reactive. NO$_2$ is taken up and distributed throughout the lung, as shown by radiotracer studies [4]. However, once absorbed, the chemical form of the radiotracer was not determined. NO$_2$ could have reacted rapidly with cellular constituents and could have been present as HNO$_2$, HNO$_3$, NO$_2^-$, NO$_3^-$, N$_2$O$_3$, N$_2$O$_4$ or a variety of organically bound compounds.

The very complexity of the lung has stymied efforts to be more precise in our understanding of mechanisms leading to these toxic effects. Every cataloging of cell types within the lung leads to a recognition of increasing numbers of them, so any experiment must be judged on the basis of its inherent limitations. In our laboratory, we have tried to integrate some of the information on the toxicity of NO$_2$ and related compounds by using a variety of tools, albeit limited in scope.

AN INTEGRATIVE INTERPRETATION OF NO$_2$ TOXICITY

To provide a perspective to NO$_2$ toxicity, a hypothetical summary of the events occurring with NO$_2$ exposure has been drawn together. Figure 1 lays out the temporal sequence of these events following a single 4-hr exposure to at least 0.25 ppm NO$_2$. Most of the events are taken from work using rats as a model. Some differences will occur between species, of course, but the sequences are likely to be similar since the major effect of NO$_2$ seems to be the killing of specific cells within the lung. Systemic toxicity also occurs, but the best documented toxicity is to the lung.

As will be discussed below, the reaction of NO$_2$ with cellular constituents is so rapid that it can be viewed as essentially instantaneous compared to the length of time required for expression of biological effects. This is shown by the shaded area marked "Chemical Reaction" in Figure 1. Cell death reaches a maximum at 24 hr postexposure, while other measures of injury, such as susceptibility to airborne microorganisms or biochemical indicators of cellular injury, occur earlier. NO$_2$ initiates a wave of mitosis to replace the type I pneumocytes and Clara cells, which have died on exposure to the toxicant. The death of pulmonary cells is not restricted to type I pneumocytes and Clara cells if the NO$_2$ concentration is greater than about 25 ppm; at these concentrations it may also affect cells higher in the airway. Because of the involvement of cells beyond the junction of the respiratory bronchiole and

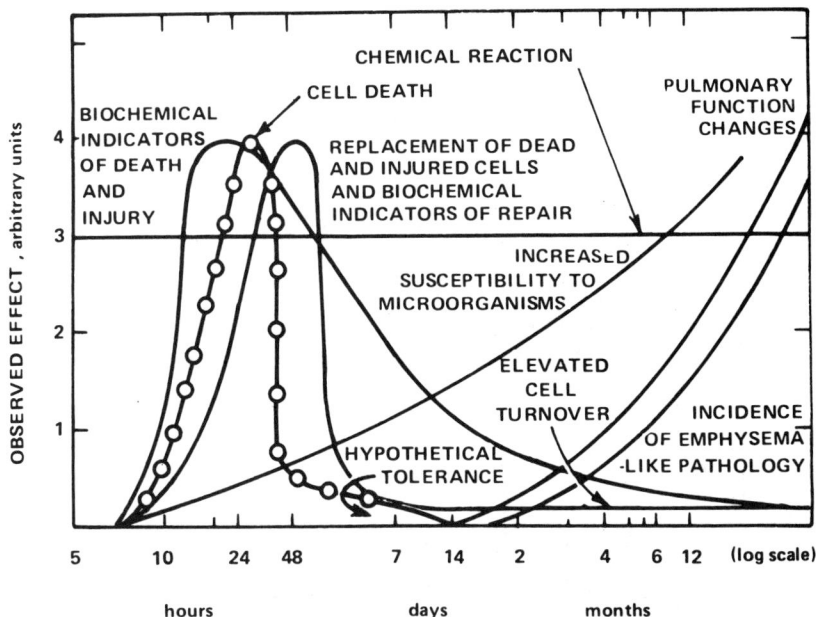

Figure 1. Temporal sequence of injury and repair hypothesized to occur in the lung following a single short-term exposure to NO_2 of less than 8 hr.

the alveolus, the toxicity of high concentrations of NO_2 may not be as pertinent to the effects of NO_2 on humans at ambient concentrations. Nonetheless, the replacement of dead and injured cells reaches a maximum at about 48 hr postexposure and declines slowly thereafter as repair proceeds. Presumably, damage from a single exposure to NO_2 concentrations of <25 ppm will be completely repaired.

The toxic effect of increasing concentrations of NO_2 is hypothesized in Figure 2. Again, chemical reaction is rapid and essentially complete by the time a single short-term exposure has ceased. As the concentration of NO_2 increases, so does the number of dying cells, probably proportionally to the log of the NO_2 concentration. Eventually, injury is so massive that pulmonary edema occurs and the exposed animals die of pulmonary insufficiency. Such a sequence of events could explain the greater effect of concentration, rather than of duration of exposure, on the toxicity of NO_2 using the bacterial infectivity model [5,6].

The temporal sequence following continuous exposure to NO_2 is shown in Figure 3. The chemical reaction of NO_2 with cellular constituents is shown as the continuous solid line. Depletion of reactants does not appear to occur with NO_2 levels of <50 ppm. Again, the reaction is rapid and continuous.

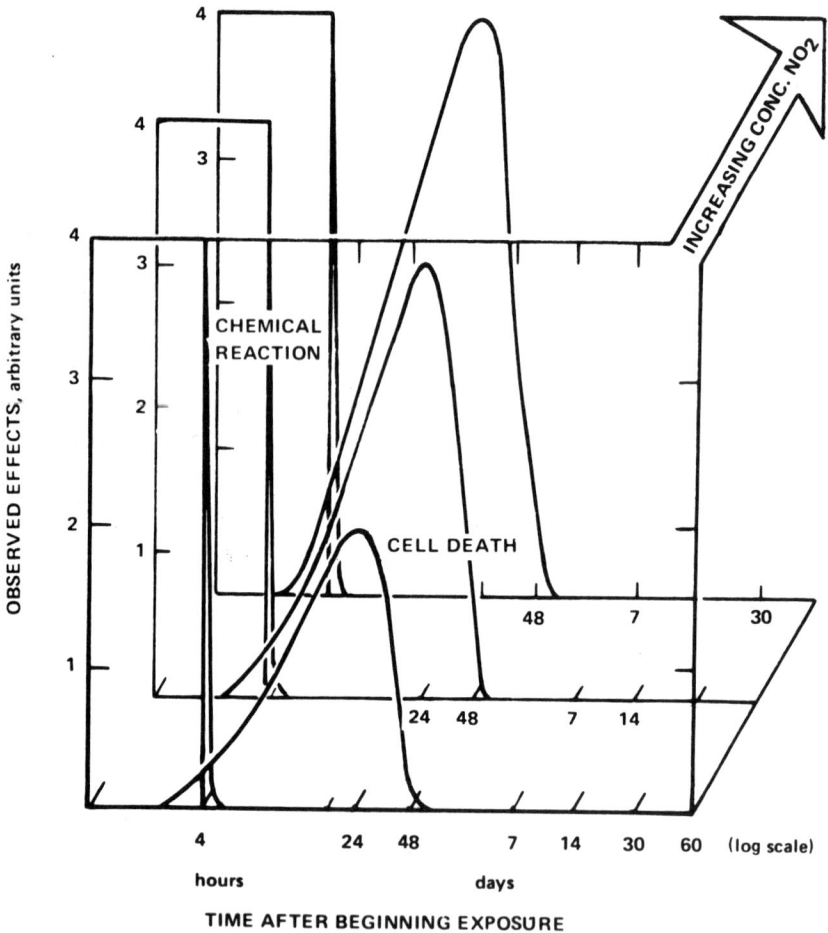

Figure 2. Proportionality between the toxic effect (cell death) and concentration of NO_2 during a constant exposure period of 4 hr. The maximum in cell death is reached at about 18 hr after exposure, and the total number of cells dying is proportional to the log of the concentration of NO_2.

Biochemical indicators of injury to, or death of, specific cells rise rapidly, with initial exposure reaching a peak near 18 hr after exposure. Cellular death, as judged from morphological studies, reaches a maximum at 24 hr, with repair, as judged by mitosis, reaching a maximum at 48 hr. Increased susceptibility to airborne microorganisms continues to increase as exposure is continued for six months or more. The susceptibility increases, suggesting continued injury by some cumulative mechanism. Pulmonary function changes are observable at higher concentrations after about two months of exposure, and emphysema occurs more frequently after six months of exposure. In such a scheme, the replacement of naïve cells with resistant cells might occur after 1–2 weeks, resulting in the hypothetical tolerance to continued exposure. Biochemical indicators that depend on cellular injury are likely to decrease in magnitude as the injured cells are replaced and may reach levels so low that differentiation between the exposed and control groups may be impossible. In fact, this seems to be the general trend of the biochemical data.

From this perspective, I would like to explore some evidence to support the hypothesis and its relationship to NO_2 toxicity to man.

CHEMICAL REACTIONS OF NO_2, NITRIC OXIDE (NO) AND RELATED POLYMERS

The chemistry of nitrogen oxides is relatively poorly understood. Much effort has been expended on the photochemistry and reactions in the atmosphere, but very little information has been gained about the reactions in solution of trace concentrations of nitrogen oxides. Thus, much of the chemistry proposed is theoretical and awaiting proof.

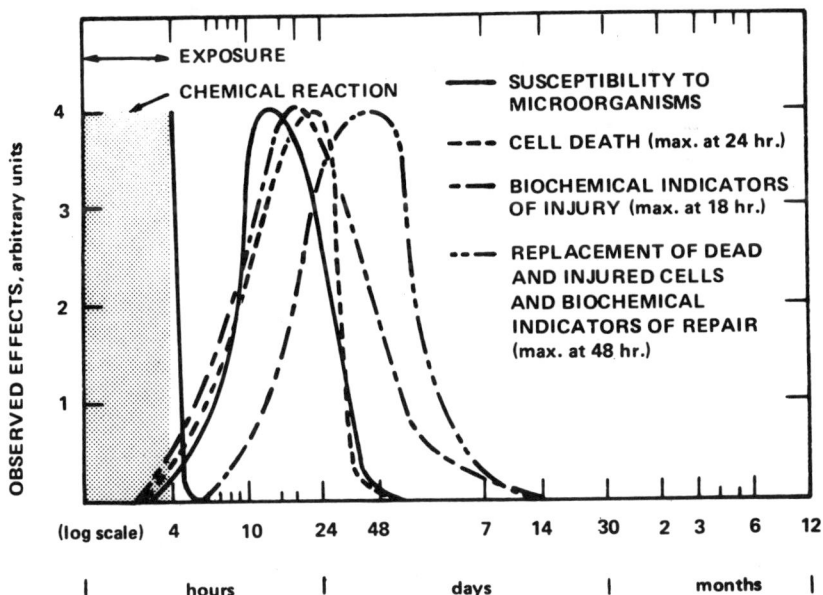

Figure 3. Temporal sequence of injury and repair hypothesized to occur in the lung on continuous exposure to NO_2.

While NO_2 is the anhydride of nitrous acid (HNO_2), the biological effects of NO_2 inhalation do not resemble those of HNO_2. In fact, the protection of animals by vitamin E and other antioxidants [7-9] and the alterations in lung fatty acid composition seen on NO_2 exposure argue that HNO_2 is not a primary reactant in NO_2 toxicity [9]. That is not to say that other soluble products of NO_2 are not significant; NO_2 dimers such as N_2O_4 and mixed dimers with NO such as N_2O_3 may play a very significant role in the long-term hazards to man of NO_2 inhalation. NO_2 is a free radical and, as such, undergoes rapid reaction without evoking solubilization.

Reaction 1 depicts the abstraction of hydrogen from a fatty acid. Nitrous acid is a stable end product. The resultant alkyl free radical will have a very short life in the lung (Reaction 2), since the reaction of alkyl radical with oxygen is controlled by their rates of diffusion. Only at very low oxygen concentrations, far below those likely to occur within the lung tissue, would the oxygen concentration become a rate-limiting factor. The hydroperoxyl free radical has a much longer life because it is less reactive. Further, lipid peroxidation could be promoted by the abstraction of another hydrogen from another molecule of fatty acid leading to a short-chain reaction (Reaction 3).

Alternatively, NO_2 could add directly across an unsaturation to give rise to the nitroperoxide radical, again via scavenging by oxygen of the alkyl-free radical generated by NO_2 addition (Reactions 4-6). Hydrogen abstraction leads to another molecule of fatty acid alkyl-free radical, which, in turn, promotes lipid peroxidation via a short-chain reaction. In both mechanisms, peroxidation of unsaturated lipids proceeds once initiation has occurred and corresponds to the peroxidation kinetics reported previously by us with model systems of fatty acid emulsions and thin films [7,9].

A role for vitamin E and other antioxidants as protectors occurs via the reaction of phenols with peroxyl free radicals. This reaction is facile and proceeds rapidly according to Reaction 7. The reaction forms a hydroperoxide and tocopherol-free radical. Tocopherol-free radical is relatively stable and dimerizes only slowly. The principal metabolite of α-tocopherol in mammals arises from such a dimer. α-Tocopherol can terminate free radical-mediated chain reactions intiated by NO_2, but has no effect on the reaction of NO_2 with unsaturated fatty acids or other cellular constituents.

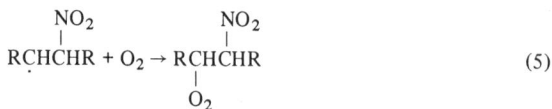

$$RH + \cdot O-N=O \rightarrow R\cdot + HONO \tag{1}$$

$$R\cdot + O_2 \rightarrow RO_2\cdot \tag{2}$$

$$RO_2\cdot + RH \rightarrow RO_2H + R\cdot \tag{3}$$

$$RCH=CHR + NO_2 \rightarrow R\overset{NO_2}{\underset{|}{C}HCHR} \tag{4}$$

$$R\overset{NO_2}{\underset{|}{C}}HCHR + O_2 \rightarrow R\overset{NO_2}{\underset{|}{C}}HC\underset{\underset{\cdot O_2}{|}}{H}R \tag{5}$$

$$RO_2{\cdot} + RH \rightarrow RO_2H + R{\cdot} \tag{6}$$

$$ROO{\cdot} + TOC{-}OH \rightarrow ROOH + TOC{-}O{:} \tag{7}$$

$$RO_2{\cdot} \rightarrow \underset{H}{\overset{R}{\diagdown}}\!\!\overset{O}{\triangle}\!\!\underset{H}{\overset{R}{\diagup}} + \underset{H}{\overset{R}{\diagdown}}\!\!\overset{O{-}O}{\triangle}\!\!\underset{H}{\overset{R}{\diagup}} + \underset{H}{\overset{R}{\diagdown}}\!\!\overset{O{-}O}{\diagup}\!\!\underset{H}{\overset{R}{\diagdown}} + RO{\cdot} + R{\cdot} \tag{8}$$

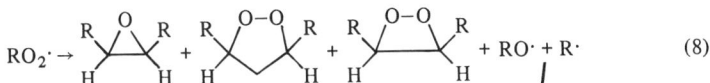

Potential Tissue Alkylation

Once peroxidation of cellular lipids occurs, complex reactions are possible. The net result is the formation of trace amounts of peroxides having a bicyclic configuration similar to naturally occurring prostaglandins H_2 and G_2 (PGH_2 and PGG_2). Vitamin E reduces the amount of these compounds formed through its termination of the chain reaction, leading to peroxidation and prevention of cyclization; but it has no effect on the cyclic peroxides once they are formed.

Peroxyl-free radical can also lead to epoxides, alkoxyl-free radicals, hydroxyl-free radicals and monocyclic peroxides. Both epoxides and alkoxyl-free radicals are potential alkylating agents. Another potential alkylating agent can be formed from nitrogen oxides by reaction of N_2O_3 or N_2O_4 with secondary and tertiary amines. Reaction leads to the formation of nitrosamines and nitramines, which can be metabolically activated to alkylating agents. Presumably, alkylation of DNA leads to mutagenesis and oncogenesis.

Peroxidation is often accompanied by alkylation and polymerization. In an early study of the polymerization of ribonuclease during the peroxidation of pure unsaturated fatty acids, we proposed that malonaldehyde was the primary agent responsible for polymerization [10]. Malonaldehyde arises from the hydrolysis of bicyclic peroxides and could be produced secondarily to peroxidation of unsaturated fatty acids initiated by NO_2. Malonaldehyde is a weak mutagen in the *Salmonella* assay.

NONRESPIRATORY FUNCTIONS OF THE LUNG

Recently, the nonrespiratory functions of the lung have been recognized as being vital to the normal physiological function of the intact organism. The lung serves both as an exocrine organ and as a regulatory metabolic organ, predominantly dealing with blood pressure. Since the lung receives all of the cardiac output, it is in a unique position to provide regulation of the circulating levels of vasoactive hormones. Further, the lung requires a precise and continuously changing regulation of the local perfusion of segments within the lung to match local blood perfusion with gas ventilation. A mismatch in either parameter is deleterious. In the clinical setting, mismatch of perfusion with ventilation is recognized as an early sign of pulmonary disease. Some of the regulatory functions of the lung particularly pertinent to our present discussion are set out in Table I.

Table I. Some Nonrespiratory Functions of the Pulmonary Capillary Bed

Activity	Physiological Effect
Converting Enzyme	
Bradykinin degradation (kininase II)	Inactivates bradykinin Increases blood pressure
Angiotensin I activation	Produces active angiotensin II Increases blood pressure
Amine Uptake	
5-hydroxytryptamine	Increases peripheral blood pressure Potential effects on CNS circulation
Prostaglandin Synthetase	
PGH$_2$ synthesis	Vasoconstriction in local areas Platelet aggregation
PGI$_2$ synthesis	Vasodilation in local areas Platelet disaggregation
Prostaglandin Degradation	
Uptake of prostaglandins	Return blood pressure to preexisting values
Metabolism of prostaglandins	Return blood pressure to preexisting values

One of the most potent blood pressure regulating systems is the renin–angiotensin system. For appropriate function, the conversion of angiotensin from its inactive form (angiotensin I) to its active form (angiotensin II) occurs in the lung. No other vascular bed is as effective as the lung in this conversion. The hydrolysis is carried out by converting enzyme, which is located in specialized caveoli on the luminal surface of the pulmonary capillary endothelial cells [11]. Converting enzyme is also kininase II or the enzyme responsible for the degradation of bradykinin to inactive peptides. The converting enzyme system is highly efficient, so a single pass through the pulmonary circulation will result in the complete inactivation of circulating levels of bradykinin and complete activation of circulating angiotensin I to angiotensin II. The net result of the passage of blood through the lung is an increase in peripheral blood pressure by a single circuit through the pulmonary vasculature.

The lung also possesses an active uptake system for the removal of circulating amines [12]. 5-Hydroxytryptamine and serontonin are rapidly removed from the circulation and metabolized to inactive compounds. In addition, drugs having the amine function are removed and metabolized prior to efflux from the lung. Since serotonin is a general vasodilator, its removal from the circulation by the lung results in an increase in peripheral blood pressure or a return to values existing prior to its release into the circulation, as in anaphylaxis.

The lung is one of the most active metabolic organs for the production of prostaglandins and their congeners. Circulating prostaglandins are removed from the blood by a carrier-mediated system, which recognizes the free carboxylic acid group as well as the presence of an oxygen function (alcohol or ketone) at position 11 and an alcohol at position 15 of the prostaglandin molecule [13]. Once absorbed, the prostaglandin is metabolized to inactive compounds, as discussed in more detail below.

The lung also produces a continuous supply of prostacyclin, or PGI_2. Prostacyclin could arise directly in the lung or indirectly in the aorta, where prostacyclin synthetase is present in large amounts. It is a potent platelet disaggregating agent acting via cyclic AMP and may be a defense mechanism preventing the aggregation and margination of platelets in the major vessels. Prostacyclin could then be the key to atherosclerosis and other vascular diseases resulting from vessel damage through platelet aggregation.

The nonrespiratory functions of the lung impinge on the regulation of blood pressure, both within the lung and peripherally. The ability of the lung to regulate local blood flow by regulation of vessel caliber is essential to the maintenance of proper respiratory function as well. The overall effect of the removal or metabolism of circulating prostaglandins from the blood by the lung is to lower blood pressure. Since the pulmonary vascular bed is a high-flow, low-pressure system, local flow is highly sensitive to small changes in vessel caliber. Flow and ventilation are exquisitely matched in the human lung. Toxicants interfering with this system could produce serious consequences not easily detected by conventional pulmonary function tests. Further, prostaglandins are not stored, but rather are synthesized and act at their site of synthesis. The molecular potency of prostaglandins ranges from 10^{-12} to 10^{-9} M, making regulation of circulating prostaglandin levels physiologically important. Studies of the mechanisms regulating local blood pressure could be highly sensitive indicators of physiological damage. To this end we have studied a number of membrane-associated transport systems. These include the active transport of phenol red from the airways to the capillary, the uptake of 5-hydroxytryptamine from the capillaries to the intracellular space, and the ion permeability of the lung in both directions.

EFFECTS OF NITROGEN DIOXIDE EXPOSURE ON PROSTAGLANDIN E$_2$ METABOLISM BY THE LUNG

Using the isolated, ventilated and perfused lung preparation, we have investigated the ability of the lung to both synthesize and metabolize prostaglandin. Early in the studies, we found that ozone inhibited the synthesis of prostaglandins [14]. Ozone inhibited prostaglandin synthesis noncompetitively, suggesting destruction of the prostaglandin synthetase. In the destruction of cellular membranes, ozone is more potent than NO_2. Membrane lipid peroxidation was most likely responsible for the inhibition. Cyclooxygenase, the key enzyme in the synthesis of prostaglandins, is self-inhibited in a noncompetitive fashion, probably by the destruction of membrane lipid by reaction with the prostaglandin peroxides PGG_2 and PGH_2.

Prostaglandin metabolism was investigated using PGE_2 as the model prostaglandin. PGE_2 is a natural product of the lung and acts to constrict the vasculature. It is an excellent substrate for both the prostaglandin active transport system and the prostaglandin degradative enzymes within the lung. The scheme for the metabolism of PGE_2 is shown in Figure 4.

Specialized techniques are necessary to measure the metabolism of PGE_2 by the lung [15]. The perfusion pathways in the lung differ significantly from each other. These naturally different pathlengths may represent the collateral circulation always present in the lung. As such, the differing pathlengths behave as though compounds perfused into the lung were mixed with a large volume of compound-free medium, making estimates of unidirectional

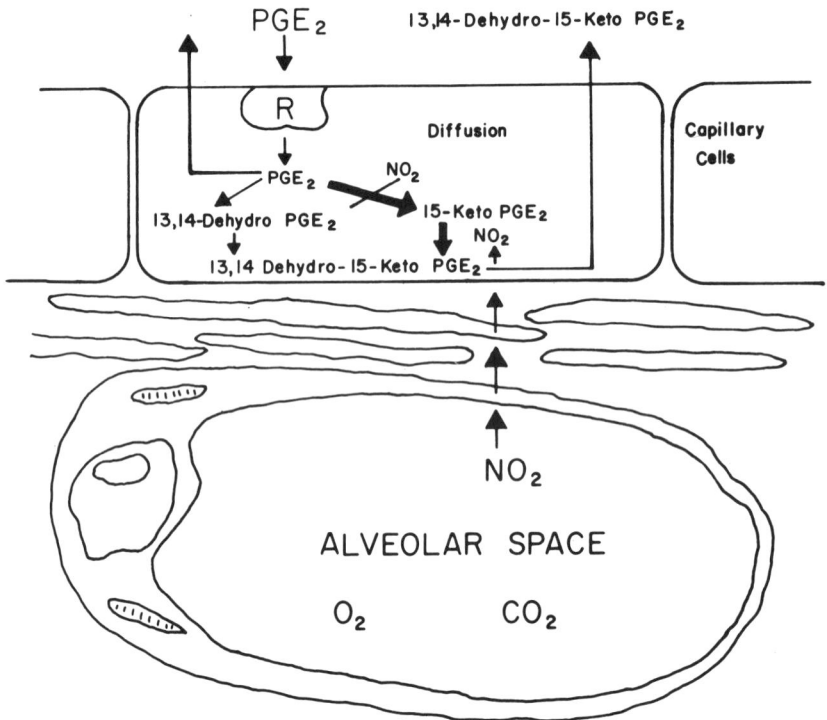

Figure 4. A schematic representation of the uptake, metabolism and release of PGE_2 by the lung. Inhaled NO_2 inhibits the metabolism of PGE_2, primarily by prevention of the conversion of PGE_2 to its inactive 15-keto metabolites.

flux difficult. To overcome these natural characteristics of the lung, Anderson and Eling [13] adapted Iverson's technique to the lung. Discontinuous gradients of both ^{3}H-PGE$_2$ and ^{14}C-dextran were applied to the lung simultaneously. The lung was prepared as previously described [15] and ventilated by negative pressure applied to an artificial thorax. A diagram of the perfusion system is shown in Figure 5. Fractions of the perfusate were collected and assayed for ^{3}H and ^{14}C content by liquid scintillation counting.

Figure 6 represents a typical curve generated by a computer reduction of the radioactivity of each fraction. Dextran is unlikely to leave the vascular space because it is of sufficiently large molecular weight. The plot of ^{14}C radioactivity (dextran) is then a plot of the average vascular transit of the perfusate, while the plot of ^{3}H radioactivity represents that PGE$_2$ not taken up by the lung. Knowing the concentration of the starting ^{14}C-dextran, the point of complete perfusion of the lung can be ascertained. This point is shown as the arrow in Figure 7 and represents the point at which maximum unidirectional flux of PGE$_2$ into the lung has occurred. Unidirectional fluxes calculated in this manner illustrate a saturable process having a K_t of 18.9 nmol/sec and a V_{max} of 0.278 nmol/sec at 2.0 ml/min (25°C) using Krebs-Hensleitt medium (pH 7.40) saturated with 5% CO_2:95% O_2.

Rats were exposed either to clean air or to 0.2, 2.0 or 19 ppm NO_2 for 3 hr. They were then killed at varying intervals starting immediately after exposure, as indicated in the figures and tables. As will be seen later, the maximum NO_2 effect was observed 18 hr postexposure, but not immediately afterwards. In Table II there was a trend toward lower uptake of PGE$_2$ by the lungs of rats exposed to higher NO_2 concentrations, but the difference between the clean air and NO_2-exposed preparations was not statistically significant. Uptake values at 1 μg PGE$_2$/ml ranged from 39.7–44.3 pmol/sec.

A biexponential rate of release of ^{3}H from the perfused lung was resolved by the method of residuals. The fastest efflux rates measured in these preparations showed some differences with prior exposure to NO_2 (Table III). The half-lives are also listed in the table to aid in conceptualization. No difference was found between the clean air and 0.2 ppm-exposed groups, while the rate of efflux was faster from the 2.0 ppm-exposed group ($t_{1/2}$ of 1.72 min vs 1.39 min). The efflux of ^{3}H was slower from the lungs of rats exposed to 19 ppm ($t_{1/2}$ of 1.72 min vs 1.96 min). These results are difficult to intercept directly, since the efflux of ^{3}H represents both ^{3}H-PGE$_2$ and ^{3}H-13,14-dehydro-15-keto PGE$_2$. As shown below, the conversion of ^{3}H-PGE$_2$ to its metabolite, ^{3}H-13,14-dehydro-15-keto PGE$_2$, is inhibited by NO_2 exposure. The ^{3}H metabolite concentration is greater in the later than in the early portions of the efflux curve and, conversely, for ^{3}H-PGE$_2$. We have chosen to lump both compounds together since we could not resolve the individual rates for the two compounds. One explanation could be that exposure to a relatively low NO_2 dose of 2.0 ppm for 3 hr increases the permeability of the lung to PGE$_2$. No clear evidence of edema in either PGE$_2$ or medium-perfused lungs could be seen in animals exposed to this level of NO_2 (Table IV). At higher concentrations of NO_2, such as 19 ppm, edema most likely occurs, increasing the extravascular space and slowing efflux.

The major effect of NO_2 exposure is, however, the inhibition of conversion of ^3H-PGE_2 to its metabolites. Thin-layer chromatography of the radiolabeled compounds extracted from the lung effluent was used to measure conversion of PGE_2 to its metabolites. The two keto metabolites, 13,14-dehydro-15-keto PGE_2 and 15-keto PGE_2, migrated with the same R_f in this system. No

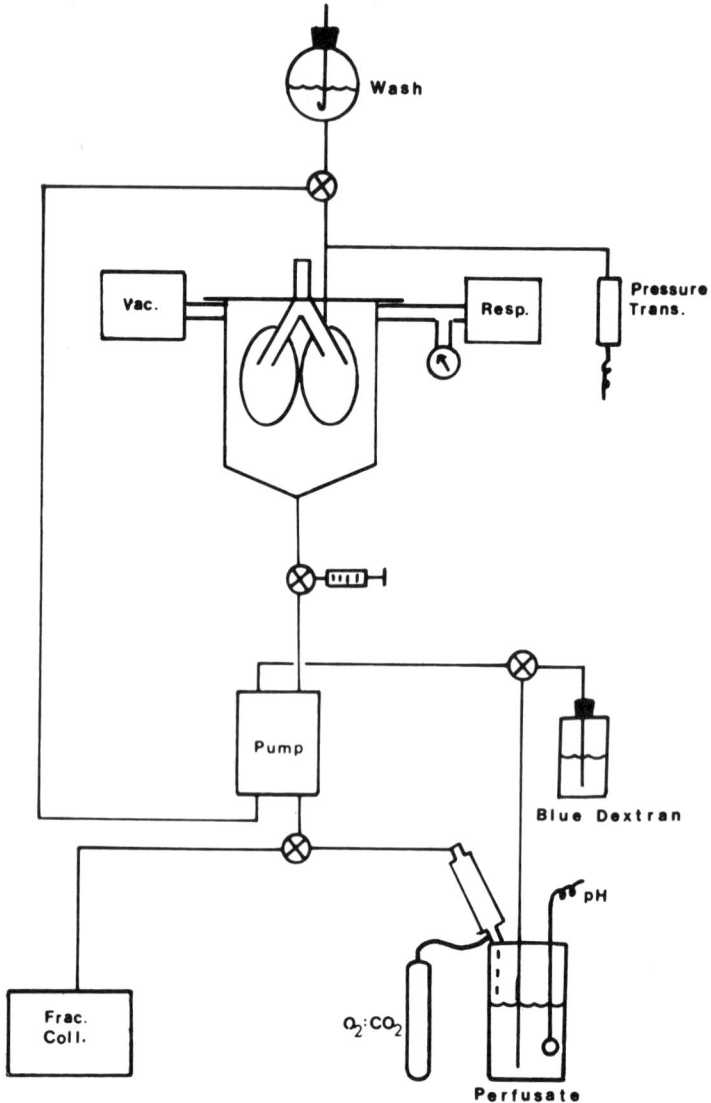

Figure 5. Block diagram of the lung perfusion apparatus.

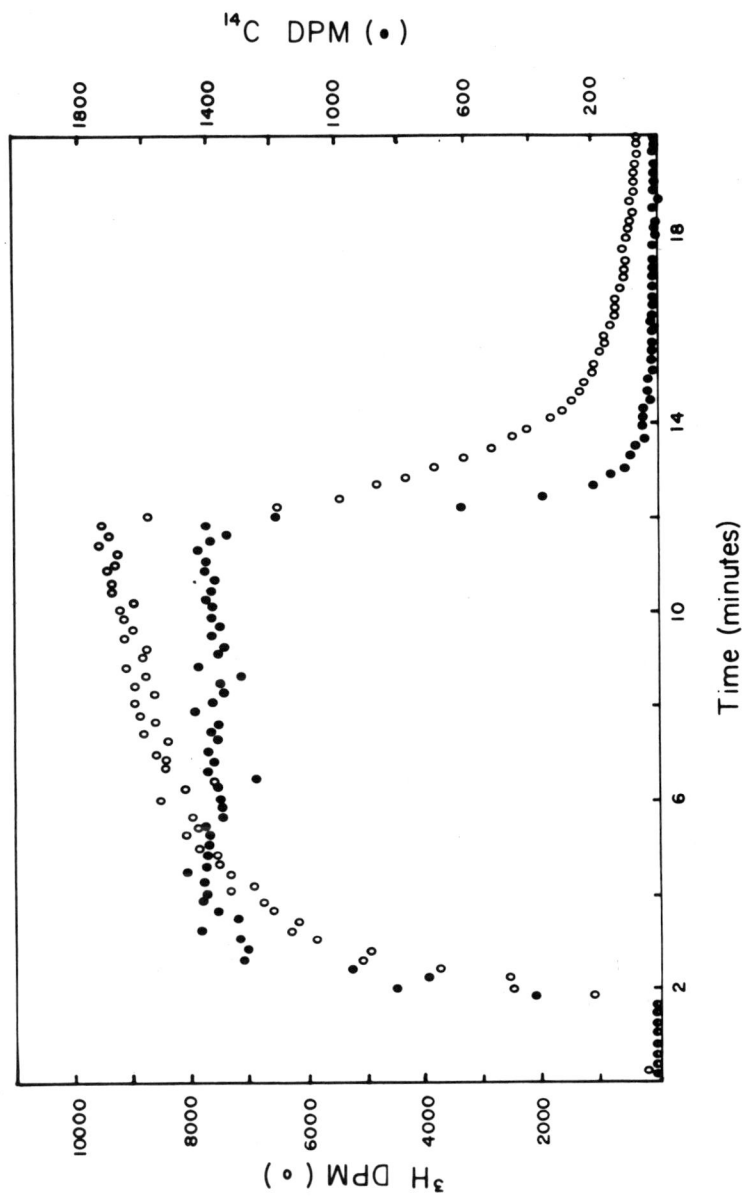

Figure 6. The appearance of ^{14}C-dextran and ^{3}H-PGE$_2$ and its ^{3}H-metabolites in the effluent from a perfused lung of an NO$_2$-exposed rat. The ^{3}H-radioactivity is indicated by open circles and the ^{14}C-radioactivity by solid circles. The perfusion period was 10 min.

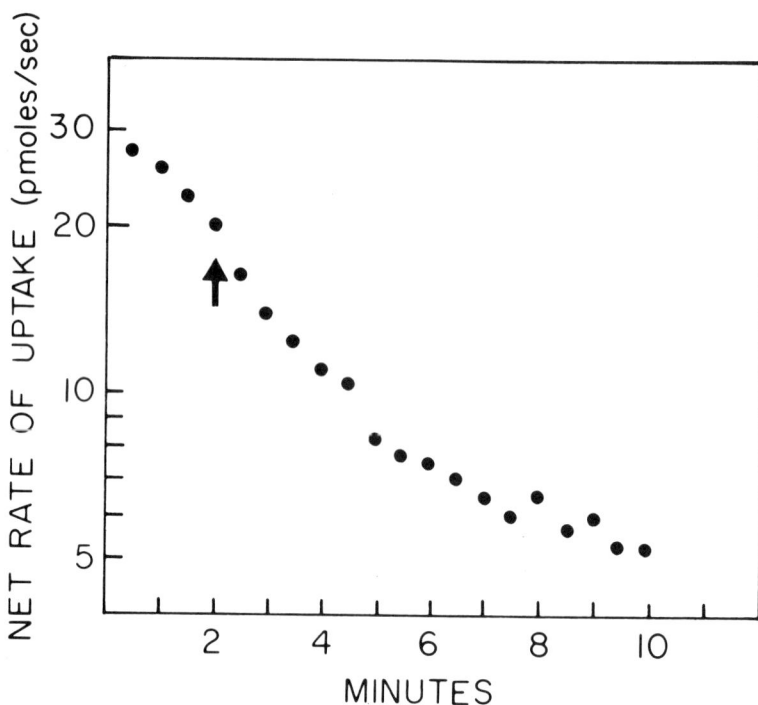

Figure 7. Net rate of uptake of PGE$_2$ calculated from the efflux of ^{3}H from the perfused lung. The unidirectional flux was estimated by computer by least squares linear regression of the curve shown, and the value taken was calculated at the point of maximum perfusion as indicated by the arrow.

13,14-dehydro PGE$_2$ was found. The inhibition of metabolism of PGE$_2$ was most likely at the conversion to 15-keto PGE$_2$. The inhibition of ^{3}H-PGE$_2$ metabolism was both dose- and time-dependent, as shown in Figure 8. The hatched bar represents the range of clean air control preparations. Inhibition was maximal at 18 hr postexposure and was greatest for the lungs of rats exposed to 19 ppm. It was the same for lungs from rats exposed to either 0.2 or 2.0 ppm NO$_2$. Inhibition declined most rapidly in rats exposed to 0.2 ppm, intermediately in those exposed to 2.0 ppm, and slowest in those exposed to 19 ppm.

CONCLUSIONS

If one assumes the temporal sequence of NO$_2$ effects on the lung outlined above, the observations on the prostaglandin metabolism by the lung can be explained easily. Since the prostaglandin transport system occurs on the

Table II. Effect of NO_2 Exposure on the Unidirectional Uptake
of ^3H-PGE_2 at 18 Hours Postexposure

	Air[a]	NO₂ Exposed (ppm)[b]		
		0.2	2.0	19
Uptake (pmole/sec)	44.3 ± 4.0	39.7 ± 3.4	40.2 ± 1.6	37.7 ± 3.5
N	8	7	5	6

[a]Mean ± S.E.
[b]None of the values is significantly different from the air control value.

Table III. Efflux of ^3H-PGE_2 and its Metabolites from NO_2-Exposed
Rat Lungs at 18 Hours Postexposure

	Air Exposed	NO₂ Exposed		
		0.2 ppm	2.0 ppm	19 ppm
k (min⁻¹)	0.40 ± 0.02	0.41 ± 0.01 NS	0.50 ± 0.03 p < 0.025	0.35 ± 0.02 p < 0.1
t₁/₂ (min)	1.72	1.68	1.39	1.96
N	8	7	5	6

Table IV. Lack of Edema in Lungs Exposed to NO_2 or Air
and Perfused with PGE_2

		Dry Weight[a] (%)
Air-Exposed, Perfused with Tyrode's (N = 4)		16.58 ± 2.96
Air-Exposed, Perfused with 1 μg PGE₂/ml (N = 3)		16.05 ± 0.27
NO₂-Exposed, Perfused with 1 μg PGE₂/ml (N = 6) 18 hr postexposure 90 hr postexposure 138 hr postexposure	14.08 15.86 15.42	15.12 ± 0.49

[a]No significant differences.

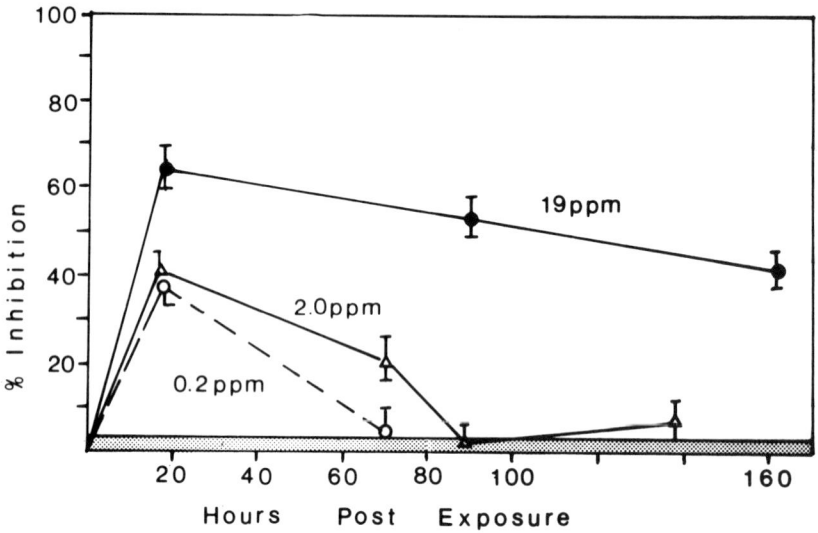

Figure 8. Inhibition of the metabolism of PGE$_2$ by NO$_2$ exposure. All animals were exposed to NO$_2$ at the indicated concentrations for 3 hr and assayed at the indicated times after exposure as shown. The shaded area represents the standard error for the control assays conducted simultaneously with the exposed groups. The bars indicate the standard error of each of the exposed groups.

capillary lumen, NO$_2$ or reaction products must have penetrated to this site to produce inhibition. The major effects of NO$_2$ exposure are not on the uptake or efflux of PGE$_2$ or its metabolites, but on the metabolism of PGE$_2$ to its inactive congeners. The length of time required for restoration of this activity suggests that the inhibition of PGE$_2$ metabolism occurred by a major disruption of the cells metabolizing prostaglandins. Such a disruption could be an early hallmark of the death of these cells. While no difference existed between the maximal inhibition of groups exposed to 0.2 and 2.0 ppm NO$_2$, a more rapid recovery was seen with the 0.2 ppm-exposed group than with the 2.0 ppm-exposed group. Similarly, the 19 ppm-exposed group recovered the slowest. This is consistent with the scheme proposed in Figure 2, where the magnitude of the NO$_2$ effect increases with dose, probably as a result of the injury or death of a greater number of susceptible cells.

These results suggest that near ambient concentrations of NO$_2$ may affect the metabolism of vasoactive compounds by the lung. PGE$_2$ is particularly important, since metabolic dysfunction may occur in asthmatic patients and since PGE$_2$ regulates local blood flow in the lung. The long time required for recovery to normal function suggests that single exposures to NO$_2$ may, in fact, be cumulative. Human exposures in urban areas follow a cyclical pattern similar to the 3-hr exposures carried out in these experiments. Peak exposures occur roughly every 12 hr, but recovery is not complete until 60 hr at the earliest, with near ambient exposures of 0.2 ppm. These results are similar to those observed using susceptibility to airborne bacterial infection as an endpoint.

If the present hypothesis of NO_2 toxicity is correct and inhibition of PGE_2 metabolism by the lung is an early reflection of death of pulmonary capillary cells, then the sequence is likely to be the same for man, provided that the pulmonary capillary cell dose is the same as the dose for rats. Adjustment or scaling of the rat exposure dose to higher or lower doses for man may be required because of the existence of anatomical differences. If these alterations in the metabolism of PGE_2 persist on chronic exposure, a significant pathophysiological consequence might be a mismatch of pulmonary blood perfusion with pulmonary ventilation. These effects are likely to exacerbate obstructive pulmonary disease and may serve as an initiating factor. In any event, these experiments support the consideration of the human health hazard of exposure to short-term high levels of NO_2 over periods less than 24 hr.

ACKNOWLEDGMENTS

The close collaboration of Drs. J. Roycroft, Jr. and J. M. Charles during the early phases of this study was essential to its successful outcome. Ms. E. D. Smolko, Mrs. K. Holden and Mr. W. B. Gunter provided excellent assistance throughout. This work was supported in part by the National Institutes of Health (NIH) Grant HL16264 and by a contract from the U.S. Environmental Protection Agency (EPA).

REFERENCES

1. Freeman, G., S. C. Cran, R. J. Stephens and N. J. Furiosi. "Pathogenesis of the Nitrogen Dioxide-Induced Lesion in the Rat Lung. A Review and Presentation of New Observations," *Am. Rev. Resp. Dis.* 98:429–443 (1968).
2. Evans, M. J., L. V. Johnson, R. J. Stephens and G. Freeman. "Renewal of the Terminal Bronchiolar Epithelium in the Rat Following Exposure to NO_2 or O_3," *Lab. Invest.* 35:246–257 (1976).
3. Miller, F. J., D. B. Menzel and D. L. Coffin. "Similarity Between Man and Laboratory Animals in Regional Pulmonary Deposition of Ozone," *Environ. Res.* 17:84–101 (1978).
4. Goldstein, E., N. F. Peele, N. J. Parks, H. H. Hines, E. P. Steffey and B. Tarkington. "Fate and Distribution of Inhaled Nitrogen Dioxide in Rhesus Monkeys," *Am. Rev. Resp. Dis.* 115:403–412 (1977).
5. Gardner, D. E., D. L. Coffin, M. A. Pinigin and G. I. Sidorenko. "Role of Time as a Factor in the Toxicity of Chemical Compounds in Intermittent and Continuous Exposures. Part I. Effects of Continuous Exposure," *J. Toxicol. Environ. Health* 3:811–820 (1977).

6. Coffin, D. L., D. E. Gardner, G. I. Sidorenko and M. A. Pinigin. "Role of Time as a Factor in the Toxicity of Chemical Compounds in Intermittent and Continuous Exposures. Part II. Effects of Intermittent Exposure," *J. Toxicol. Environ. Health* 3:821–828 (1977).
7. Menzel, D. B., J. N. Roehm and S. D. Lee. "Vitamin E: The Biological and Environmental Antioxidant," *J. Agric. Food Chem.* 20:481–486 (1972).
8. Fletcher, B. L., and A. L. Tappel. "Protective Effects of Dietary α-Tocopherol in Rats Exposed to Ozone and Nitrogen Dioxide," *Environ. Res.* 6:165–175 (1973).
9. Menzel, D. B. In: *Free Radicals in Biology*, W. Pryor, Ed. (New York: Academic Press, Inc., 1976), p. 181.
10. Menzel, D. B. "Reaction of Oxidizing Lipids with Ribonuclease," *Lipids* 2:83–84 (1967).
11. Ryan, U. S., J. W. Ryan, C. Whitaker and A. Chiu. "Localization of Angiotensin Converting Enzyme (Kininase II). II. Immunocytochemistry and Immunofluorescence," *Tissue and Cell* 8:125–145 (1976).
12. Steinberg, H., D. J. P. Bassett and A. B. Fisher. "Depression of Pulmonary 5-Hydroxytryptamine Uptake by Metabolic Inhibitors," *Am. J. Physiol.* 228:1298–1303 (1975).
13. Anderson, M. W., and T. E. Eling. "Prostaglandin Removal and Metabolism by Isolated Perfused Rat Lung," *Prostaglandins* 11:645–677 (1976).
14. Menzel, D. B., W. G. Anderson and M. B. Abou-Donia. "Ozone Exposure Modifies Prostaglandin Biosynthesis in Perfused Rat Lungs," *Res. Comm. Chem. Path. Pharm.* 15:135–147 (1976).
15. Charles, J. M., M. B. Abou-Donia and D. B. Menzel. "Absorption of Paraquat and Diquat from the Airways of the Perfused Rat Lung," *Toxicology* 9:59–67 (1978).

CHAPTER 14

DOES NITROGEN DIOXIDE MODIFY
THE RESPIRATORY EFFECTS OF OZONE?

Eiji Yokoyama and Isamu Ichikawa

Department of Industrial Health
The Institute of Public Health
Tokyo, Japan

Kiyoyuki Kawai

Department of Experimental Toxicology
National Institute of Industrial Health
Kawasaki, Japan

INTRODUCTION

An appreciable concentration of oxides of nitrogen (NO_x) always appears in the ambient air before the development of photochemical oxidants [1]. A preexposure or concurrent exposure of the population to NO_x may modify the respiratory effects caused by oxidants, but experimental evidence is lacking, although "cross-tolerance" has been demonstrated between nitrogen dioxide (NO_2) and ozone (O_3) at higher concentrations [2]. Ambient conditions were simulated with sequential exposure to NO_2 and O_3 in rats.

METHODS

Each of four groups of male Wistar rats were exposed 7 days a week to one of the following exposure regimens: NO_2 for 3 hours (10:00 A.M.–1:00 P.M.) then O_3 for 3 hours (1:00 P.M.–4:00 P.M.); one group was exposed to NO_2 for 3 hours; one group was exposed to O_3 for 3 hours; and a control group was exposed to filtered air. A cubic, stainless-steel chamber was supplied with filtered air at a rate of 20 air changes per hour. NO_2 was added to the air at the rate of 1% balanced with nitrogen. Ozone was produced by passing pure oxygen through a silent discharge O_3 generator. This was added to the air stream at the entrance to the chamber. Concentrations of NO_2 or O_3 in the chamber were monitored with chemiluminescent NO-NO_x analyzer (Toshiba-Beckman, Tokyo, model 951A) or a Mast O_3 analyzer, respectively, and recorded on an ink-writing recorder. The NO_2 instrument was calibrated using the Saltzman reagent method [3], and the O_3 instrument using the neutral-buffered KI method. All animals were deprived of food and water daily during the length of all gas exposure (10:00 A.M.–4:00 P.M.).

Two experiments were carried out, and the age of animal, concentration of the gases and duration of exposure were similar. The recording of temperature and relative humidity inside the chamber were incompletely recorded in Experiment 1, but ranged around 25°C. Relative humidity ranged between 48% and 65%. In Experiment 2, the mean was $27.0 \pm 0.9°C$ and $50.7 \pm 7.5\%$, respectively.

Immediately after the final exposure, animals were killed by abdominal exsanguination following intraperitoneal injection of Na-pentobarbital (50 mg/kg). A polyethylene tube, 1.5 cm long and 2.3 mm i.d., was inserted into the trachea at a point about 1 cm below the larynx. A pneumothorax was produced by cutting the diaphragm, and the chest wall opened.

MECHANICAL PROPERTIES OF LUNGS

The lungs were left intact in situ for measurements of ventilatory mechanics. Details of method are described elsewhere [4]. In brief, a static deflation volume-pressure (V/P) curve was obtained by recording the air volume retained in the lungs at fixed levels of elastic recoil pressure (P_L) during step-wise deflation from 30 cm H_2O of P_L. The lung volume defined at this pressure is total lung capacity (TLC). When measuring pulmonary flow resistance (R_1) and flow-volume curve ($\dot{V}/V$), the animals were placed on their backs in a body plethysmograph. R_1 was measured with the oscillation technique at different levels of P_L: sinusoidal flow (6 cycles/sec, 10 ml/sec of peak-to-peak flow) was applied via a pneumotachograph (Fleisch #00) to lungs, the volume of which was adjusted as to keep a fixed P_L (5.0 – 0.6 cm H_2O) by changing the negative pressure in the plethysmograph. A ratio of the magnitude of transpulmonary pressure corresponding to the peak-to-peak flow rate was calculated from which the resistance of the tracheal cannula was subtracted. R_1 was expressed as flow divided by TLC to negate the difference in lung size. The $\dot{V}/V$ curve during forced exhalation was obtained by venting the lungs maintained at TLC by applying the pressure through the tracheal cannula to a bottle negatively charged at 35–40 cm H_2O. In this case, flow was measured by the pneumotachograph attached to the body plethysmograph, and volume was obtained by the electronic integration of the flow signal. Lung volume changed during exhalation was defined as vital capacity (VC). The flow divided by VC was presented at 25% VC and 50% VC. In experiment 1, only the V/P study was done. In experiment 2, V/P curve and R_1 were successively measured in about half the animals in each group, and V/V curve was obtained in the remainder to minimize an increase in trapped air at 0 cm H_2O of P_L, which might be produced if applied repeatedly with these techniques.

Enzyme Activities in Subcellular Fractions of Lung Tissue

In experiment 1, the entire lung was removed from the chest after the completion of the V/P study. It was weighed, homogenized and separated into subcellular (mitochondrial, microsomal and supernatant) fractions by differential centrifugation. The details of the methods used to measure the enzyme activities in the above subcellular fractions are described elsewhere [5]. The activities of phospholipase A_1 and A_2 [6] in microsomal and supernatant fractions were obtained by counting the radioactivity of 2-^{14}C-oleolyl-lysolecithin (lysoPC) or ^{14}C-oleic acid, which were produced from 2-^{14}-oleoyl-lecithin-(PC) [7] incubated with each fraction. The activity of lysolecithin acyltransferase [8,9] in microsomal fraction was obtained by counting the radioactivity of PC formed when incubating this fraction with 1-acyl-LysoPC and ^{14}C-fatty acid (palmitic acid- or linoleic acid- [1-^{14}C]). The activity of lysolecithin-lysolecithin acyltransferease [8,10] in the supernatant fraction was obtained by counting the radioactivity of PC formed when incubating it with 1-^{14}C-palmitoyl-LysoPC [11].

Histology of Lungs

In experiment 2, the lungs of animals exposed for 30 days were prepared for light-microscopic examination after the V̇/V recording. The left and right lungs were separated by cutting the main bronchi at the bifurcation, and the left lung was fixed in 10% buffered formalin. The right lung was inflated by the intratracheal injection of 10% buffered formalin. The right lungs were fixed in an inflated state (calculated to be 65% of TLC), and the left lungs were fixed in an uninflated state. Paraffin sections (ca. 5 μ) of all lung lobes were prepared using standard histological techniques. Routinely sections were stained using hematoxiline-eosine, Mallory, Weigert's and PAS-alcian blue. All sections were read using light microscopy.

RESULTS

Experiment 1

Daily exposure to mean concentrations of 5.5 ($\pm$0.3 SD) ppm of NO_2 and/or 1.1 ($\pm$0.1 SD) ppm of O_3 were carried out in 8-week-old rats during 14 consecutive days and in 9-week-old rats during 7 consecutive days (6 in each group). No significant change in body weight, lung weight or TLC was observed in experiment 1 (Table I). Their V/P curves, with volume expressed as % TLC, were quite similar to those of control animals (Figure 1). As

Table I. Effects on Body Weight, Lung Weight and Total Lung Capacity of
Rats of Daily 3-Hour Exposure to NO_2 and O_3 and Sequential Exposure
to These Gases for 7 and 14 Consecutive Days

Group	Number	Body Weight (g)	Lung Weight (g)	LW/BW × 10^3	TLC/LW (ml/g)
		(Mean ± SD)			
Control	6	322 ± 25	1.20 ± 0.09	3.70 ± 0.11	9.6 ± 1.3
7 Days					
5.5 ppm NO_2, 3 hr	4	339 ± 19	1.20 ± 0.06	3.58 ± 0.11	10.1 ± 1.5
1.1 ppm O_3, 3 hr	4	328 ± 23	1.16 ± 0.07	3.53 ± 0.15	9.2 ± 1.0
5.5 ppm NO_2, 3 hr– 1.1 ppm O_3, 3 hr	4	312 ± 12	1.20 + 0.05	3.84 ± 0.10	9.6 ± 0.6
14 Days					
5.5 ppm NO_2, 3 hr	5	325 ± 11	1.16 ± 0.02	3.57 ± 0.10	10.2 ± 1.5
1.1 ppm O_3, 3 hr	5	328 ± 16	1.17 ± 0.10	3.57 ± 0.14	9.7 ± 1.1
5.5 ppm NO_2, 3 hr– 1.1 ppm O_3, 3 hr	5	326 ± 19	1.22 ± 0.05	3.74 ± 0.12	9.8 ± 1.1

shown in Table II, no detectable change in phospholipase A_1 activity was observed in these animals, but phospholipase A_2 activity in mitochondrial fraction tended to increase in animals exposed to either O_3 or NO_2–O_3 for 14 days, compared with that in control animals. The increase in the NO_2–O_3 exposed group was significant ($p < 0.05$). Significant change in the activity of lysolecithin acyltransferase in microsomal fraction was not detected in the lungs of rats exposed to the gases when using saturated (palmitic) acid as a substrate, although there was a tendency for it to decrease in animals exposed for 14 days. On the other hand, the activity of the enzyme using unsaturated (linoleic) acid as a substrate was decreased significantly ($p < 0.05$ or 0.01) in the same animals. It tended to be lower in animals exposed for 14 days than for 7 days, regardless of the gases exposed, although the differences were not significant. The activity of lysolecithin-lysolecithin acyltransferase in supernatant fraction was relatively low in animals exposed to either O_3 or NO_2–O_3 for 14 days compared with that in the control: the difference between the control and the latter group was significant ($p < 0.01$).

Experiment 2

Rats were 7 weeks old, weighing about 220 g, at the beginning of exposure. Daily exposure to mean concentrations of 5.4 (±0.9 SD) ppm of NO_2 and/or 1.0 (±0.1 SD) ppm of O_3 were performed during 14 (16 animals in each group) and 30 consecutive days (12 animals in each group). Figure 2 shows the curves of growth of animals exposed for 30 days, in which no

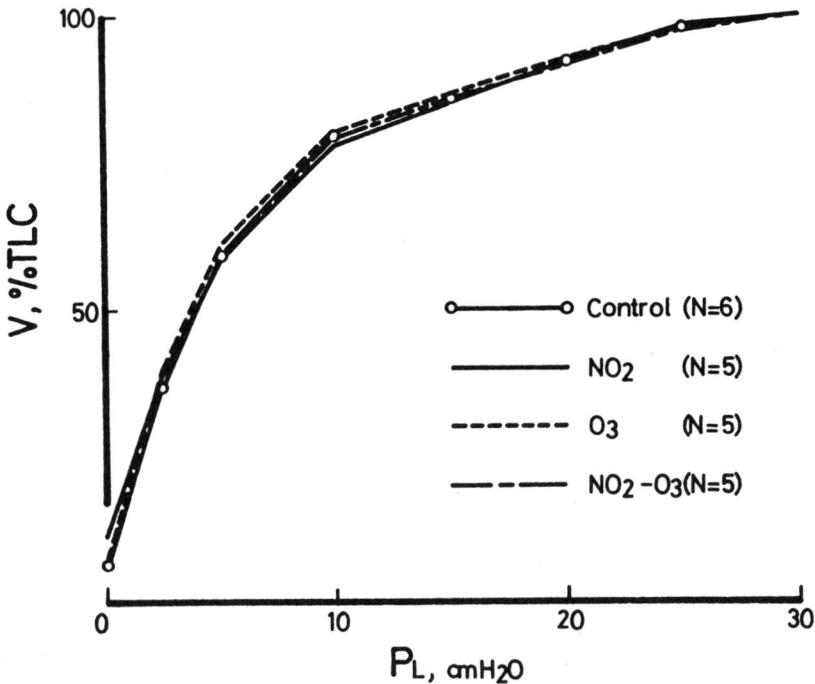

Figure 1. Static deflation volume-pressure curves of lungs of rats exposed to NO_2 (5.5 ppm) or O_3 (1.1 ppm) for 3 hr/day and sequentially exposed to these gases for 14 days. Volume is expressed as a percentage of total lung capacity.

significant difference was observed between the groups. At the termination of exposure, significant difference in a ratio of lung weight to body weight was not detected between the groups, but a ratio of TLC to lung weight in animals exposed to O_3 was slightly ($p < 0.05$), although significantly, smaller than that in the control as shown in Table III. TLC in both the NO_2 and NO_2-O_3 exposed groups did not differ from that in the control. Figures 3 and 4 show mean values of ventilatory mechanics (V/P curve with volume expressed as % TLC and R_1 at P_L below 5 cm H_2O, and the flows at 50% VC and 25% VC obtained by forced deflation) in animals exposed for 14 and 30 days, respectively. The shaded area represents 95% confidence limit of control. No significant change in these measurements was observed in any groups exposed for 14 days. However, the curves of R_1 expressed as a function of P_L in both the O_3- and NO_2-O_3-exposed groups could be superimposed and shifted upwards parallel with the curve in the control. The increase in R_1 was significant ($p < 0.05$) at nearly all P_L. R_1 in the NO_2-exposed animals did not differ from the control. Neither V/P curves nor the flows at 50% VC and 25% VC at the forced deflation in these animals showed

Table II. Effects of Daily Sequential 3-hr Exposure to NO_2 and O_3, During Consecutive 7- and 14-Day Periods on the Activities of Phospholipase A and Acyltransferases of Lecithin in Lungs of Rats (mean ± SD nmol/mg protein/hr)

| Group | Number | Phospholipase A | | | | Lysolecithin Acyltransferase—Microsome | | Lysolecithin-Lysolecithin Acyltransferase—Supernatant |
| | | Mitochondria | | Supernatant | | ^{14}C-Palmitic Acid | ^{14}C-Linoleic Acid | |
		A_1	A_2	A_1	A_2			
Control	5	22.0 ± 3.9	78.3 ± 29.6	3.0 ± 1.5	25.8 ± 9.7	7.5 ± 3.7	38.6 ± 2.8	57.9 ± 6.2
7 Days								
5.5 ppm NO₂	5	22.6 ± 8.3	75.3 ± 29.0	2.8 ± 2.0	24.4 ± 3.2	8.6 ± 4.0	24.5 ± 6.7[a]	71.9 ± 13.1
1.1 ppm O₃	5	20.6 ± 11.8	83.9 ± 41.3	2.7 ± 1.6	23.9 ± 4.2	7.8 ± 5.0	24.7 ± 7.5[a]	58.8 ± 19.9
5.5 ppm NO₂ 1.1 ppm O₃	5	21.1 ± 8.8	92.7 ± 25.8	2.7 ± 0.7	24.5 ± 6.8	7.1 ± 1.4	24.6 ± 8.8[a]	61.6 ± 24.0
14 Days								
5.5 ppm NO₂	5	23.8 ± 3.1	98.8 ± 19.0	3.2 ± 1.5	25.5 ± 1.7	5.9 ± 3.3	18.5 ± 16.1[b]	52.1 ± 34.8
1.1 ppm O₃	5	22.3 ± 1.6	109.9 ± 10.2	2.8 ± 1.5	25.9 ± 1.8	4.4 ± 1.9	14.8 ± 10.0[a]	42.9 ± 23.1
5.5 ppm NO₂ 1.1 ppm O₃	5	21.6 ± 4.0	114.2 ± 6.8[b]	3.0 ± 1.3	25.1 ± 2.3	6.2 ± 3.6	15.4 ± 5.7[a]	38.5 ± 7.2[b]

[a]Significantly different from the control; $p < 0.01$.
[b]Significantly different from the control; $p < 0.05$.

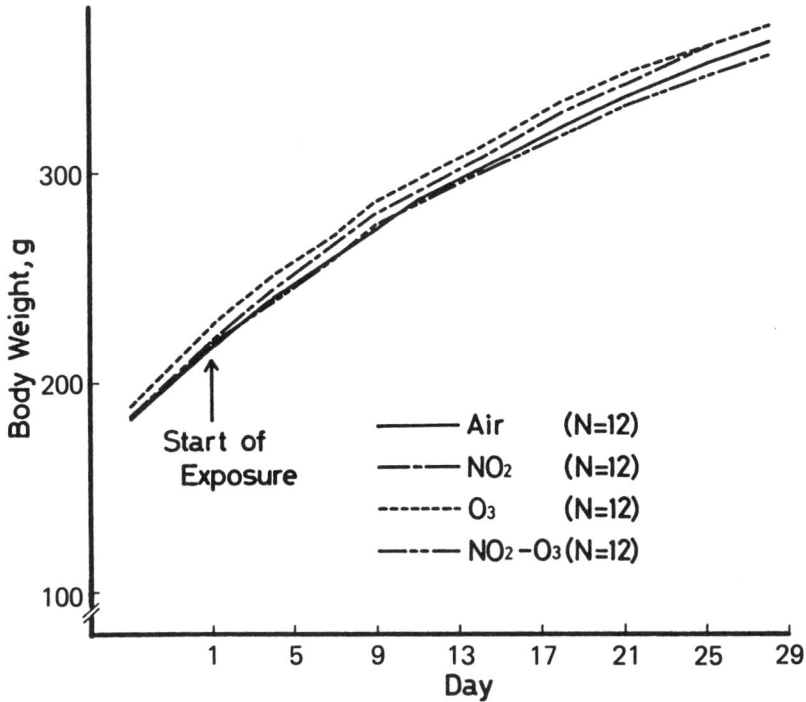

Figure 2. Growth curves of rats during 30 days of exposure to NO_2 (5.4 ppm) or O_3 (1.0 ppm) for 3 hr/day and sequential exposure to these gases. The exposure was started at day 1.

detectable differences from those in control animals. A somewhat larger volume was retained in the lungs at P_L of 2.5 cm H_2O in the NO_2-exposed animals, but this increase did not attain the statistical significance. Microscopically, lungs of the control group were entirely normal, except for sparse cell infiltration in the periarterial connective tissue. In animals exposed to NO_2 alone, the nonciliated epithelia in peripheral bronchioli increased in number, showing slight hypertrophy and basophilia of their cytoplasm. In some animals, similar cells were found along the wall of alveolar ducts and adjacent alveoli. Lung injury in animals exposed to O_3 was more marked, compared with NO_2-exposed animals. Marked formation of the mucosal folding was observed over wide regions of airways, particularly at the region from larger to medium-sized bronchi and tentatively interpreted as evidence of persistent existence of airway constriction. The epithelia lining the top of the projected mucosal folding was hypetrophic and the cells cytoplasm was basophilic. Irregularity and loss of cilia was also noticed. Those changes extended to respiratory bronchioles and alveolar ducts, which showed slight thickening of the wall accompanied by the formation of delicate collagen

Table III. Effects on Body Weight, Lung Weight and Total Lung Capacity of Rats of Daily 3-Hour Exposure to NO_2 and O_3 and Sequential Exposure to These Gases for 14 and 30 Consecutive Days

Group	Number	Body Weight (g)	Lung Weight (g)	$LW/BW \times 10^3$	TLC/LW (ml/g)
		(Mean ± SD)			
Control	8	323 ± 17	1.09 ± 0.09	3.38 ± 0.27	10.6 ± 0.9
14 Days					
5.4 ppm NO_2, 3 hr	8	312 ± 16	1.09 ± 0.08	3.48 ± 0.24	10.5 ± 1.5
1.0 ppm O_3, 3 hr	8	311 ± 25	1.11 ± 0.13	3.57 ± 0.30	10.1 ± 1.4
5.4 ppm NO_2, 3 hr– 1.0 ppm O_3, 3 hr	8	303 + 25	1.08 ± 0.09	3.58 ± 0.27	10.1 ± 1.1
Control	6	382 ± 6	1.20 ± 0.07	3.14 ± 0.21	11.3 ± 0.6
30 Days					
5.4 ppm NO_2, 3 hr	6	381 ± 10	1.08 ± 0.11	2.83 ± 0.26	12.4 ± 1.8
1.0 ppm O_3, 3 hr	6	393 ± 18	1.24 ± 0.06	3.17 ± 0.16	10.2 ± 0.6[a]
5.4 ppm NO_2, 3 hr– 1.0 ppm O_3, 3 hr	6	367 ± 8	1.20 ± 0.07	3.27 ± 0.22	10.9 ± 0.8

[a]Significantly different from the control, $p < 0.05$.

fibers. Alveolar structure was regular, as seen on the reinflated lung, and no appreciable changes could be demonstrated under the light microscope. Lung injuries observed in rats exposed to NO_2-O_3 were essentially the same as those of animals exposed to O_3 alone, although slight epithelial necrosis in medium-sized bronchi was noticed in a few animals. Changes at the bronchiolo-pulmonary junction appeared to be somewhat more marked than with O_3 alone.

DISCUSSION AND CONCLUSIONS

Concentrations of NO_2 and O_3 are changed in the ambient atmosphere when photochemical oxidants are formed. It is difficult to reproduce this pattern of concentration changes in experimental chambers. Preliminarily selected sequential exposure to constant concentrations of NO_2 and O_3 was done to mimic these changes. The concentrations used in the experiments were selected because it was observed that a single 3-hour exposure to 5 ppm of NO_2 did not cause the change in activities of enzymes catalyzing the phospholipid metabolism in rat lung [12], and that the 3-hour exposure to 1 ppm of O_3 did not cause the change in V/P curve of rat lung [13].

Goldstein [14] reported additive effects on lipid peroxidation of an in vitro (human red cell) system sequentially exposed to NO_2 and O_3. Fukase

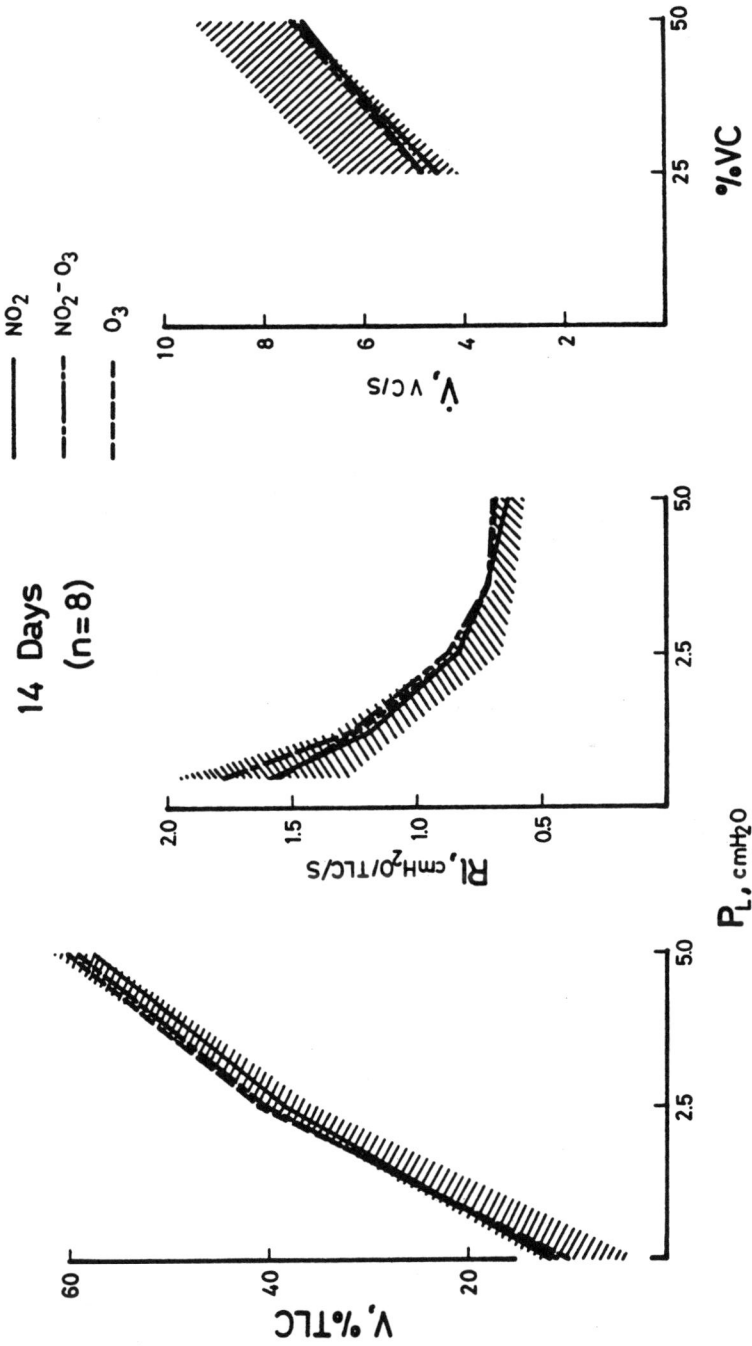

Figure 3. Ventilatory mechanics of lungs of rats exposed to NO_2 (5.4 ppm) or O_3 (1 ppm) for 3 hr/day and sequentially exposed to these gases for 14 days. Shaded area represents 95% confidence limit.

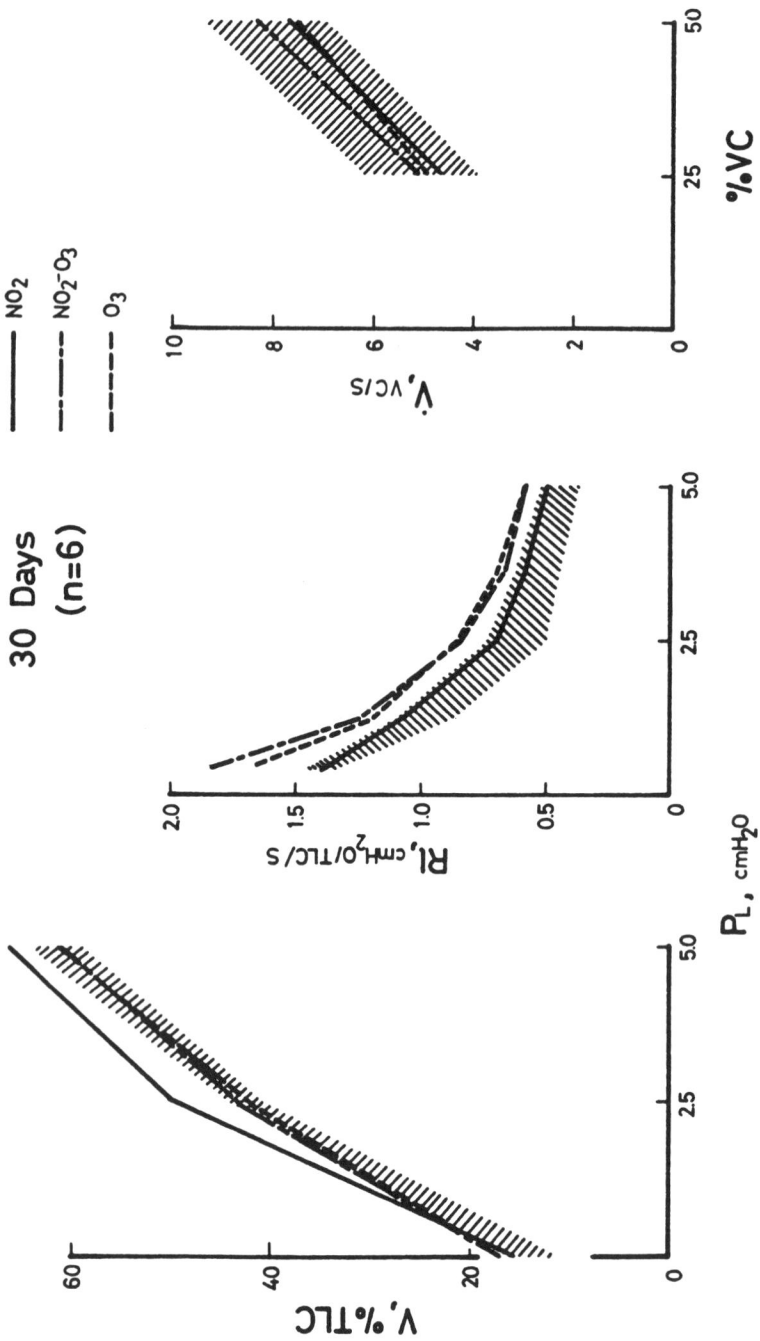

Figure 4. Ventilatory mechanics of lungs of rats exposed to NO_2 (5.4 ppm) or O_3 (1 ppm) for 3 hr/day and sequentially exposed to these gases for 30 days. Shaded area represents 95% confidence limit.

et al. [15] performed the sequential exposure of mice similar to ours: mice were exposed daily to NO_2 (3–15 ppm) for 3 hours, followed by 1 ppm of O_3 for 3 hours for 7 days, and additive effects on GSH level of lungs were found. Tsunoda et al. [16] did not observe significant changes in ventilatory function of healthy human subjects after exposure to 0.5 ppm of NO_2 or to O_3 alone for 3 hours, or their sequential exposure.

Effects of experimental exposure to air pollutants on the lipid content and its fractional distribution in lungs have been studied extensively [17–19], but relatively little information is available concerning behavior of enzymes catalyzing its metabolisms. We reported [6,12] the elevated activity of phospholipase A in rat lungs acutely exposed to 2 ppm of O_3. We also noticed that daily exposure to 0.45 ppm of O_3 for 6 hours for 45 days did not cause significant changes in activities of phospholipase A or of acyltransferases for lecithin in rat lungs [5]. In experiment 1 we observed detectable, although not necessarily significant, changes in the activities of these enzymes. The mechanisms are discussed in detail elsewhere [5,6].

An increasing trend in the activity of phospholipase A_2 in mitochondrial fraction and a decreasing trend in the activity of lysolecithin–lysolecithin acyltransferase in supernatant fraction were evident in the lungs of rats exposed to O_3 alone and rats sequentially exposed to NO_2 and O_3 for 14 days. There was practically no difference in the activities between the groups. However, the changes were statistically significant only in the latter group, largely due to less variation in measured values. If individual response is to be limited in a narrower range with stronger stimuli, the above result may mean that NO_2 acted as an additional stimulus on the enzyme activity.

The response of activity of lysolecithin acyltransferase in microsomal fraction appeared to depend on the kind of fatty acid used as a substrate. When unsaturated acid was used (linoleic), its activity decreased proportionally to the duration of exposure, regardless of the gases exposed. We presently cannot explain the vulnerability of this enzyme, except for its chemical structure (SH-enzyme). By contrast, no significant change was noted in the activity of the enzyme when saturated fatty acid was used as a substrate under any condition of exposure, although it tended to decrease after 14 days. It is hard to draw a definite conclusion because of the relatively wide variation of the measured values, but this may mean that mechanisms for maintaining the level of saturated lecithin or dipalmitoyllecithin (DPL) in lungs are relatively resistant to these pollutant gases.

The results obtained in experiment 2 were somewhat different from those of experiment 1, which indicated that the sequential exposure to NO_2 and O_3 for 30 days did not produce effects different from those of O_3 alone (which causes the changes in mechanical properties of rat lungs). Also, the histological findings suggestive of the change in bronchial size were essentially the same in both groups, although our findings suggested more marked effects of sequential NO_2–O_3 exposure. However, it is possible to obtain different results in living animals because of the limits in our present technique using postmortum lungs for mechanical measurements.

Reduction in TLC accompanied by no detectable change in V/P curve with volume expressed as % TLC suggests a reduction of ventilated units as one of the most probable causes, but its extent was minor. We presently interpret the increase of R_1 over range of P_L below 5 cm H_2O to indicate the narrowing of the large airways because the tendency that R_1 increased more at lower P_L was never observed. This may be compatible with the histological changes of lungs in these groups. However, we cannot explain why the present technique did not disclose any changes in peripheral airway resistance, despite the histological changes which included thickening of the bronchiolar walls and collagen formation.

Before these studies were done, it was anticipated that NO_2 might lessen the effects of O_3 in sequential exposures. The present preliminary experiments do not permit us to speculate on the problems encountered in an actual atmosphere, as more sophisticated modes of exposure are needed. However, it is demonstrated from the present results that physiological, structural or biochemical responses are modified under complex combinations of air pollutants.

ACKNOWLEDGMENTS

This study was supported in part by Environmental Agency of Japan. We are grateful for the careful technical assistance of Mrs. H. Arakawa, and our sincere thanks is extended to Dr. John Orthoefer of the U.S. Environmental Protection Agency for his kind help with this chapter.

REFERENCES

1. "Air Quality Criteria for Photochemical Oxidants," U.S. Department of Health, Education and Welfare, Public Health Service, Environmental Health Service, National Air Pollution Control Administration, Washington, D.C. (1970).
2. Stokinger, H. E., and L. D. Scheel. "Ozone Toxicity. Immunochemical and Tolerance-Producing Aspects," *Arch. Environ. Health* 4:117-124 (1962).
3. Saltzman, B. E. "Colorimetric Microdetermination of NO_2 in the Atmosphere," *Anal. Chem.* 20:1949-1954 (1964).
4. Yokoyama, E., H. Arakawa, Y. Otsu and K. Iizuka. "Methods to Measure Ventilatory Mechanics of Rat Lung," *Bull. Inst. Public Health* 27:64-77 (1978).
5. Yokoyama, E., and I. Ichikawa. "Respiratory Effects on Rats of Intermittent Exposure to Ozone." Unpublished results.
6. Ichikawa, I., and E. Yokoyama. "Phospholipase A Activity in Lung of Rats Acutely Exposed to Ozone," *J. Japan Soc. Biol. Interface* 5:37-41 (1974).
7. Robertson, A. F., and W. E. M. Lands. "Metabolism of Phospholipids in Normal and Spherocytic Human Erythrocytes," *J. Lipid Res.* 5:88-93 (1964).

8. Abe, M., T. Akino and K. Ohno. "The Formation of Lecithin From Lysolecithin in Rat Lung Supernatant," *Biochim. Biophys. Acta* 280: 275–280 (1972).
9. Lands, W. E. M. "Metabolism of Glycerolipidases: A Comparison of Lecithin and Triglycerides Syntheses," *J. Biol. Chem.* 231:883–888 (1958).
10. Erbland, J. F., and G. V. Marinetti. "The Enzymatic Acylation and Hydrolysis of Lysolecithin," *Biochim. Biophys. Acta* 106:128–138 (1965).
11. Akino, T., M. Abe and T. Arai. "Studies on the Biosynthetic Pathways of Molecular Species of Lecithin by Rat Lung Slices," *Biochim. Biophys. Acta* 248:274–281 (1971).
12. Ichikawa, I., and E. Yokoyama. "Phospholipase A Activity in O_3- and NO_2-Exposed Rat Lung," *J. Japan Soc. Air Poll.* 9:333 (1974).
13. Yokoyama, E. "O_3 Exposure and Ventilatory Functions of Rat," *Japan J. Thorac. Dis.* 16(Suppl.):91 (1978).
14. Goldstein, B. D. "Combined Exposure to Ozone and Nitrogen Dioxide," *Environ. Health Persp.* 13:107–110 (1976).
15. Fukase, O., K. Isomura and H. Watanabe. "Combination Effects of Ozone and Gaseous Pollutants on Mice," *Proc. 19th Meeting of Japan Society of Air Pollution* (1978), p. 329.
16. Tsunoda, T., T. Toyama and M. Nakaza. "Respiratory Effects of Exposure to NO_2 and O_3," *Proc. 51st Meeting of Japan Association of Industrial Health* 1978), pp. 478, 479.
17. Roehm, J. N., G. Hadly and D. B. Menzel. "The Influence of Vitamin E on the Lung Fatty Acids of Rats Exposed to Ozone," *Arch. Environ. Health* 24:237–242 (1972).
18. Shimasaki, H., T. Takatori, W. R. Anderson, H. L. Horten and O. S. Privett. "Alteration of Lung Lipids in Ozone Exposed Rats," *Biochem. Biophys. Res. Commun.* 68:1256–1262 (1976).
19. Blank, M. L., W. Dalbey, P. Nettesheim, J. Price, D. Creasia and F. Snyder. "Sequential Changes in Phospholipid Composition and Synthesis in Lungs Exposed to Nitrogen Dioxide," *Am. Rev. Resp. Dis.* 117:273–280 (1978).

CHAPTER 15

EFFECTS ON PULMONARY FUNCTION OF LOW-LEVEL NITROGEN DIOXIDE EXPOSURE

J. R. Gillespie and J. B. Berry
Department of Physiological Sciences

L. L. White
Family Nurse Practitioner Program

P. Lindsay
Pulmonary Section, Department of Medicine
The University of California at Davis
Davis, California 95616

INTRODUCTION

A special committee on the Environmental and Occupational Health Assembly of the American Thoracic Society (ATS) [1] recently prepared a comprehensive review of the health effects of air pollution, which included a section on nitrogen dioxide (NO_2). This document summarized the published data on the major health effects associated with inhalation of the NO_2, which are shown in Table I. There are few data on the health effects of NO_2. Nevertheless, data in the ATS review, including those from the studies of Freeman and co-workers [2,3] and Haydon et al. [4] strongly suggest that NO_2 is damaging to the lung parenchyma. At relatively high experimental exposure levels (about or greater than 1000 $\mu g/m^3$), NO_2 causes pulmonary parenchymal damage in rodents; may complicate or predispose animals or human subjects to airway infection; and is associated with increased airway reactivity in some asthmatic individuals [5,6].

There have been few studies on the effects of NO_2 exposure at levels near the World Health Organization task committee recommended maximum level, 190–320 $\mu g/m^3$. Shy et al. [7] and Pearlman et al. [8] found increased

Table I. Summary of Health Effects of NO_2 Exposure at Ambient or Near Ambient Concentrations [1]

Concentration ($\mu g/m^3$)	Duration of Exposure	Effect
Animal Studies		
3760 (2 ppm)	Continuous	Alterations in alveoli and respiratory bronchioles of laboratory animals; type I replaced by type II pneumocytes; wave of cell replication and loss of ciliated cells.
940 (0.5 ppm)	Continuous for 12 months	Development of emphysematous lesions in mice: ciliary loss, alveolar cell disruption, expanded alveoli and progressive pulmonary parenchymal damage.
2860 (1.5 ppm) 6580 (3.5 ppm)	Continuous for 16 days or more Intermittent exposure	Increased mortality in mice exposed to airborne infection of bacterial pathogens.
Human Studies		
3000–3760 (1.6–2.0 ppm)	15 min	Increased airway resistance in experimentally exposed normal subjects and persons with chronic bronchitis
210 (0.11 ppm)	1 hr	Increase in the airway response of asthmatics to a broncho-constricting agent.
150–182 (0.08–0.10 ppm) in Combination with Other Pollutants	Several years	Increased incidence of acute respiratory disease in school children and their parents and in children 1–3 years of age.
Estimated Peak Exposures of 940–880 (0.5–1.0 ppm)	Daily for undetermined intervals	Increased incidence of lower respiratory infections among children residing in homes using gas versus electricity for cooking. Illness rates adjusted for social class, age, sex, crowding and latitude.

incidence of acute respiratory disease in infants, school children and their parents living in a polluted air environment that contained 150–182 $\mu g/m^3$ NO_2 and other air pollutants. Orehek et al. [6] reported an increase in the airway response of some asthmatics to a bronchoconstricting agent, carbachol, when their subjects breathed 210 $\mu g/m^3$ NO_2 for 1 hr. The mechanisms underlying these effects and the potential long-term effects of exposure to ambient levels of NO_2 have not yet been described.

As part of a large interdisciplinary study, we measured the pulmonary function of dogs exposed for 68 months to low levels of NO_2 and NO. Our aims were to evaluate their pulmonary function during exposure for evidence of dysfunction and after exposure for evidence of residual effects of the air pollutants. Results of pulmonary and cardiovascular studies done during and near the end (TE) of the exposure were provided by the investigators in Cincinnati and are reported elsewhere [9–11]. This chapter emphasizes our physiological findings in the NO_2-exposed dogs and contrasts the NO_2-exposed dogs' physiology with that of the other groups.

METHODS

Table II shows the exposure levels of the various air pollutants and Figure 1 shows the schedule followed for the physiology studies. The exposure was at the National Environmental Research Center in Cincinnati. The air pollutant exposure protocol and technique have been described in detail [5,12,13]. The female beagle dogs studied were six months old at the beginning of the exposure. A control group of dogs breathed filtered, temperature- and humidity-conditioned clean air. The rest were divided into seven treatment groups and exposed 16 hr/day for 68 months to one of the following pollutants or mixtures: raw auto exhaust (R); irradiated auto exhaust (I); SO_x; R + SO_x; I + SO_x; NO low concentration + NO_2 high concentration (NO_2 high); NO high concentration + NO_2 low concentration (NO_2 low) (Table II).

Approximately six months after the termination of exposure, 86 of the dogs were transported to Davis, California where they were housed in outdoor kennels. The environment in Davis, which is a rural area, lacked measurable quantities of the pollutants to which the dogs had previously been exposed.

The pulmonary function techniques employed in this study have been reported in detail [14–16], so only a brief account of each test will be given here. We studied the dogs in a random order, with the investigators having no knowledge of the subject's treatment group.

We anesthetized the dogs with intravenous thiamylol sodium (approximately 8 mg/kg) and incubated them with the largest possible diameter endotracheal tube (8–9 mm i.d.). The depth of anesthesia was carefully monitored and maintained near minimum anesthetic concentration I [17,18] as judged by pedal responses. After giving the dogs anesthetic, but before

Table II. Atmospheric Mean Concentrations and Their Standard Deviations Administered from 0800 to 2400 Each Day (mg/m^3)

Group	Atmosphere	Abbreviation	Pollutant						
			CO	HC (as CH$_4$)	NO$_2$	NO	O$_X$ (as O$_3$)	SO$_2$	H$_2$SO$_4$
1	Control air	CA	---	---	---	---	---	---	---
2	Nonirradiated auto exhaust	R	112.1 ± 11.5	18.0 ± 2.9	0.09 ± 0.04	1.78 ± 0.52	---	---	---
3	Irradiated auto exhaust	I	108.6 ± 22.5	15.6 ± 4.0	1.77 ± 0.68	0.23 ± 0.36	0.39 ± 0.18	---	---
4	SO$_2$ + H$_2$SO$_4$	SO$_X$	---	---	---	---	---	1.10 ± 0.57	0.09 ± 0.04
5	Nonirradiated auto exhaust + SO$_2$ + H$_2$SO$_4$	R + SO$_X$	113.1 ± 15.9	17.9 ± 2.8	0.09 ± 0.06	1.86 ± 0.54	---	1.27 ± 0.61	0.09 ± 0.04
6	Irradiated auto exhaust + SO$_2$ + H$_2$SO$_4$	I + SO$_X$	109.0 ± 22.8	15.6 ± 3.9	1.68 ± 0.68	0.23 ± 0.36	0.39 ± 0.16	1.10 ± 0.56	0.11 ± 0.04
7	Nitrogen oxides, 2	NO$_L$[a] + NO$_2$H[b]	---	---	1.21 ± 0.22	0.31 ± 0.08	---	---	---
8	Nitrogen oxides, 2	NO$_H$ + NO$_2$L	---	---	0.26 ± 0.62	2.05 ± 0.26	---	---	---

[a] L = low concentration.
[b] H = high concentration.

TIME OF STUDIES

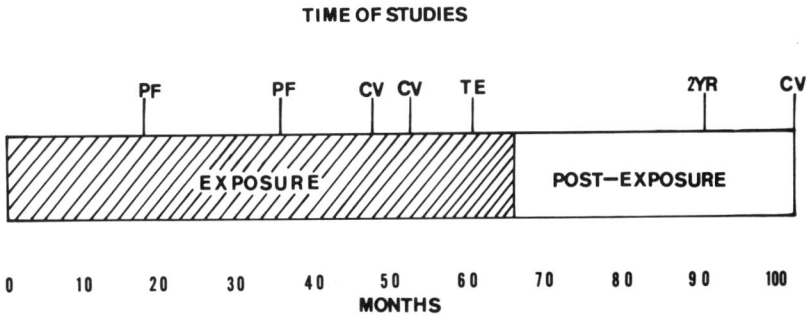

Figure 1. Schedule followed for the physiology studies: pulmonary function studies (PF); cardiovascular studies (CE); pulmonary function studies near termination of exposure period (TE); and pulmonary function studies at 2 years after exposure (2 YR.).

mechanically ventilating them, data were collected for the calculation of physiological dead space (V_D), dead space ventilation ($\dot{V}_D$) and dead space tidal volume ratios (V_D/V_T).

The dogs were ventilated with a fixed volume ventilator. The tidal volume was adjusted to give a transpulmonary (P_{TP}) change of 5–10 cm H_2O with each breath and the respiratory rate to give an end tidal carbon dioxide (CO_2) concentration of 5%.

We obtained total pulmonary resistance (R_{pul}), total respiratory resistance (R_{rs}) and chest wall resistance (R_{cw}) using the method of Mead and Wittenberger [19]. Sinusoidal flow oscillations were generated by a speaker system attached to the endotracheal tube. We oscillated the airway at 2–5 Hz when the lungs were at their resting end-expired volume (FRC), following a deep breath. Recordings of transpulmonary, airway and pleural pressures versus flow on an X-Y oscilloscope provided R_{pul}, R_{rs} and R_{cw}, respectively, after subtraction of endotracheal tube resistance.

We measured pulmonary diffusing capacity (D_{LCO}) with the single-breath carbon monoxide test [16,20], calculating its value using the formula of Ogilvie and co-workers [21]. This test was repeated after a period of 10–15 minutes, during which the dogs received three breaths of oxygen-rich gas. From these measurements pulmonary capillary blood volume was calculated using the technique of Roughton and Forester [22].

Inspiratory and expiratory lung compliance and chest wall compliance (C_{L_I}, C_{L_E} and C_{cw}, respectively) were obtained from records of quasistatic inspiratory and expiratory lung and chest wall pressure–volume curves, using techniques previously reported [14,15]. Residual volume (RV) was taken to be the volume at which the lung–pressure–volume trace became nearly horizontal (P_{TP} of −10 to −15 cm H_2O).

We measured FRC by the Boyle's law technique (FRC_{BL}) in the body plethysmograph and by the nitrogen equilibrium technique (FRC_{N2}) [14,15]. Vital capacity (VC), inspiratory capacity (IC) and expiratory reserve volume (ERV) were calculated from TLC, FRC and RV values.

We assessed the frequency dependence of compliance by plotting dynamic compliance (C_{dyn}) measured at 10, 20, 30, 40 and 50 breaths/min against respiratory frequency (f). Since larger changes in C_{dyn}/f are associated with greater frequency dependence of lung compliance, the slope of the line joining the points of this graph was used to quantify the measurement.

RESULTS

There was no significant difference in body weight, PCV or Hb of the eight groups of dogs at 24 months postexposure, although all lost weight during the postexposure period because of less restrictive caging.

A comparison of pulmonary physiological values between the control group and external control dogs from Davis previously studied in this laboratory [14,15] showed no significant difference. This confirmed our view that this group represented a satisfactory and healthy experimental control group.

The Pa_{CO_2}, V_D and $\dot{V}_D$ of the NO_2-low group was significantly greater than the control group. Its R_{rs}, R_{pul} and P_{cw} were not different from control.

The NO_2-high group had the lowest $D_{L_{CO}}$ value of all exposure groups, and the $D_{L_{CO}}/TLC$ ratio was significantly less than control for both NO_2-exposure groups. Their pulmonary capillary blood volume was not different from the control group. The lung and chest wall compliances were not different from the control group. However, their dynamic compliances at low frequencies were significantly greater than those of the controls. The NO_2-high group had a significantly greater TLC, RV and RV/TLC than the controls. Their FRC was not different from the controls.

The data were also analyzed by comparing each dog's values at the end of the exposure period (TE) to its value two years after exposure (2-YR).

The control group values for C_{L_I}, C_{L_I}/FRC and R_{pul} were not significantly different between TE and 2-YR. Control group values for $D_{L_{CO}}$ and $D_{L_{CO}}/TLC$ were significantly greater at 2-YR than at TE, as predicted by our previous studies on healthy dogs [16]. The control group's lung volumes increased between TE and 2-YR, as predicted by Robinson and Gillespie [14], and the RV/TLC ratio did not change significantly.

The C_{L_I} increased significantly between TE and 2-YR in the NO_2-high group. The individual values are shown in Figure 2. This group had a significant decrease in $D_{L_{CO}}$ and $D_{L_{CO}}/TLC$ between TE and 2-YR. All lung volumes except ERV increased significantly between TE and 2-YR. The TLC, FRC and RV tended to increase more in the exposed groups than in the control group; the average increases for TLC, FRC and RV, respectively,

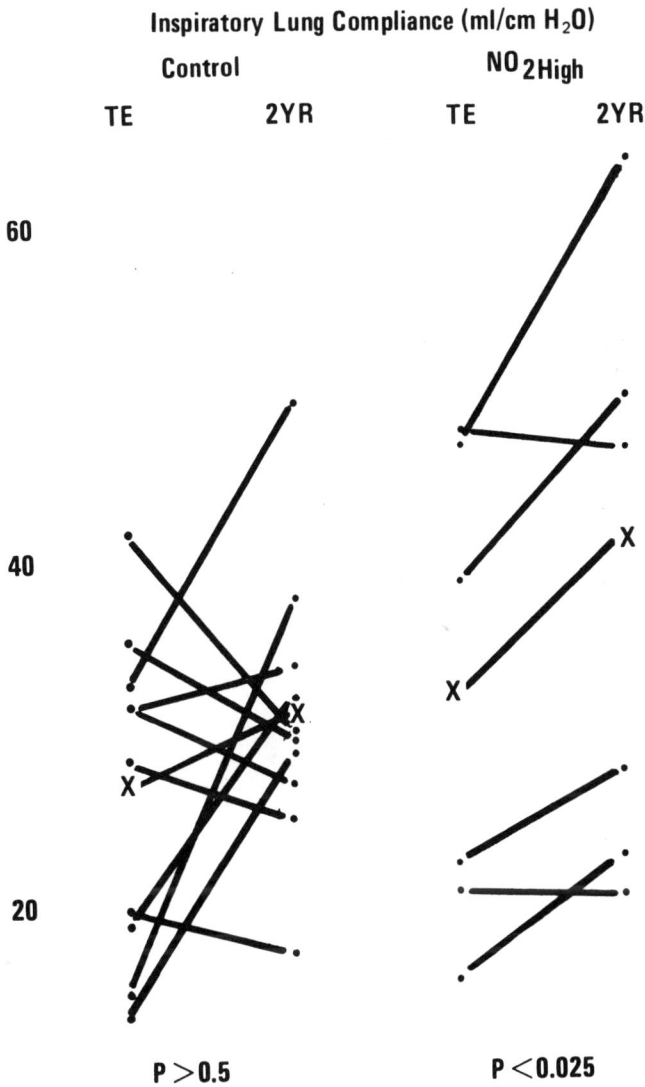

Figure 2. Inspiratory lung compliance of each dog in the control and NO_2-high group at TE and 2 YR. Mean values for each group are shown by X's.

were 229 ± 32, 88 ± 14 and 73 ± 12 ml for the exposed dogs, compared to 116-, 15- and 15-ml increases for the control dogs. Figure 3 shows the changes in RV between TE and 2-YR. The RV/TLC ratio increased significantly between TE and 2-YR in NO_2-high and NO_2-low groups.

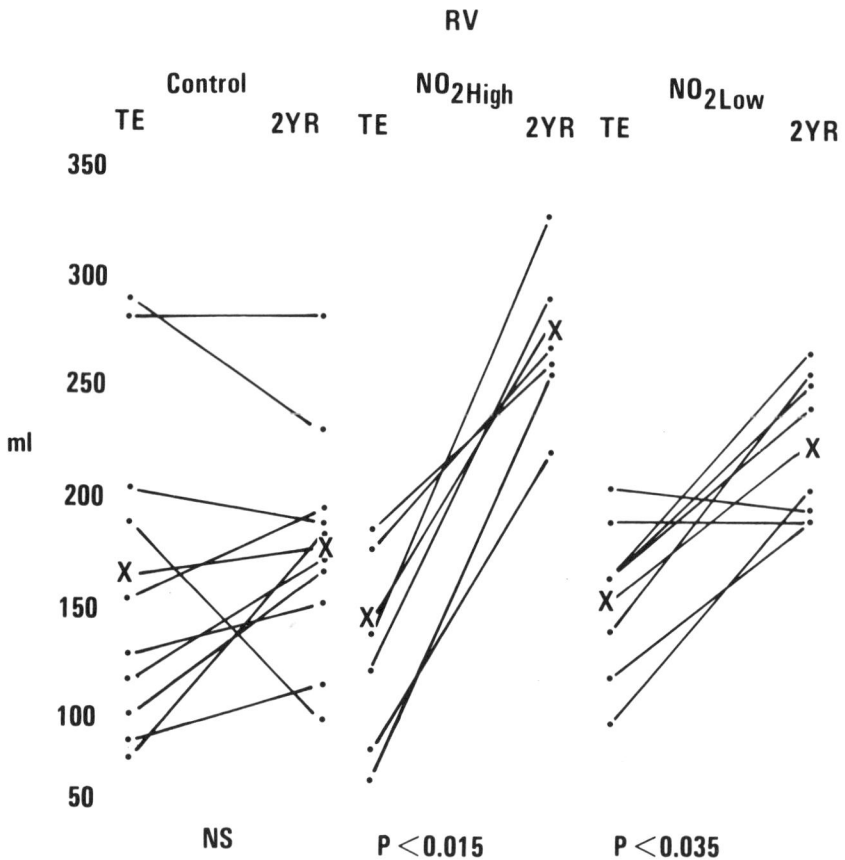

Figure 3. Residual volume (RV) of each dog in the control, NO_2-high and NO_2-low group at TE and 2 YR. Mean values for each group are shown by X's.

DISCUSSION AND CONCLUSIONS

We have concluded that pulmonary injury continues after termination of long-term, low-level NO_2 exposure. Our studies show greater pulmonary function changes following the 2-year recovery period, during which the dogs breathed clean air, than at the end of 68 months of exposure.

The first differences in the pulmonary function values between the exposed and control dogs were measured by Lewis et al. [11] in Cincinnati after 36 months of exposure. They found small differences in D_{LCO} between those dogs receiving NO and NO_2 and those of the controls. Additional changes at 68 months (TE) were increased RV in groups NO_2-low and -high.

Two years after exposure there were differences between the pulmonary function values of all groups exposed to air pollutants compared to: (1) those of the control group; (2) those of an external control group; and (3) their own values near the termination of exposure (TE).

The pulmonary function values of the control group appeared to remain nearly the same during the 2-yr postexposure period and were similar to those of other healthy dogs previously studied in our laboratory [14–16]. There were no differences in breed, body weight or age among the eight groups of dogs in this study, and we concluded that these factors were not responsible for the differences in pulmonary function we measured between control and exposed groups. There were no consistent or widespread pathological lesions seen on examination of the lungs of the control group [23,24]. Based on the above, we concluded that the control group was an acceptable representation of the healthy dog population, and that it was a satisfactory control for the study of the effects of air pollutants on our seven exposure groups.

In general, we believe our findings to be consistent with parenchymal injury in dogs exposed to NO, and NO_2. The histological evidence for parenchymal damage was increased alveolar size, increased number and size of alveolar pores, and tissue destruction [23,24]. This type of structural abnormality was most severe in the NO_2-high and -low groups.

We believe that the abnormalities seen in the pulmonary function measurements can be explained by the observed structural damage. The high C_{dyn} at low f in the NO_2-high group at 2-YR might have been because of the high lung compliance of several individuals in this group caused by pulmonary parenchymal destruction. Lung compliance increased significantly between TE and 2-YR in groups NO_2-high and -low. The decreased D_{LCO}/TLC values of all of the exposure groups at 2-YR compared to the control group also suggests parenchymal damage. The trend toward larger TLC, RV and RV/TLC values at 2-YR could have been due to parenchymal damage.

The increases in RV suggest air trapping because RV in dogs is set by airway closure [15]. Early airway closure may be caused by peribronchiolar tissue destruction, which was seen in groups NO_2-high and -low. These lesions may also have been the cause of hypoventilation and/or uneven ventilation, resulting in the ventilatory differences (Pa_{CO_2}, V_D, V_D/V_T, $\dot{V}_D$) observed in group NO_2-low compared to control values.

It is difficult to assess the measured differences in Pa_{CO_2} and V_D between the control group and exposure groups because of possible inconsistent effects of anesthesia. However, all dogs were given about the same dosage of anesthetic and were in light surgical anesthesia, as judged by their weak pedal pain response [17,18]. Differences seen in ventilatory values may be the result of lung damage resulting from exposure to air pollutants. This view is supported by histological evidence of parenchymal changes in the exposed lungs [23,24] and by differences in pulmonary volume and mechanics values between the control and NO_2-high and -low groups, which are not affected by anesthesia.

The correlation of our results from NO and NO_2 exposures and those of others [2-4] is not clear, due in part to differences in: species studied; level, method and duration of exposure; and methods of evaluation. Experimental exposures by others have been of shorter duration and usually at higher concentrations than we employed. Our studies are also unique in that they included study of the animals two years after the exposure period. Freeman et al. [2,3] and Haydon et al. [4] reviewed the studies on effects of NO_2 and described the parenchymal damage that they and others have seen in animals following exposure. Our dogs showed similar lung damage on exposure to comparatively low levels of NO and NO_2. We believe that these data provide substantial evidence that ambient levels of NO_2 are harmful to the lung parenchyma and that termination of exposure will not necessarily halt lung damage.

In summary, we have presented evidence that supports the conclusions of Lewis et al. [11] that there is pulmonary injury in dogs resulting from exposure to high ambient levels of NO_2 and NO. We also showed that there was a continuation of pulmonary function loss during a 2-yr postexposure period, during which the previously exposed dogs breathed clean air. NO_2 mainly caused a progressive emphysematous lesion.

REFERENCES

1. American Thoracic Society. "Health Effects of Air Pollution," *Am. Thor. Soc. News* 4(2):22–63 (1978).
2. Freeman, G., S. C. Crane, R. J. Stephens and N. J. Furiosi. "Pathogenesis of the Nitrogen Dioxide-Induced Lesion in the Rat Lung: A Review and Presentation of New Observation," *Am. Rev. Resp. Dis.* 98:429 (1968).
3. Freeman, G., S. C. Crane, N. J. Furiosi, R. J. Stephens, M. J. Evans and W. D. Moore. "Covert Reduction in Ventilatory Surface in Rats During Prolonged Exposure to Subacute Nitrogen Dioxide," *Am. Rev. Resp. Dis.* 106:563 (1972).
4. Haydon, G. B., G. Freeman and N. J. Furiosi. "Covert Pathogenesis of NO_2 Induced Emphysema in the Rat," *Arch. Environ. Health* 11:776 (1965).
5. Hinners, R. G., J. K. Burkart and G. L. Contner. "Animal Exposure Chambers in Air Pollutant Studies," *Arch. Environ. Health* 13:209 (1966).
6. Orehek, J., J. P. Massari, P. Gayrard, C. Crimaud and J. Charpin. "Effect of Short-Term, Low-Level Nitrogen Dioxide Exposure on Bronchial Sensitivity of Asthmatic Patients," *J. Clin. Invest.* 57:301 (1976).
7. Shy, C. M., J. P. Creason, M. E. Pearlman, K. E. McClain, F. B. Benson and M. M. Young. "The Chattanooga Schoolchildren Study: I Methods, Description of Pollutant Exposure, and Results of Ventilatory Function Testing," *J. Air Poll. Control Assoc.* 20:539 (1970).

8. Pearlman, M. E., J. F. Finklea, J. P. Creason, C. M. Shy, M. M. Young and R. J. Horton. "Nitrogen Dioxide and Lower Respiratory Illness," *Pediatrics* 47:391 (1971).
9. Block, W. N., Jr., T. R. Lewis, K. A. Busch, J. G. Othoefer and J. F. Stara. "Cardiovascular Status of Female Beagles Exposed to Air Pollutants," *Arch. Environ. Health* 24:342 (1972).
10. Vaughan, T. R., L. F. Jennelle and T. R. Lewis. "Long-Term Exposure to Low Levels of Air Pollutants. Effects on Pulmonary Function in the Beagle," *Arch. Environ. Health* 19:45 (1969).
11. Lewis, T. R., W. J. Moorman, Y. Yang and J. F. Stara. "Long-Term Exposure to Auto Exhaust and Other Pollutant Mixtures. Effects on Pulmonary Function in the Beagle," *Arch. Environ. Health* 29:102 (1974).
12. Barkley, N. P., K. A. Busch, W. L. Grider and M. Malanchuk. "The Concentration of Lead in Automobile Exhaust Exposure Chambers," *Am. Ind. Hyg. Assoc. J.* 33:678 (1972).
13. Hueter, F. G., G. L. Contner, K. A. Busch and R. G. Hinners. "Biological Effects of Atmospheres Contaminated by Auto Exhaust," *Arch. Environ. Health* 12:553 (1966).
14. Robinson, N. E., J. R. Gillespie, J. D. Berry and A. Simpson. "Lung Compliance, Lung Volumes, and Single-Breath Diffusing Capacity in Dogs," *J. Appl. Physiol.* 33:808 (1972).
15. Robinson, N. E., and J. R. Gillespie. "Lung Volumes in Aging Beagle Dogs," *J. Appl. Physiol.* 35:317 (1973).
16. Robinson, N. E., and J. R. Gillespie. "Pulmonary Diffusing Capacity and Capillary Blood Volume in Aging Dogs," *J. Appl. Physiol.* 38:647 (1975).
17. Steffey, E. P., J. R. Gillespie, J. D. Berry, E. I. Eger II and E. A. Rhode. "Circulatory Effects of Halothane and Halothane-Nitrous Oxide Anesthesia in the Dog: Spontaneous Ventilation," *Am. J. Vet. Res.* 36:197 (1975).
18. Steffey, E. P., J. R. Gillespie, J. D. Berry, E. I. Eger II and E. A. Rhode. "Circulatory Effects of Halothane and Halothane-Nitrous Oxide Anesthesia in the Dog: Controlled Ventilation," *Am. J. Vet Res.* 35:1289 (1974).
19. Mead, J., and J. L. Whittenberger. "Physical Properties of Human Lungs Measured During Spontaneous Respiration," *J. Appl. Physiol.* 5:769 (1953).
20. Karp, R. B., P. D. Graf and J. A. Nadel. "Regulation of Pulmonary Capillary Blood Volume by Pulmonary Arterial and Left Atrial Pressures," *Circulation Res.* 22:1 (1968).
21. Ogilvie, C. M., R. E. Forester, W. S. Blakemore and J. W. Morton. "A Standardized Breath Holding Technique for the Clinical Measurement of Diffusing Capacity of the Lung for Carbon Monozide," *J. Clin. Invest.* 36:1 (1957).
22. Roughton, F. J. W., and R. E. Forster. "Relative Importance of Diffusion and Chemical Reaction Rates in Determining Rate of Exchange in Gases in the Human Lung, with Special Reference to True Diffusing Capacity of Pulmonary Membrane and Volume of Blood in the Lung Capillaries," *J. Appl. Physiol.* 11:290 (1957).

23. Hyde, D., D. Hallberg, A. Wiggins, W. Tyler, D. Dungworth and J. Orthoefer. "Morphometric Evaluation of Lungs Using Automated Image Analysis," in *Proc. 6th Conf. on Environmental Toxicology*, Dayton, OH (1975).
24. Hyde, D., J. Orthoefer, D. Dungworth, W. Tyler, R. Carter and H. Lun. "Morphometric and Morphologic Evaluation of Pulmonary Lesions in Beagle Dogs Chronically Exposed to High Ambient Levels of Air Pollutants," *Lab. Invest.* 38(19):455 (1978).

CHAPTER 16

MORPHOLOGICAL AND PATHOLOGICAL EFFECTS OF NO$_2$ ON THE RAT LUNG

Michael J. Evans and Gustave Freeman

Medical Sciences Laboratory
Life Sciences Division
SRI International
Menlo Park, California 94025

INTRODUCTION

The effects of nitrogen dioxide (NO$_2$) on pulmonary tissue have been under investigation for many years. The original studies were on NO$_2$-polluted occupational environments, the generation of NO$_2$ in cigarette smoke, and the combustion of nitrogen-containing fuels [1]—conditions that have been associated with an increasing prevalence of emphysema in humans. Early studies demonstrated acute injury, mainly to the epithelium, and severe pulmonary edema on exposure to NO$_2$. More chronic phases included changes in the small airways, the tissue framework and the lung parenchyma, leading to aggregation of macrophages and, ultimately, to an emphysema-like disease in the rat [2]. Common to all phases of injury by NO$_2$ was the area of the bronchioles, alveolar ducts and adjacent alveoli.

Subsequent detailed studies described the early stages of injury and the mechanism of epithelial repair. Factors that affect the repair process, such as age and dietary antioxidants, were examined also. Chronic, long-range exposure studies have revealed the gross morphological changes and the changes in the epithelium, the connective tissue framework of the airways and the parenchyma that are associated with the emphysema-like disease. However, relatively little is known about the detailed structural events that account for the loss of alveolar walls that define the experimental emphysema induced by NO$_2$.

This overview of the morphological effects of NO_2 on the rat's lung is addressed principally to the quantitative dynamic aspects of injury to cells and their renewal and, secondarily, to the order of events during prolonged exposure that lead to the irreversible emphysema-like disease. Both near-ambient peak concentrations of NO_2 and the high subacute levels that exist in tobacco smoke are considered. Some functional aspects of respiration are also discussed.

INJURY OF THE PULMONARY EPITHELIUM

The small airways of the normal lung are lined with a single layer of ciliated and nonciliated cells [3,4]. Nonciliated cells usually protrude above adjacent ciliated cells and contain secretory granules (Clara cells). In the alveoli, large, flat squamous cells (Type 1) cover 97% of the alveolar surface. The remaining 3% of the gas-exchanging area is covered by Type 2 cells, which characteristically contain lamellar bodies [5].

When rats are exposed continuously to sublethal concentrations of NO_2 (2–17 ppm), the epithelial cells lining the airways and alveoli are damaged [6,7]. In the bronchiolar regions, the first changes are noted in ciliated cells. After four hours exposure, a loss of cilia begins. By eight hours the loss is obvious, and sloughing of dying and necrotic ciliated cells is observed (Figures 1–3). After 16 hrs exposure, the ciliated and nonciliated cells of the bronchiolar epithelium become uniform in height. Although Clara cells are relatively resistant to injury, they lose secretory granules during the first 24 hours of exposure [8]. By the second day of exposure, cellular hypertrophy and hyperplasia are evident in the bronchiolar epithelium, and foci of multilayered cells become obvious. From the second through the fifteenth days, the epithelium tends to maintain this cellular appearance with a distinct absence of cilia. During this period, numerous ciliated vacuoles are observed (Figure 4), along with rod-like structures in nonciliated cells (Figures 5 and 6). When the animals are allowed to recover in air, normal ciliated and nonciliated cells appear again by the seventh day of recovery (Figure 7) [8,9].

In alveolar regions, the first morphological changes are observed also by the fourth hour of exposure [6,7]. At this time, Type 1 cells near the opening of the terminal bronchioles exhibit swelling, and occasional areas of alveoli have lost the covering Type 1 cells, leaving basement membrane exposed. These changes increase in intensity, so that after a day of continuous exposure the larger areas of alveoli have lost their damaged Type 1 cells (Figures 8 and 9). Cellular debris and some interstitial and alveolar edema become apparent. Type 2 cells appear to be relatively resistant to NO_2, as do alveolar macrophages; endothelial cells appear not to be affected.

After 48 hours of continuous exposure, areas previously devoid of Type 1 cells have been covered by a low cuboidal epithelium (Figures 10–12). These cells are derived from Type 2 cells, but lack lamellar bodies. Cellular debris, mucoid material and macrophages are present near the openings of the

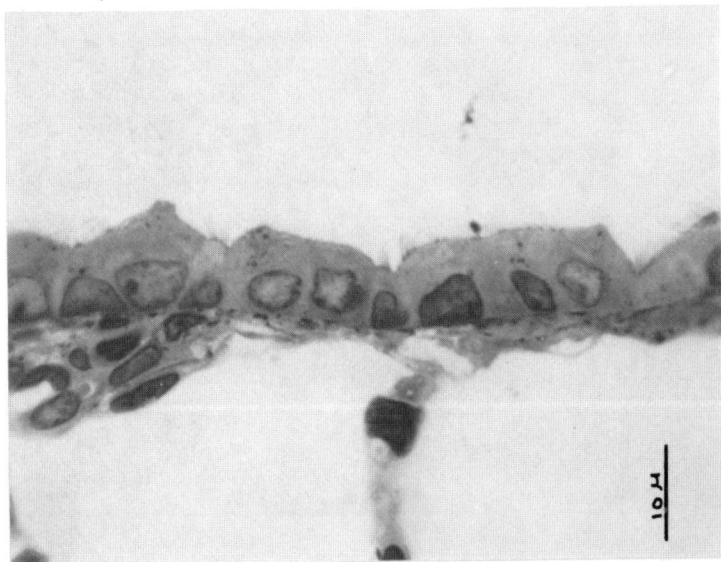

Figure 2. Bronchiolar epithelium after 24 hours of exposure to 15 ppm NO_2. Note the lack of cilia (1200X) [9].

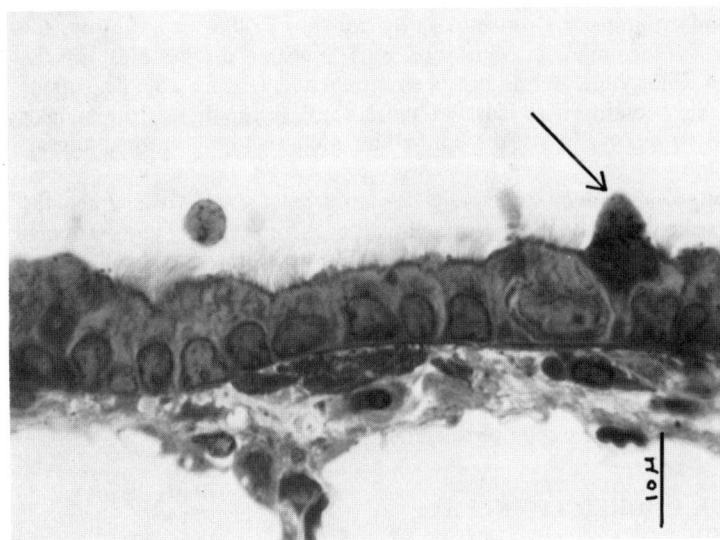

Figure 1. Portion of bronchiolar epithelium from a control rat demonstrating a Clara cell protruding above adjacent ciliated cells (1200X) [9].

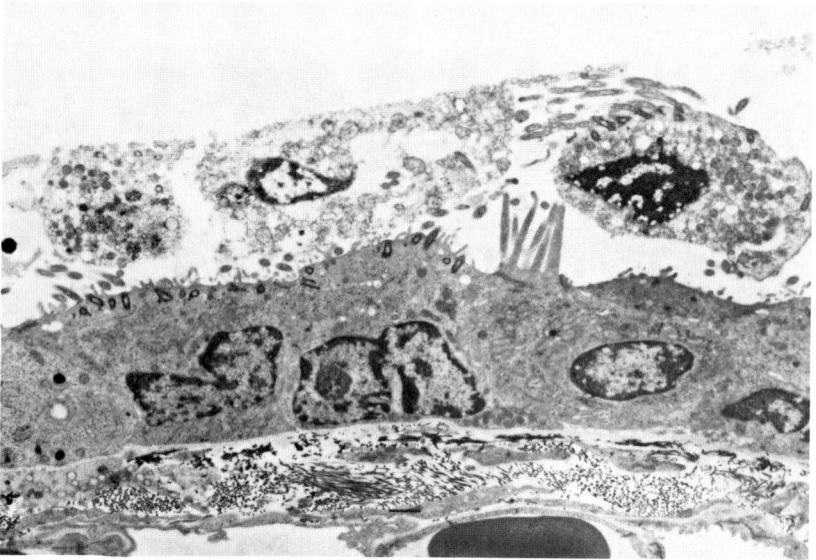

Figure 3. Electron micrograph of bronchiolar epithelium after 8 hr of exposure to 14 ppm NO_2, demonstrating necrotic cells that had sloughed off the bronchiolar wall (4176×) [7].

terminal and respiratory bronchioles. By the fourth day of exposure, the epithelium has assumed a more squamous appearance, and most of the cellular and amorphous debris has been eliminated (Figure 13). The tissue remains in this condition at least 15 more days (Figure 14). If the animals are allowed to recover in air, the epithelium soon assumes a normal appearance [10].

Three important conclusions were derived from these studies. First, the cells most vulnerable to NO_2 are the ciliated bronchiolar and Type 1 alveolar cells. Both cell types have a much larger surface area exposed to the NO_2 than the more resistant types (Type 2 and Clara cells). Second, the epithelium can repair itself in the presence of sublethal levels of NO_2. Third, the new cells are relatively resistant to continuous exposure to the same concentration of NO_2. These observations led to new studies directed at describing the mechanism for repair of injured epithelium; the factors that affect the amount of tissue damaged; and the nature of tolerance of the epithelium.

REPAIR OF INJURED EPITHELIUM

After injury to the bronchiolar and alveolar epithelium, cellular proliferation leads to replacement of damaged cells. In the alveoli of young rats

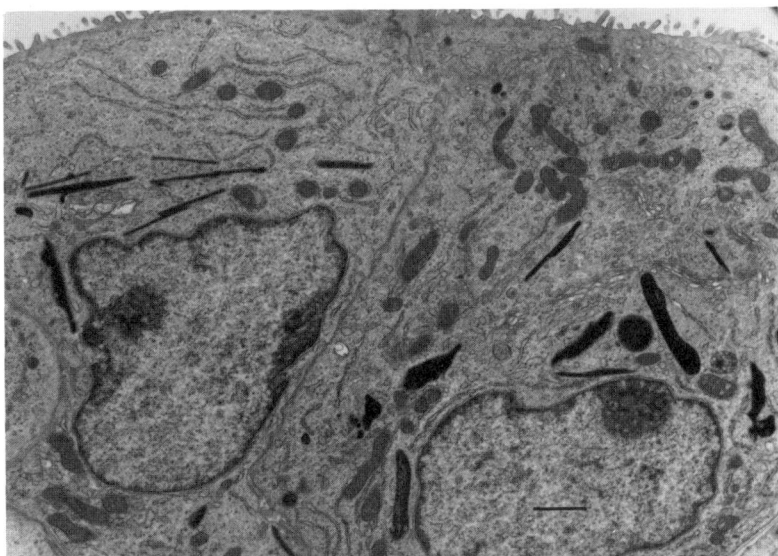

Figure 4. Rodlike structures in nonciliated cells after five days of exposure to 12 ppm NO_2 (7056×)[7].

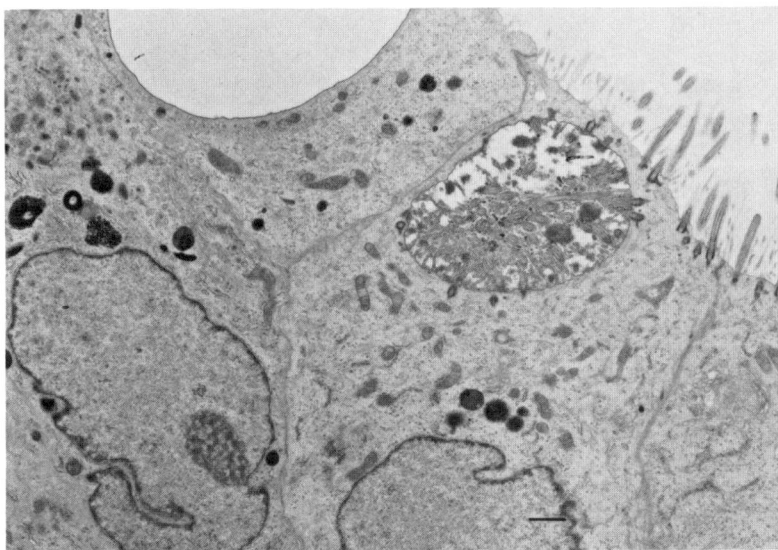

Figure 5. Ciliated vacuole in a ciliated cell after 15 days of exposure to 12 ppm NO_2 (4368×)[7].

Figure 7. Bronchiolar epithelium in a rat exposed to 15 ppm NO_2 for 24 hours and allowed to recover in air for 7 days. Compare with Figure 1 and note the return of a normal ciliary bed (1337.5×).

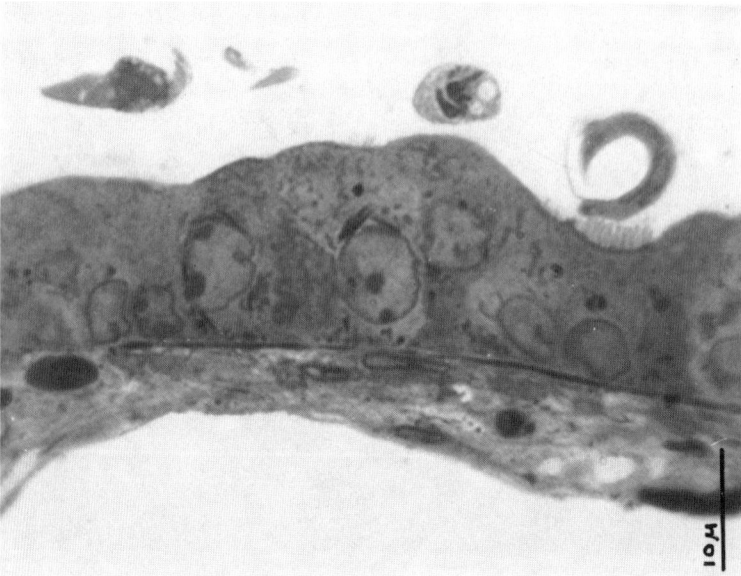

Figure 6. Bronchiolar epithelium after 15 days of exposure to 17 ppm NO_2 (1337.5×).

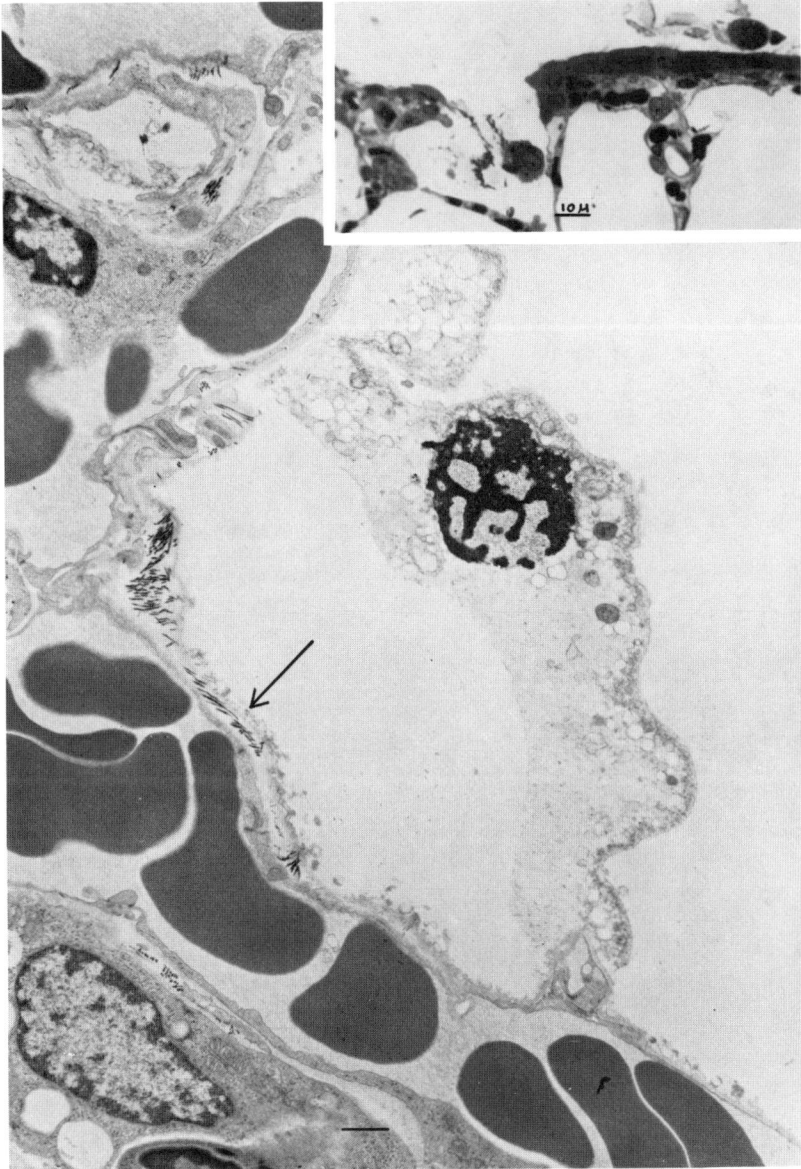

Figure 8. Example of injured alveoli near the opening of the terminal bronchiole after 12 hours of exposure to 15 ppm NO_2 (433×).

Figure 9. Necrotic Type 1 cell sloughing off the alveolar wall after 24 hours of exposure to 14 ppm NO_2. Arrow indicates area of basement membrane (5572×) [7].

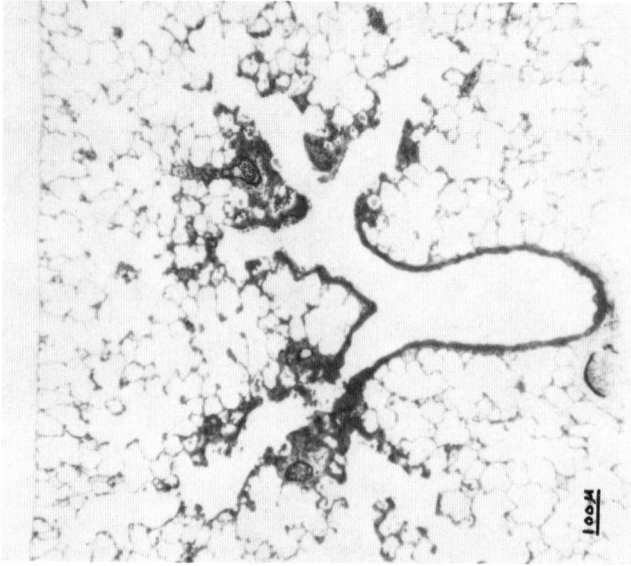

Figure 11. Area similar to that in Figure 10 after 48 hours of exposure to 17 ppm NO_2. The increased density of alveoli near the opening of terminal bronchioles is caused by proliferation of cells and accumulation of macrophages associated with the damage caused during the first 24 hours of exposure and its repair (55×) [7].

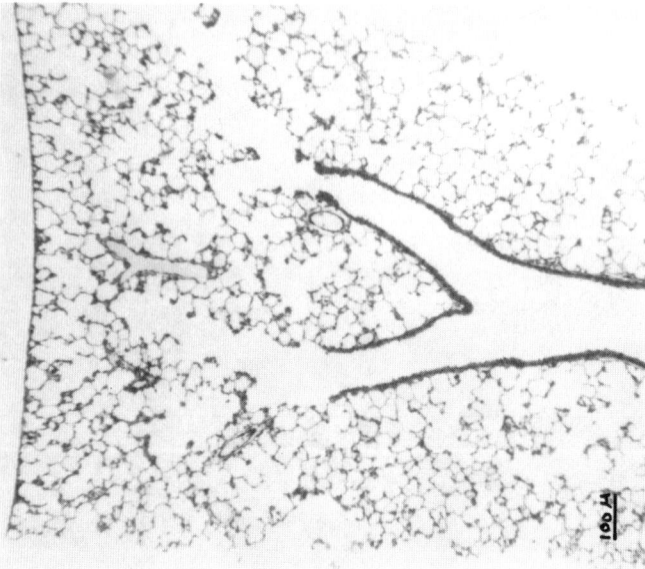

Figure 10. Tissue architecture in a control rat (55×) [7].

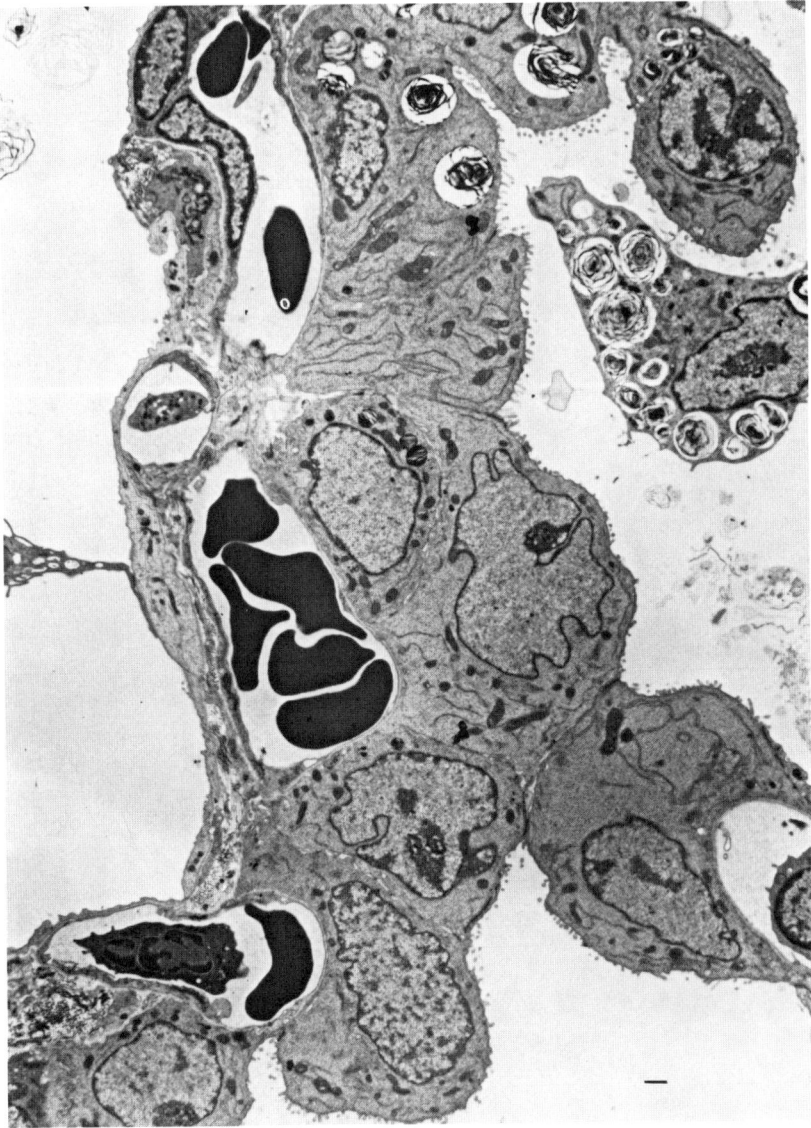

Figure 12. Electron micrograph of the cuboidal epithelium that results from proliferation of Type 2 cells for renewal of injured Type 1 epithelium (2960×) [7].

Figure 14. Tissue architecture after 15 days of exposure to 17 ppm NO$_2$ (55×) [7].

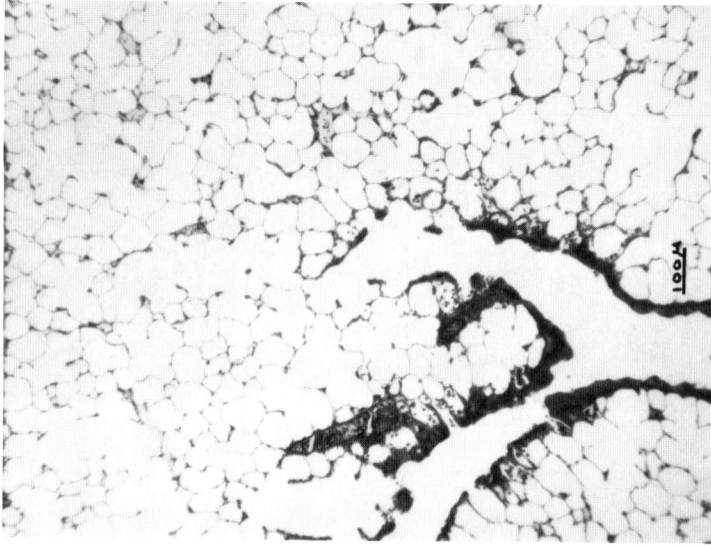

Figure 13. Tissue architecture after 4 days of exposure to 17 ppm NO$_2$, demonstrating the return of the epithelium to a more squamous type (55×) [7].

exposed continuously to NO_2, Type 2 cells begin dividing within the first 24 hours (Figure 15) [11]. Maximal cell proliferation is reached by the second or third day of exposure and then declines, so that by the fifth day it approximates control levels. Proliferation remains at these low levels despite continued exposure to the gas, and no further epithelial injury is observed. If the exposure period is brief (5 hours), the proliferative response reaches a peak within 24 hours and then declines to near control levels by the third day of recovery [12].

During maximum Type 2 cell proliferation (second and third days), proximal alveoli appear to be covered by a layer of cuboidal epithelium. When proliferation ceases (on the fourth and fifth days), the epithelium assumes a more normal squamous appearance. Earlier studies of the kinetics of Type 2 cell proliferation in animals recovering in air revealed that, after division, a cuboidal intermediate form flattened to become a typical squamous Type 1 cell (Figure 16). The cuboidal intermediate cell described in those studies [10,13] is similar to the type seen lining previously injured alveoli on the second and third days of exposure. When it differentiates into a Type 1 cell at 4 and 5 days, the more squamous epithelium predominates. This pattern of alveolar epithelial repair is not unique to NO_2; it has been observed under a variety of conditions involving injury to the alveoli [14].

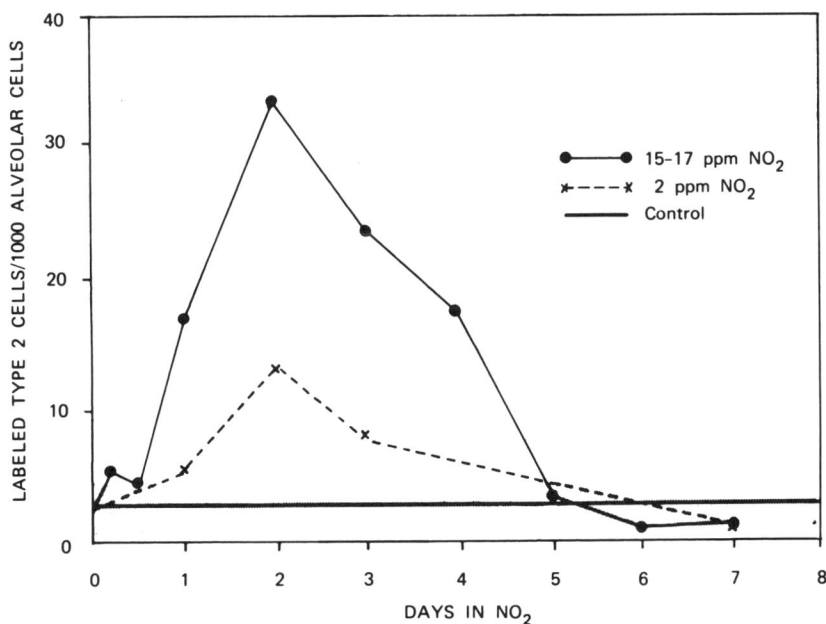

Figure 15. The proliferative response of Type 2 cells during continuous exposure to NO_2, as demonstrated by labeling with tritiated thymidine [1].

Although the epithelium assumes a squamous appearance by the fifth day of exposure, the number of cells is greater than in controls [13,15]. In a recent study, the number of Type 1 and 2 cells in a prescribed area of injured alveoli was counted after 5 and 15 days of exposure to NO_2 [16]. In control rats, the average numbers of Type 1, 2 and intermediate cells in that area were 156, 258 and 0, respectively. However, after 5 and 15 days of exposure, the respective numbers of Type 1 cells were 145 and 138; of Type 2 cells, 267 and 330; and of intermediate cells, 82 and 109. The increased cellularity has not been explained, but it may be associated incidentally with the development of tolerance to NO_2.

Other cells in the alveoli also multiply during exposure to NO_2 [11]. The number of dividing macrophages increases by the third day of exposure. Whether this is in response to the presence of cellular debris or to hormonal stimulation associated with Type 2 cell division is not clear. In the capillaries,

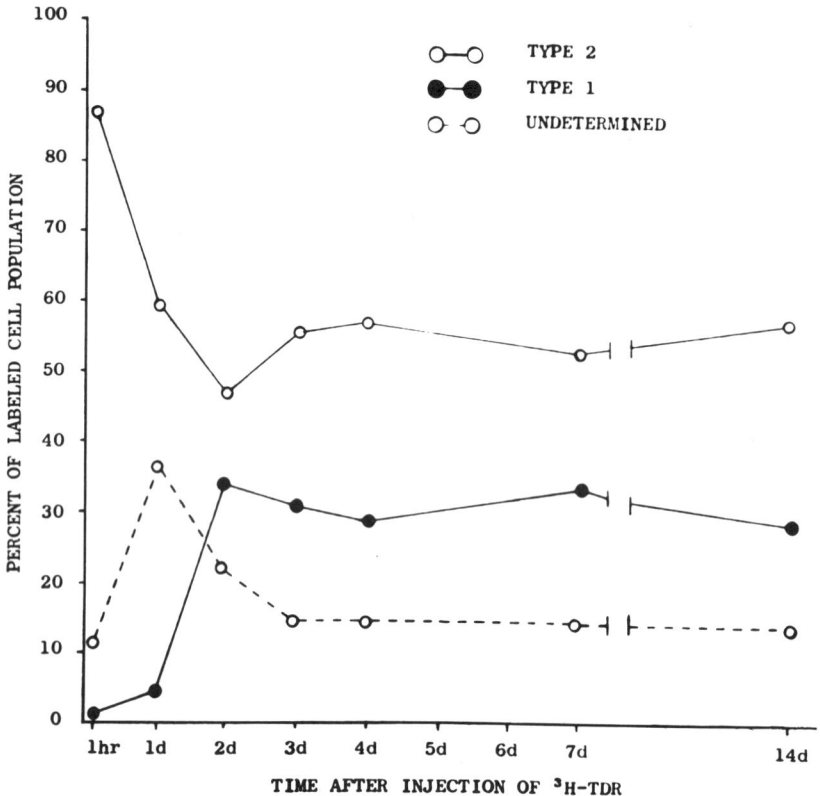

Figure 16. The differentiation of Type 2 cells into Type 1 cells during recovery from 48 hours of exposure to 15 ppm NO_2 [10].

circulating mononuclear cells are often observed to divide on the third and fourth days of exposure. This may be a hematopoietic response for recruitment of alveolar macrophages for the lung.

A proliferative response to injury was also observed in the terminal bronchioles [11]. However, in contrast to the alveolar reaction, the nonciliated Clara cells reach maximum proliferation during the first day of exposure and then decline rapidly during the second (Figure 17). Control levels are approached by the fifth day. The kinetics have been described in detail for animals recovering in air (Figure 18) [8,9]. First, the ciliated cell is injured and sloughs off; then Clara cells divide, resulting in an increased

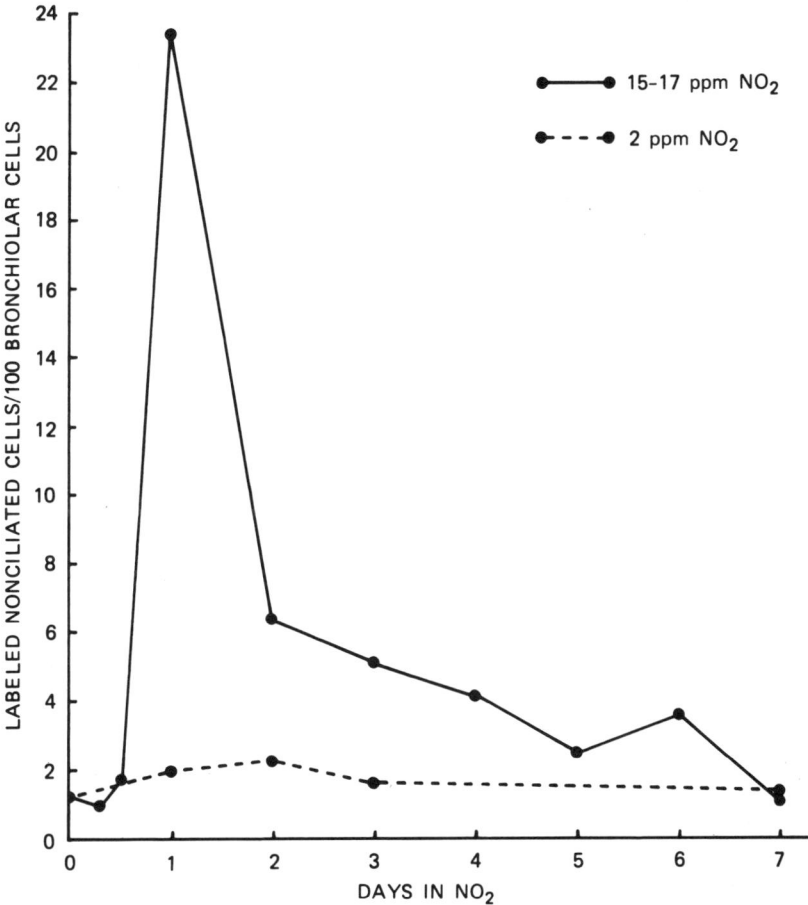

Figure 17. The proliferative response of nonciliated bronchiolar cells during continuous exposure to NO_2, as demonstrated by labeling with tritiated thymidine.

number of nonciliated cells in the lining of the bronchiolar epithelium. The nonciliated cells differentiate into mature Clara cells and then into ciliated cells. This process has not been studied during continuous exposure but the sequence is probably the same, although ciliated cells tend not to form normal cilia. This is evidenced by the sparseness of lumenal cilia and the presence of intracellular ciliated vacuoles (Figures 5 and 6). In addition, Clara cells of normal appearance are lacking. Instead, numerous nonciliated cells are present that contain large rod-shaped crystals (Figure 4). Again, the basis for this phenomenon is not known [6,7,17].

The magnitude of the proliferative response to injury of the alveoli was shown recently to correlate directly with the amount of epithelial tissue injured during the first 24 hours of exposure to NO_2 [18]. Previous studies had indicated that the magnitude was related to the dose; that is, labeling indexes for Type 2 cells were higher with higher levels of NO_2 [13,19,20]. In a recent study, we measured alveolar injury by determining the extent of exposed basement membrane after Type 1 cells had been damaged and desquamated. This value was compared with the proliferative response of Type 2 cells after the injury (Figure 19). A very close correlation was obtained ($r = 0.90$), indicating that Type 2 cell division is controlled by a

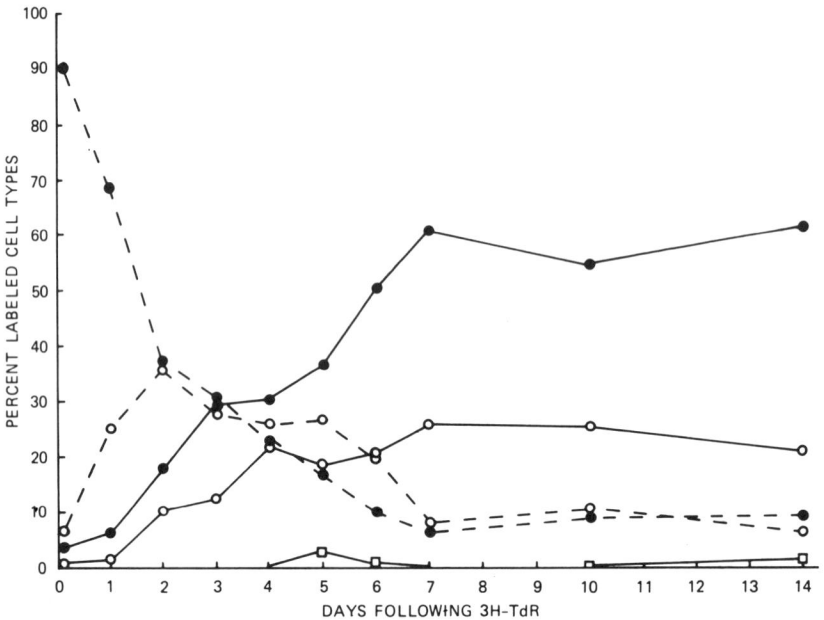

Figure 18. The differentiation of nonciliated cells into mature Clara cells and ciliated cells during recovery from 24 hours of exposure to 14 ppm NO_2: •---•, Type A cells; o---o, Type B cells; •—•, Clara cells; o—o, ciliated cells; □—□, brush cells [8].

negative feedback inhibition mechanism similar to that observed for other epithelial tissues [21,22]. Thus, monitoring cell proliferation could be useful as a means of quantitating sublethal injury to the alveolar epithelium. Similar measurements have not been made in the bronchiolar epithelium. However, the dose–response effect appears to be similar to that seen in the alveoli. Research is currently underway to determine whether bronchiolar injury also could be estimated by the proliferative response.

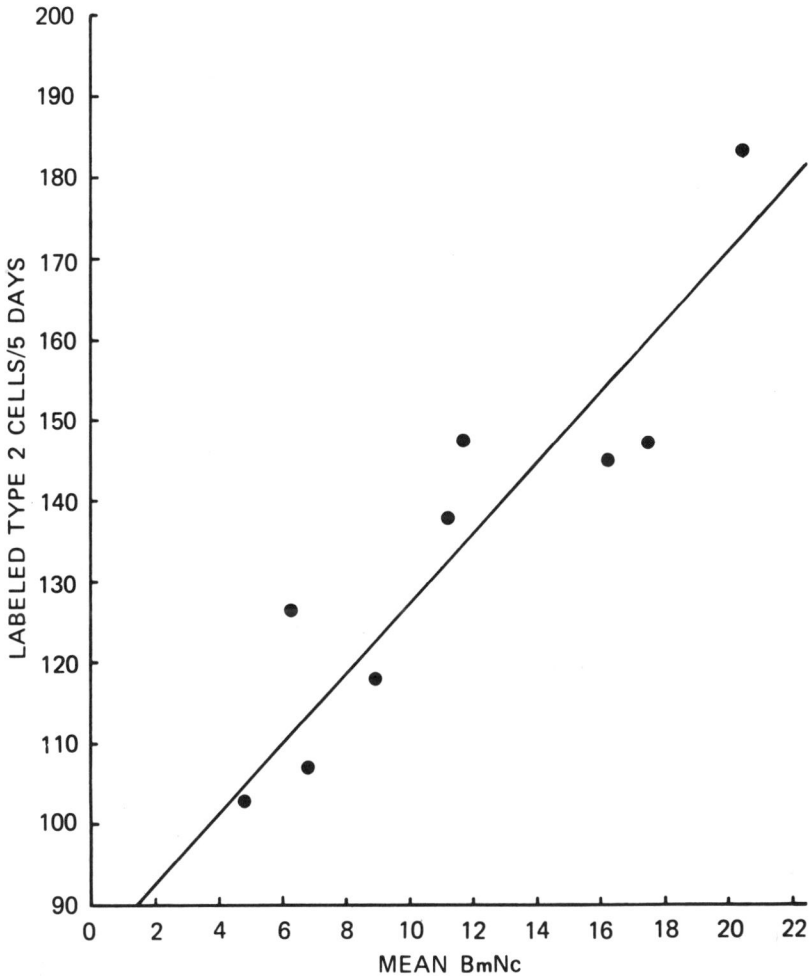

Figure 19. Relationship between the total proliferative response of Type 2 cells (as demonstrated by the curve in Figure 15) and the proportion of basement membrane not covered by Type 1 cells (BmNC), after 24 hours of exposure to NO_2 (as demonstrated in Figure 9). The correlation coefficient r = 0.090 [18].

EFFECT OF CONCENTRATION ON DEGREE OF INJURY

High concentrations of NO_2 are clearly more injurious than lower levels. Morphologically, the larger the number of alveoli damaged, the greater the injury, mainly because of the concentration gradient of NO_2 during breathing. Thus, the areas most susceptible to injury are the terminal bronchioles, alveolar ducts, adjacent alveoli and, finally, the more peripheral alveoli [6,7,23]. The concentration gradient effect is most likely affected by the aerodynamics, solubility, reactivity and dilution of NO_2 in alveolar air as it permeates more distal alveoli.

The larger airways appear much less affected by NO_2 than the terminal and respiratory bronchioles. However, with relatively high concentrations, cells of the larger airways also become damaged. Although this has not been studied in detail, evidence is provided mainly by the labeling studies of Kleinerman [24] and Creasia et al. [12]. Creasia et al. [12] showed a proliferative response in the main bronchi and bronchioles (but not in the trachea) of hamsters exposed to 10 ppm NO_2. Using higher concentrations, Kleinerman [24] was able to demonstrate an effect at all airway levels. Epithelial repair in the large airways is sufficiently different from that in bronchioles and alveoli that comparing data directly is not possible with respect to injury and repair [14]. However, the data indicate that cell proliferation does occur, starting in the bronchioles and involving the larger airways with increasing concentrations of NO_2. If injury in the large airways is followed by cell replication, a concentration-dependent effect should be demonstrable also in the epithelium of particular airway levels.

FACTORS THAT AFFECT THE AMOUNT OF TISSUE INJURY

Factors other than concentration affect the degree of tissue injury during exposure to NO_2. A study of aging [7,20] revealed that relatively old rats exposed to 14 ppm NO_2 suffered greater injury than young rats exposed to 17 ppm NO_2. Morphologically, the nature of the injury observed in the airways and alveoli was the same for both age groups, but a greater area of damage was observed in the alveoli in the older animals [7]. This was associated with increased alveolar edema. Further studies showed that initiation of repair of damaged epithelium in older animals was delayed by approximately one day [20]. Both factors—the greater extent of tissue damage and the retarded onset of cell proliferation—are likely to be responsible for the edema and, therefore, the higher mortality rate in the older animals. Also, in these studies 12% of 11-month-old rats died after exposure to about 14 ppm NO_2, whereas 36% of the 19-month-old rats died after exposure to the same concentration. Furiosi and Freeman [25] obtained similar mortality data over a number of years. An interesting note, however, is that old animals that survived the first three days of exposure were able to repair the damage, as did young rats, despite continuous exposure to NO_2. The reason for the difference in initial susceptibility to injury was not determined.

In contrast, Stephens et al. [26] observed resistance to high concentrations of NO_2 in rats exposed from birth until they were about 30 days old. In this study, loss of cilia was reported in the terminal bronchioles of nursing rats from 8 to 20 days of age. In the alveoli, no evidence was seen of damage to Type 1 cells until the period of observation from 21 to 35 days of age. The basis for the refractory nature of the alveolar epithelium up to 20 days was not apparent. Burri [27] observed that alveoli expanded in 20-day-old rats, after growth-related cellular proliferation had ceased [28]. Geometrically, this would result in spreading and thinning of Type 1 cells to cover the increased alveolar surface area. Conceivably, the larger surface area of individual cells may increase their susceptibility to injury to a noxious gas [16].

Another factor that may affect oxidant damage is the tissue level of the natural antioxidants vitamin E and selenium [29,30]. In a series of studies with exposed rats fed diets containing various amounts of vitamin E, Fletcher and Tappel [31] found a delay in mortality in vitamin E-supplemented animals compared with those fed vitamin E-deficient diets. Death was not averted, however, in animals supplemented with vitamin E during exposure to NO_2.

Recently, we have been studying the effect of dietary vitamin E and selenium as protection against injury of the alveolar epithelium by NO_2, using the proliferative response of Type 2 cells to quantitate damage [18]. Rats fed diets deficient in vitamin E and selenium and rats receiving presumably adequate levels of these antioxidants were exposed continuously to 14 ppm NO_2 for six days. Comparison of the proliferative responses at that time revealed no difference. Because effective levels of vitamin E and selenium might have become depleted with such continuous exposure, we exposed a similar group of animals for only six hours and followed the proliferative response during five days of recovery. Preliminary results indicate that the group on diets lacking vitamin E and selenium sustained 50% greater damage to Type 1 cells than the group fed supplemented diets. These studies confirm earlier evidence that biological antioxidants affect the extent of damage sustained by alveolar epithelium during exposure to NO_2 [31].

TOLERANCE AFTER EXPOSURE TO NO_2

Although NO_2 injures pulmonary cells of rats on initial exposure, the tissue becomes relatively tolerant to the same concentration during continuous exposure. Experimentally, this is demonstrated clearly in Figures 15 and 17, which show that nonciliated and Type 2 cell proliferation fall to near control levels by the fifth day of exposure and remain at this level despite continuous exposure to the same oxidant for up to two years [11]. If the animals are removed from the NO_2-containing environment, the tolerance decreases. Creasia et al. [12] used intermittent exposures and showed that tolerance developed to 10 ppm NO_2 persisted for about seven days before reexposure to the same concentration again caused significant injury. We obtained similar results in rats exposed to ozone [16].

The biological components defining tolerance have not been clearly established. Mechanisms have been proposed whereby activation of biochemical antioxidant systems in the cell may detoxify the free radicals and peroxides resulting from exposure to oxidants [32]. Another factor conferring tolerance may be the surface area of individual cells. Observations on rats suggest that tolerance may be related to the reduced surface area of cuboidal Type 1 cells after reparative regeneration [16]. This suggests that the surface area requiring protection by antioxidant systems may be significant in the development of tolerance. However, to date, the mechanism by which cells become tolerant to NO_2 has not been explained completely.

THE EFFECTS OF CHRONIC EXPOSURE

Early pathological events during and after limited initial exposure of rats to NO_2 are reparable, as described, but persistent exposure to subacute concentrations of at least 14 ppm for several months leads to permanent distortion of the architecture of the rats' lungs (Figures 20 and 21) and an emphysema-like disease (Figure 22) with loss of alveolar tissue [2,33]. Lower concentrations affect the lungs also, but 1 or 2 ppm—even for the lifetime of the rat—fail to induce parenchymal destruction [34,35], although respiratory rates and lung volumes become elevated and metabolic changes are seen in the bronchiolar epithelium [36].

The characteristic paucity of cilia in the airways in chronically exposed animals and the relatively uniform hypertrophy of the epithelium of the small airways, with their occasional foci of hyperplasia, indicate a much reduced, but continuing, response of the lining cells to the persistent presence of NO_2. From the onset of exposure, the most severely affected tissues of the airways and parenchyma—mainly the terminal bronchioles, alveolar ducts and adjacent alveoli—are infiltrated with inflammatory cells. These are largely macrophages, both within the alveoli and the small airways (where they often fill the former close to the small airways and ducts) and in the interstitial tissue. Other types of mononuclear cells and occasionally granulocytic cells are included. Mast cells are often seen dispersed throughout the lung.

The ubiquitous network of connective tissue elements that permeates the alveolar walls and small airways reveals changes in staining properties of both the elastic tissue and the collagen strands in long-exposed rats [37]. Besides appearing to become thicker with time, the elastic strands tend to fracture, and some of the collagen fibers appear as thick, ropelike structures [38]. Simultaneously, the basement membranes supporting both the alveolar epithelium and the capillary endothelium become broader in proximal alveoli and contrast with the unusually thin, attenuated alveolar walls of more peripheral alveoli. In general, pulmonary blood volume per gram of tissue is reduced in the heavy, voluminous, nonedematous lungs that develop during exposure to high subacute concentrations of about 15 ppm NO_2 [35].

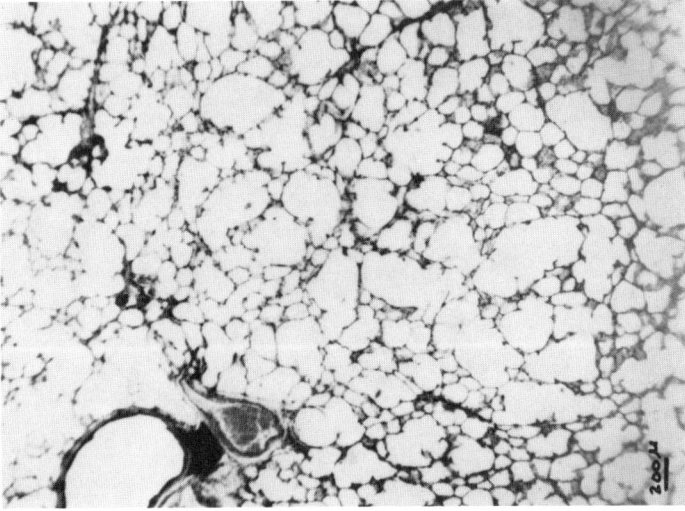

Figure 21. Parenchyma of a rat exposed at prolonged, irregular intervals to ~15 ppm NO_2 for 24 months, exhibiting a loss of alveolar surface area (24×) [33].

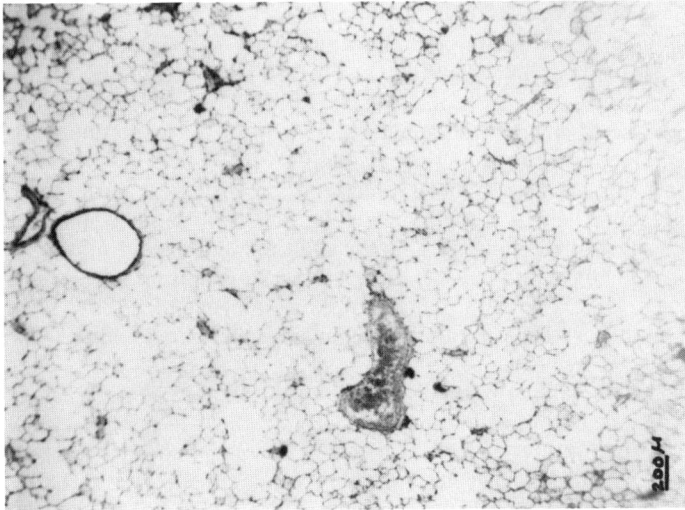

Figure 20. Parenchyma of a 24-month old control rat (24×) [33].

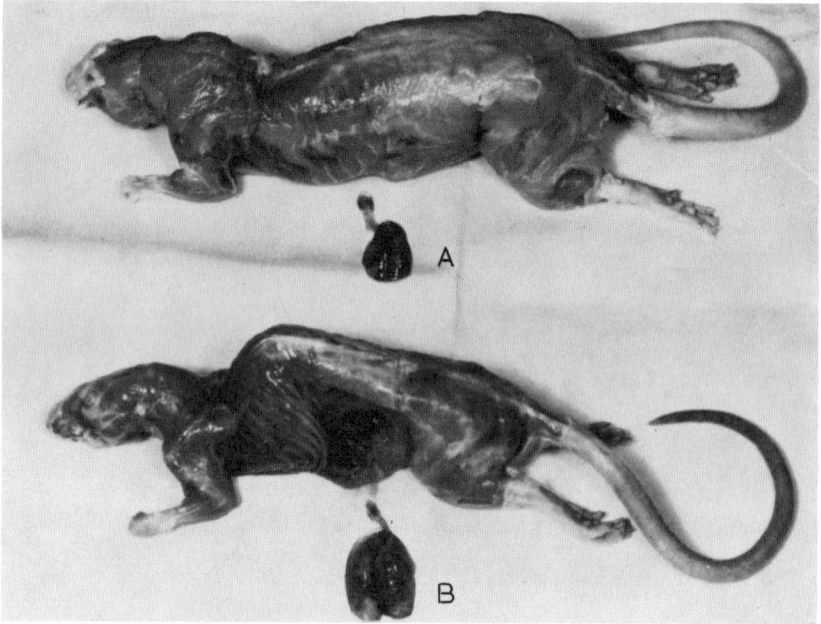

Figure 22. (a) Normal and (b) emphysematous lungs of a rat exposed continuously for 6 months to ~15 ppm NO_2 [35].

The internal diameters of terminal bronchioles narrow during prolonged exposure as their walls (composed mainly of smooth muscle, connective tissue and epithelial lining) become thicker [39]. The diameters are reduced further because of an accumulation of mucous and serous secretions from the increased number of peripherally activated "goblet" and other secreting cells of the bronchial and bronchiolar mucosa; thus, airflow is correspondingly reduced. Cellular debris, fibrinous exudate, and the attendant macrophages and other cells of the immune system are additional obstacles to flow.

More frequent breathing, the relative prolongation of the expiratory phase, the enlargement of total and residual lung volumes in the hyper-extended thorax, and the narrow bronchiolar diameters are consistent with a potential rise in resistance to expiratory airflow. Determination of the total number of alveoli and/or parenchymal air spaces shows a considerable reduction in the large, exposed lungs compared with those in control rats of the same age. This indicates a significant loss of alveolar walls and surface area, in addition to the alveolar distention in rats exposed intermittently for approximately 2 years to about 15 ppm NO_2 [37]. Such animals remain relatively inactive, presumably for want of oxygen. Their arterial pO_2 tends to become reduced and their pCO_2 elevated, reflecting a metabolic acidosis with a blood pH of about 7.3. Both the variability in bronchiolar caliber and irregular remodeling of the parenchyma lead to inhomogeneity of airflow during respiration, as suggested by changes in frequency dependence of compliance [40].

Thus, continuous exposure to 1-2 ppm NO_2 over the lifetime of the rat results in modest but real changes in the respiratory epithelium, breathing rate and erythropoiesis. However, rats exposed on a similar regimen to higher concentrations, equivalent to NO_2 levels likely to be found in fresh tobacco smoke, develop an overtly emphysema-like disease in about 5-6 months or longer, depending on the exposure schedule.

This is strong evidence that airborne NO_2 is probably a factor in the induction of chronic pulmonary disease in cigarette smokers and workers exposed to similar concentrations of NO_2. Populations exposed to much lower ambient levels may be subject to increased vulnerability to ambient pathogenic particles.

ACKNOWLEDGMENTS

We are grateful to our colleagues listed among the references for their basic contributions and to many other collaborators for excellent technical assistance.

This work was supported by National Institutes of Health, Grant Nos. HL16330 and ES00842, and by the Environmental Protection Agency, Contract No. 68-02-1944.

REFERENCES

1. World Health Organization. "Environmental Health Criteria 4: Oxides of Nitrogen," United Nations Environment Programme and the World Health Organization, Geneva, Switzerland (1977).
2. Freeman, G., and G. B. Haydon. "Emphysema After Low-Level Exposure to NO_2," *Arch. Environ. Health* 8:125-128 (1964).
3. Jeffery, P. K., and L. Reid. "New Observations of Rat Epithelium: A Quantitative and Electron Microscope Study," *J. Anat.* 120:295-320 (1975).
4. Breeze, R. G., and E. B. Wheeldon. "The Cells of the Pulmonary Airways," *Am. Rev. Resp. Dis.* 116:705-777 (1977).
5. Weibel, E. R., P. Gehr, D. Haies, J. Gil and M. Bachofen. In: *Lung Cells in Disease*, A. Bouhuys, Ed. (Amsterdam, Holland: Elsevier/North-Holland Biomedical Press, 1976), p. 1.
6. Stephens, R. J., G. Freeman and M. J. Evans. "Early Response of Lungs to Low Levels of Nitrogen Dioxide," *Arch. Environ. Health* 24:160-179 (1972).
7. Cabral-Anderson, L. J., M. J. Evans and G. Freeman. "Effects of NO_2 on the Lungs of Aging Rats. I. Morphology," *Exp. Mol. Pathol.* 27:353-365 (1977).

8. Evans, M. J., L. J. Cabral-Anderson and G. Freeman. "Role of the Clara Cell in Renewal of the Bronchiolar Epithelium," *Lab. Invest.* 38:648–655 (1978).

9. "Renewal of Terminal Bronchiolar Epithelium Following Exposure to O_3 and NO_2," *Lab. Invest.* 35:246–257 (1976).

10. Evans, M. J., L. J. Cabral, R. J. Stephens and G. Freeman. "Transformation of Alveolar Type 2 Cells to Type 1 Cells Following Exposure to NO_2," *Exp. Mol. Pathol.* 22:142–150 (1975).

11. Evans, M. J., R. J. Stephens, L. J. Cabral and G. Freeman. "Cell Renewal in the Lungs of Rats Exposed to Low Levels of NO_2," *Arch. Environ. Health* 24:180–188 (1972).

12. Creasia, D. A., P. Nettesheim and J. C. S. Kim. "Stimulation of DNA Synthesis in the Lungs of Hamsters Exposed Intermittently to Nitrogen Dioxide," *J. Toxicol. Environ. Health* 2:1173–1181 (1977).

13. Evans, M. J., L. J. Cabral, R. J. Stephens and G. Freeman. "Renewal of Alveolar Epithelium in the Rat Following Exposure to NO_2," *Am. J. Pathol.* 70:171–194 (1973).

14. Evans, M. J. In: *Mechanisms in Respiratory Toxicology*, H. Witschi and P. Nettesheim, Eds. (Cleveland: CRC Press, Inc., in press).

15. Yuen, T. G. A., and R. P. Sherwin. "Hyperplasia of Type II Pneumocytes and Nitrogen Dioxide (10 ppm): A Quantitation Based on Electron Photomicrographs," *Arch. Environ. Health* 24:178–188 (1971).

16. Evans, M. J., L. J. Cabral-Anderson, N. P. Dekker and G. Freeman. "Role of the Epithelium in the Development of Tolerance." in preparation.

17. Stephens, R. J., G. Freeman, S. C. Crane and N. J. Furiosi. "Ultrastructural Changes in the Terminal Bronchioles of the Rat During Continuous, Low-Level Exposure to NO_2 (Ciliogenesis)," *Exp. Mol. Pathol.* 14:1–19 (1971).

18. Evans, M. J., N. Dekker, L. J. Cabral-Anderson and G. Freeman. "Quantitation of Damage to the Alveolar Epithelium by Means of Type 2 Cell Proliferation," *Am. Rev. Resp. Dis.* 118:787–790 (1978).

19. Evans, M. J., L. J. Cabral-Anderson and G. Freeman. "Cell Proliferation in the Lungs of Aging Rats Exposed to NO_2," *Proc. 69th Annual Meeting of Air Poll. Control Assoc.*, No. 76-32, Portland, OR (1976).

20. "Effects of NO_2 on the Lungs of Aging Rats. II. Cell Proliferation," *Exp. Mol. Pathol.* 27:366–376 (1977).

21. Bullough, W. S. "Mitotic and Functional Homeostasis: A Speculative Review," *Cancer Res.* 25:1683–1727 (1965).

22. Oehlert, W. "Cell Proliferation in Carcinogenesis," *Cell Tissue Kinet.* 6:325–335 (1973).

23. Crapo, J. D., J. Marsh-Salin, P. Ingram and P. C. Pratt. "Tolerance and Cross-Tolerance Using NO_2 and O_2. II. Pulmonary Morphology and Morphometry," *J. Appl. Physiol.* 44:370–379 (1978).

24. Kleinerman, J. "Inhalation Carcinogenesis," M. G. Hanna, P. Nettesheim and J. R. Gilbert, Eds. (Oak Ridge, TN: U.S. Atomic Energy Commission Division of Technical Information, 1970), p. 271.

25. Furiosi, N., and G. Freeman. Unpublished results.

26. Stephens, R. J., M. F. Sloan, D. G. Growth, D. S. Negi and K. D. Lunan. "Cytologic Response of Postnatal Rat Lungs to O_3 or NO_2 Exposure," *Am. J. Pathol.* 93:183–200 (1978).

27. Burri, P. H. "The Postnatal Growth of the Rat Lung. III. Morphology," *Anat. Rec.* 180:77–98 (1974).
28. Kauffman, S. L., P. H. Burri and E. R. Weibel. "The Postnatal Growth of the Rat Lung. II. Autoradiography," *Anat. Rec.* 180:63–76 (1974).
29. Combs, G. F., T. Noguchi and M. L. Scott. "Mechanisms of Action of Selenium and Vitamin E in Protection of Biological Membranes," *Fed. Proc.* 34:2090–2095 (1975).
30. Tappel, A. L. "Lipid Peroxidation Damage to Cell Components," *Fed. Proc.* 32:1870–1874 (1973).
31. Fletcher, B. L., and A. L. Tappel. "Protective Effects of Dietary α-Tocopherol in Rats Exposed to Toxic Levels of Ozone and Nitrogen Dioxide," *Environ. Res.* 6:165–175 (1973).
32. Crapo, J. D., K. Sjostrom and R. T. Drew. "Tolerance and Cross-Tolerance Using NO_2 and O_2. I. Toxicology and Biochemistry," *J. Appl. Physiol.* 44:364–369 (1978).
33. Freeman, G., S. C. Crane, N. J. Furiosi, R. J. Stephens, M. J. Evans and W. D. Moore. "Covert Reduction in Ventilatory Surface in Rats During Prolonged Exposure to Subacute Nitrogen Dioxide," *Am. Rev. Resp. Dis.* 106:563–579 (1972).
34. Freeman, G., N. J. Furiosi and G. B. Haydon. "Effects of Continuous Exposure of 0.8 ppm NO_2 on Respiration of Rats," *Arch. Environ. Health* 13:454–456 (1966).
35. Freeman, G., S. C. Crane, R. J. Stephens and N. J. Furiosi. "Pathogenesis of the Nitrogen Dioxide-Induced Lesion in the Rat Lung: A Review and Presentation of New Observations," *Am. Rev. Resp. Dis.* 98:429–443 (1968).
36. Freeman, G., R. Stephens, S. Crane and N. Furiosi. "Lesion of the Lung in Rats Continuously Exposed to Two Parts per Million of Nitrogen Dioxide," *Arch. Environ. Health* 17:181–192 (1968).
37. Freeman, G., S. C. Crane, R. J. Stephens and N. J. Furiosi. "The Subacute Nitrogen Dioxide Induced Lesion in the Rat Lung," *Arch. Environ. Health* 18:609–612 (1969).
38. Stephens, R. J., G. Freeman and M. J. Evans. "Ultrastructural Changes in Connective Tissue in Lungs of Rats Exposed to NO_2," *Arch Intern. Med.* 127:873–883 (1971).
39. Freeman, G. Unpublished results.
40. Freeman, G., L. Juhos, N. J. Furiosi, W. Powell and R. Mussenden. "Criteria from Animals Exposed to Known Concentrations of Nitrogen Dioxide and Ozone, With Potential Use in Epidemiology," *Proc. Recent Adv. in the Assessment of Health Effects of Environ. Poll.* Vol. II, EUR 5360, World Health Organization, Commission of the European Communities, and U.S. Environmental Protection Agency (1974), pp. 833–844.

CHAPTER 17

INFLUENCE OF EXPOSURE PATTERNS OF NITROGEN DIOXIDE ON SUSCEPTIBILITY TO INFECTIOUS RESPIRATORY DISEASE

Donald E. Gardner

Inhalation Toxicology Branch
Health Effects Research Laboratory
Environmental Protection Agency
Research Triangle Park, North Carolina 27711

INTRODUCTION

Nitrogen dioxide (NO_2) concentrations in the atmosphere are subject to wide variations. The concentration of NO_2 varies with the rate of combustion, the presence of other atmospheric pollutants and various meteorological conditions, such as sunlight, wind and temperature. Closer examination of the available atmospheric data derived from monitoring the concentration of NO_2 over short time periods enables certain general conclusions to be drawn concerning the adequacy and appropriateness of using a simple annual arithmetic average for protecting against short-term health effects.

Nationwide data for 1975 on the ratio of peak hourly NO_2 concentrations to yearly arithmetic means indicate that this ratio is quite variable, ranging from 3.6 to 13.5 (Table I) [1]. The monitoring data reveal that exposure to NO_2 concentrations of 1 mg/m^3 (0.53 ppm) or greater for 1 hr was seen in areas in California. Peak 1-hr exposure equalling or exceeding 0.75 mg/m^3 (0.4 ppm) were also reported in other sites in California, Kentucky, Illinois and Ohio.

The 1-hr average NO_2 concentration pattern vs time, for periods of 3 days during which high NO_2 levels were observed, is shown in Figure 1 [2]. The Los Angeles data showed a typical diurnal profile during which the NO_2 concentration climbed steadily after a small morning peak. The sharp morning peak exemplifies the combined effects of poor atmospheric dispersion, high emissions and photochemical activity that are common for this region.

Table I. Ratio of Maximum Observed Hourly Nitrogen Dioxide Concentrations to Annual Means During 1975 for Selected Locations

| State | Location | Maximum Hourly Concentration/ Yearly Arithmetic Mean | |
		Method A[a]	Method B[b]
California	Anaheim	940/101 = 9.3	–
	Costa Mesa	658/58 = 11.3	–
	Los Angeles	1053/126 = 8.4	–
	Barstow	432/39 = 11.1	–
	Chula Vista	–	451/64 = 7.1
Illinois	Chicago	–	395/104 = 3.8
		–	395/109 = 3.6
		–	244/41 = 6.0
Kentucky	Paducah	714/66 = 10.8	–
	Ashland	895/85 = 10.5	–
Texas	Dallas	–	432/32 = 13.5

[a]Method A: Instrumental Colorimetric-Lyshkow (MOD) method, a variation of the continuous Greiss–Saltzman Method.
[b]Method B: Instrumental Chemiluminescence Method.

The data for Ashland, Kentucky exhibited a similar diurnal trend, except they were much lower on the third day. In contrast, a quite different pattern was observed in McLean, Virginia, since a major increase in NO_2 concentrations did not take place until 5:00 or 6:00 PM.

Repeated exposures to hourly peaks of this magnitude or greater can occur repeatedly over a period of several days, especially in: (1) large metropolitan areas, presumably as the combined results of mobile source and other emissions and unfavorable meteorological conditions; (2) suburban areas near large metropolitan centers, presumably as a result of local mobile source emissions and/or pollutant transport; and (3) small cities, presumably primarily as a result of local industrial emissions. Reducing monitoring data to a simple annual average concentration minimizes the sporadic pollutant peaks. Thus, this average is not very useful as an index of the potential risk for peak, short-term human exposures. Air quality standards that do not account for variation in the frequency and amplitude of such pollutant peak levels may have serious implications on the health of the population at risk.

Due to ambiguities in the literature regarding the dose–time response relationships for NO_2, it became obvious that more definitive work remains to be done to determine the relative importance of various modes of exposure and the possible influence of "thresholds," tolerance, recovery and adaptation in the host's biological response to NO_2. The Health Effects Research Laboratory of the U.S. Environmental Protection Agency has been conducting experiments specifically designed to examine and compare the influence

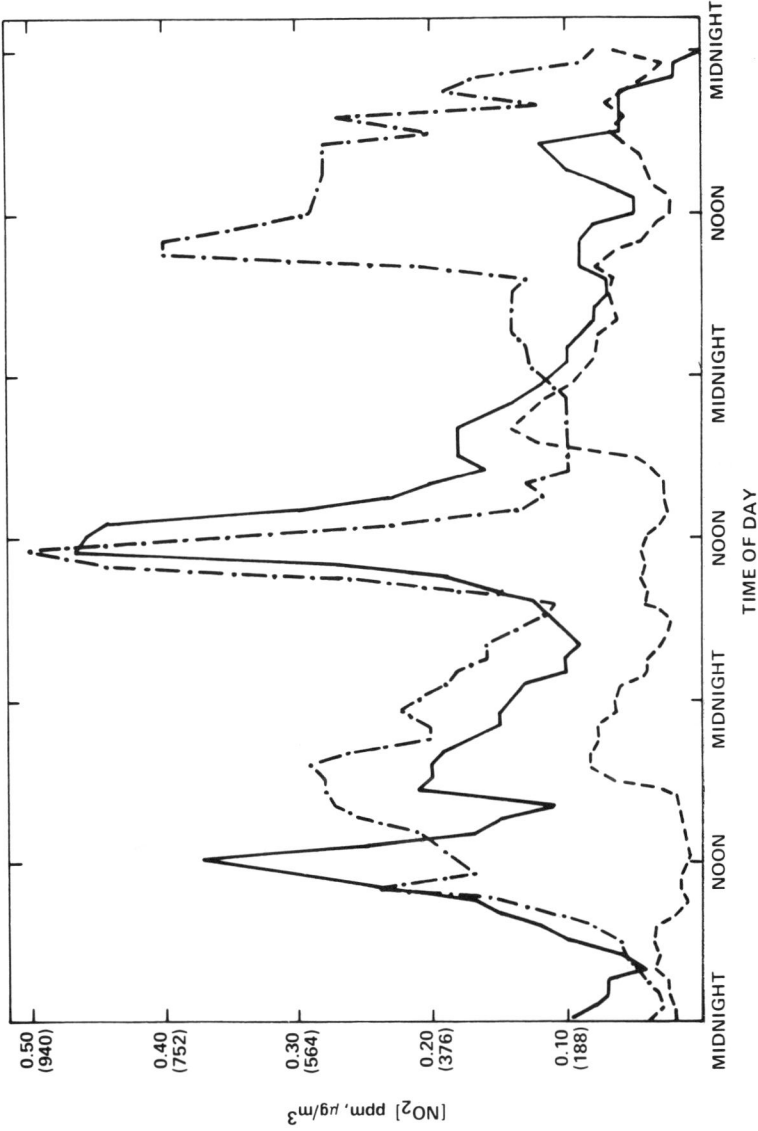

Figure 1. One-hour NO₂ concentrations in three cities: (–) Ashland, Kentucky; (-·-) Los Angeles, California; and (——) Mclean, Virginia during 1975.

of various pollutant exposure patterns on NO_2 toxicity. These studies employ a simple biological model system that measures the host resistance to respiratory infections. This infectivity model has proven to be a particularly sensitive indicator of pulmonary injury. The procedure integrates many of the defense mechanisms of the lung, thereby reflecting the overall damage that has occurred during the pollutant exposure.

The following research was designed to examine the:

1. dose response at different time intervals T (hr) and concentrations C (ppm); in these studies both concentration and time were varied while maintaining a constant C × T product;
2. dose–time relationships for various concentrations of NO_2; in these studies the concentration of NO_2 was kept constant and the length of exposure was increased;
3. comparison of intermittent and continuous exposure regimens;
4. persistence of effect for various lengths of a spiked exposure without a baseline level of NO_2;
5. persistence of effect for various lengths of a spiked exposure with a background concentration;
6. effects of multiple spikes on a continuous background concentration of NO_2; and
7. effects on the dose-response relationship when NO_2 was combined with a photochemical oxidant, ozone.

BIOLOGICAL MODEL SYSTEM

The lung has a remarkable ability to cope with inhaled microorganisms and, under normal conditions, is highly efficient in maintaining the sterility of the lower respiratory tract, even though it is frequently assaulted by potentially pathogenic organisms from the external environment [3–5]. However, there are numerous environmental chemicals, i.e., O_3, NiO, $CdCl_2$, $NiCl_2$ and MnO_2, that can alter or suppress these pulmonary defense systems, thus increasing the risk of disease [6–9]. Due to the effects of pollutants on these host defense systems, microorganisms that would normally be kept under control through these bactericidal and clearance mechanisms can multiply and cause pulmonary disease.

In the study to be described here, the animal model system tests the effect of NO_2 as an environmental modifier in the course of the experimental infection. This model system reflects the summation of various deleterious changes within the lung caused by the action of NO_2 and/or its by-products. Such subtle changes include edema, reduced phagocytic or bactericidal activity of the alveolar macrophages, altered mucociliary function, inflammation and immunosuppression. All these effects have been reported to be caused by inhalation of NO_2 exposure [10–14].

The animal model system used in these studies has been explained in detail elsewhere [15] and is shown scyematically in Figure 2. Briefly, female pathogen-free Swiss Albino Mice, CD-1 (Charles River Laboratory), weighing 20–25 g, are randomly selected to be exposed to either clean air or to various

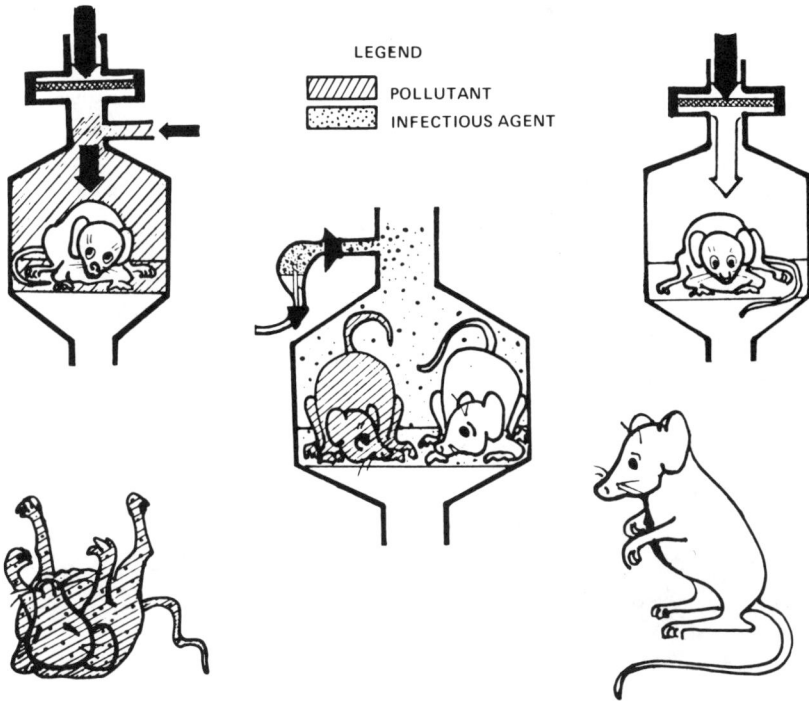

Figure 2. Schematic of mouse infectivity model showing exposure of animal to pollutant (NO$_2$) or clean air followed by challenge with strepococci pyogenes. Data is presented as difference in mortality of treated minus control animals.

exposure regimens of NO$_2$. Each animal is exposed in an individual compartment and is provided food and water ad libitum. The exposure chambers used in these studies are shown in Figure 3.

After NO$_2$ exposure, both the control and pollutant-exposed groups were combined and placed in an infectivity chamber and exposed for 15 min to an aerosol of variable microorganisms (*Streptococcus pyogenes*, group C). Other microorganisms employed included *Klebsiella pneumoniae, Diplococcus pneumoniae* and influenza virus. A flow diagram of the infectivity chamber is given in Figure 4. The number of viable microorganisms deposited in the lung immediately after the bacterial aerosol ranged from 200–4000 colony-forming units per lung. This range of variation in deposited microbes does not have any influence on the resulting mortality [16]. After receiving the bacterial aerosols, the test animals were separated into their appropriate treatment groups and were maintained in clear air for 15 days, during which time total cumulative mortality rates were determined. The data are reported

Figure 3. Animal exposure facilities for NO₂ toxicity studies.

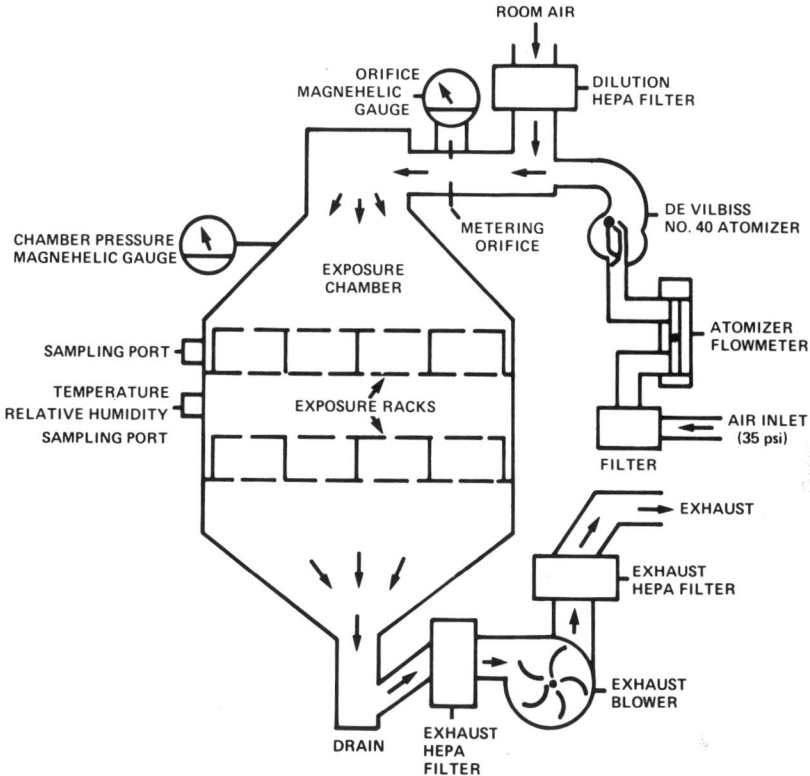

Figure 4. Schematic drawing of the exposure chamber and associated instruments for laboratory-induced pulmonary infection in mice.

as differences in percent mortality between the NO_2 test group and the control response. The mortality in the control groups is usually 15–20%, reflecting the virulence of the microorganisms and the susceptibility of the host.

NO_2 MEASUREMENTS/QUALITY ASSURANCE

The concentration of NO_2 within the chamber was continuously monitored by the *Federal Register* method of chemiluminescence [17]. In addition, to ensure proper quality assurance, all instruments used to monitor exposure parameters were calibrated on a proscribed schedule. Multipoint calibrations were performed every two weeks on the continuous chemiluminescent NO_x monitors. National Bureau of Standards certified 0.01% NO and a dynamic gas-phase calibration unit were used to calibrate the analyzers according to *Federal Register* methodology. External audits were conducted every 3–4 months. Data were recorded on strip-chart recorders that were calibrated simultaneously with their respective monitors. Instrument maintenance was performed as suggested in appropriate instrument manuals. Exposure chamber air flow was calibrated periodically to allow theoretical NO_2 concentrations to be predicted. Airflow measuring equipment was subjected to volumetric calibrations.

Objective One

One purpose of these experiments was to study the interaction between concentration and time. Toxicity is an expression of the capacity of a chemical to produce injury in a living host. Injury is the response to a dose of a substance, while dose is the total amount of a substance administered over a given period of time. A basic concept often employed by the toxicologist is to equate differences in concentration and time by assuming that the product of concentration and time are directly proportional to the toxic effect and to express this mathematically as $C \times T = K$, where K is a constant or endpoint of the response. The CT concept (Haber's Rule) is often used in forming comparisons of toxicities of different substances and in estimating either the concentration or the time of exposure when either of these conditions changes. However, this rule is not without exception, and such an assumption could have a direct impact on the assessment of the toxicity of short-term exposures to peak concentrations. Therefore, it is important to determine if one can simply use the product of $C \times T$ of the NO_2 exposure and expect a fixed response. If no interaction occurred between concentration and time, no statistical difference in response would be noted when either factor is varied, proving the product remained a constant value.

Therefore, this concept was tested using the infectivity model. The relationship between concentration and time produced significantly different mortality responses, although the $C \times T$ product remained constant (Figure 5). From these data, it is evident that the actual concentration has more influence than the duration of the exposure. If no interaction occurred between concentration and time, i.e., if NO_2 followed Haber's Rule, all the test groups should yield the same mortality, which is indicated by the parallel line at approximately 15%. Instead, the data indicate a gradient response from a high of approximately 50% to a low mortality not different from controls.

Additional evidence indicates that susceptibility to infection is influenced more by the concentration of NO_2 than by the duration of the exposure (Table II). In each case, the exposure to higher concentrations of NO_2 for brief periods resulted in a greater percentage mortality than did exposure to lower concentrations of NO_2, even though the $C \times T$ product remained constant.

Objective Two

A second objective was to study dose–time relationships for various concentrations of NO_2. The interaction of a constant concentration with time

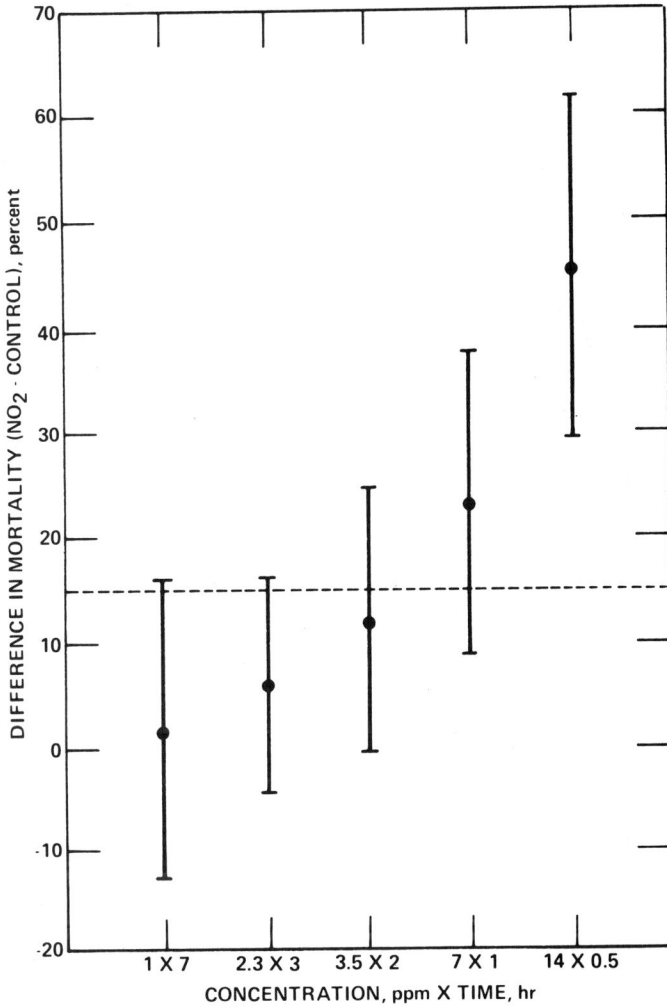

Figure 5. Enhancement of mortality from laboratory-induced infection resulting from the same concentration × time exposure of NO_2 over different exposure periods. Average percent mortality differences are shown with the 95% confidence interval. The horizontal line indicates the predicted effect if there were a direct relationship between concentration and time.

over a longer period of continuous exposure to NO_2 was also examined. In these studies [18], six different concentrations were employed ranging from 0.95–526 mg/m^3 (0.5–28 ppm) of NO_2 (Figure 6). The length of time varied from a few minutes to several months. It should be noted that the data for the 0.95 mg/m^3 (0.5 ppm) were taken from the study of Ehrlich and

Table II. The Influence of Concentration and Time on Enhancement of Mortality
Resulting from Various NO_2 Concentrations

Concentration (ppm)	Concentration X Time					
	7		14		21	
	Time (hr)	% Mortality	Time (hr)	% Mortality	Time (hr)	% Mortality
1.5	4.7	6.4	9.3	10.2	14.0	12.5
3.5	2.0	18.7	4.0	27.0	6.0	31.9
7.0	1.0	30.2	2.0	41.8	3.0	48.6
14.0	0.5	21.7	1.0	44.9	1.5	58.5
28.0	0.25	55.5	0.5	67.2	0.75	74.0

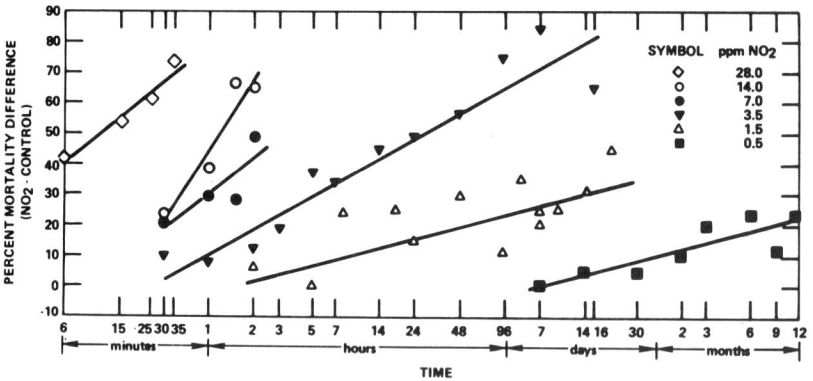

Figure 6. Percent mortality of mice versus length of continuous exposure to various concentrations of NO_2 prior to challenge with streptococci.

Henry [19]. In these studies, the microorganism used was *Klebsiella pneumoniae* instead of *Streptococcus pyogenes*, which was used in the other treatment groups. A positive linear dose response (p < 0.05) was observed for each concentration, and with increasing concentration the slope of the regression line becomes steeper.

Utilizing these curves, a concentration–time curve was prepared to predict the length of the exposure needed to produce a given enhancement in mortality with this model system. Figure 7 illustrates this point by plotting the

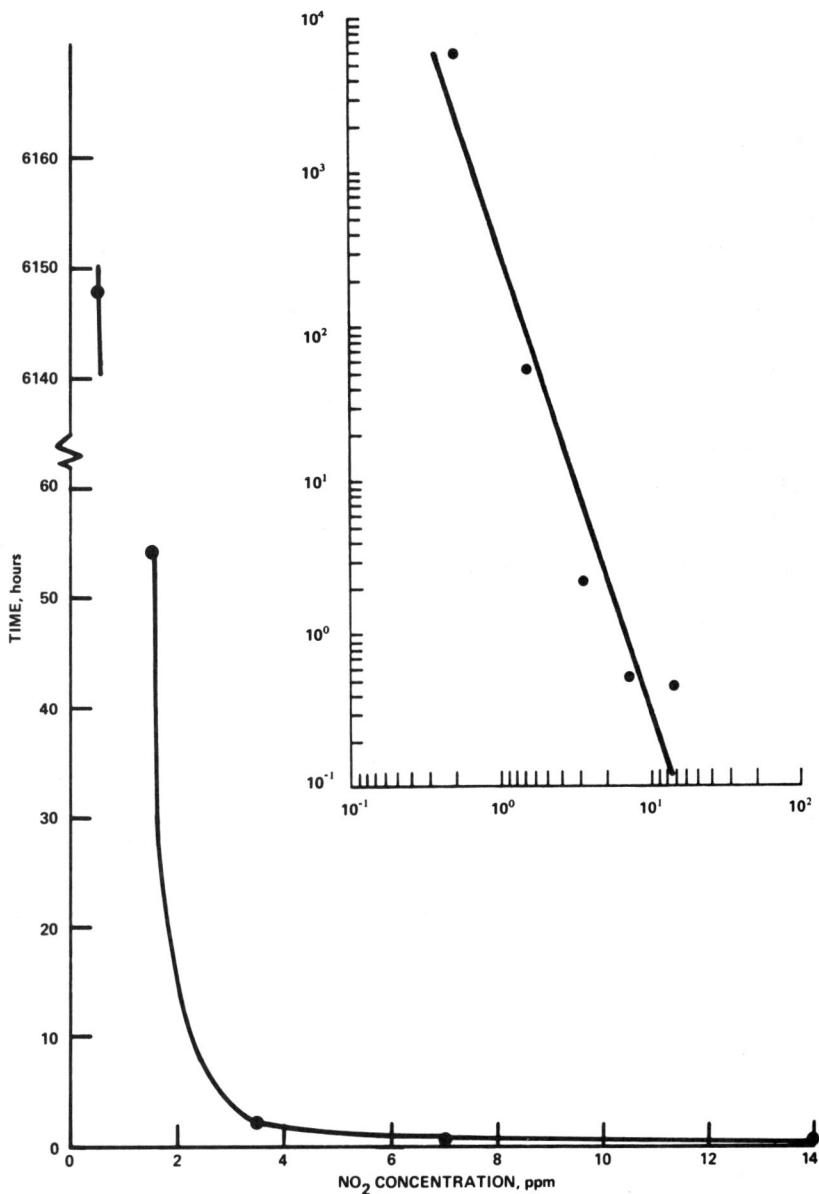

Figure 7. Time required to elicit a 20% mortality in mice vs concentration of NO_2.

time necessary to elicit at 20% mortality (difference from control) vs concentration of NO_2. This figure shows that the concentration–time relationship of mortality resulting from exposure to NO_2 can be approximated by a straight line on a log–log scale.

Objective Three

A third objective was to compare intermittent and continuous exposure modes. Because of the wide fluctuations in concentration and duration of community air pollutants, it is important for environmental scientists to study and compare the effects on health of intermittent, as well as continuous, exposures. Such information improves the understanding of the processes governing the interaction between chemical compounds and the host, such as adaptation, tolerance and recovery.

To date, two concentrations have been studied in our laboratory [20]. Figure 8 illustrates the results of exposure to 6.6 mg/m^3 (3.5 ppm) from continuous (24 hr/day, 7 days/wk) and intermittent (7 hr/day, 7 days/wk) exposure for periods up to 15 days. In each group there was a noticeable increase in response with time; however, the effect observed for the two exposure modes was not significantly different at any time period tested. This indicates that the damage to the host from the 7-hr repeated exposure was sufficient to prevent complete recovery during the 17-hr intervals between exposures when the animals breathed clean air. Table III compares the effects of concentration $\times$ time and the continuous vs intermittent exposure when the concentration of NO_2 was 6.6 mg/m^3 (3.5 ppm). When these data were adjusted for actual exposure time and concentration (C $\times$ T), the percent mortality to exposure modes was essentially the same.

At 2.8 mg/m^3 (1.5 ppm), similar effects were seen at this lower concentration of NO_2, but only after a 14-day exposure (Figure 9). Prior to this, the continuous exposure was always significantly enhanced over the 7 hr/day intermittent exposure.

Why the first four exposure periods were not statistically different from the controls can be elucidated by reexamining the family of curves (Figure 6) and applying these data to the problem. A significant response to 2.8 mg/m^3 (1.5 ppm) would not be expected until the target organ had been exposed to the gas for approximately 30 hr (C $\times$ T = 45). Not until after this critical length of exposure would one expect to see any enhancement of mortality over controls. In this intermittent study, the first four periods of exposure had an accumulative C $\times$ T of approximately 42.

Thus, in this intermittent exposure, where the animals are exposed 7 hr/day, the total gas delivered to the target site during the first few days would be below this threshold compared with the continuous exposure groups, where this threshold for effect would have been exceeded within the first 24 hr of exposure. However, the seventh intermittent exposure is significant, and the level of effect begins to approach that observed with continuous exposure. With subsequent exposures, the response was similar regardless of the mode of exposure.

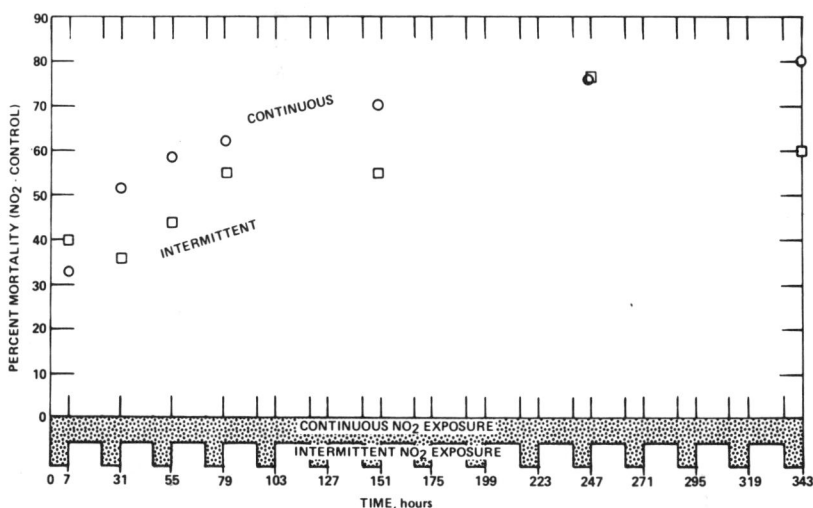

Figure 8. Percent mortality of mice vs the length of either continuous or intermittent exposure to 3.5 ppm NO_2 prior to challenge with streptococci.

Table III. A Comparison of the Effects of C × T and Mode of Exposure to 3.5 ppm NO_2 on Percent Mortality

Concentration X Time	Exposure Mode	
	Intermittent (% mortality)	Continuous (% mortality)
49.0	37	42
73.5	43	47
98.0	55	50
171.5	55	57
269.5	75	62
367.5	60	66

Objective Four

To achieve persistence of the effect of various lengths of a spiked exposure with no NO_2 background was the fourth objective.

To determine whether such spiked levels have any serious implications to the health of the population at risk, a study was designed to determine the

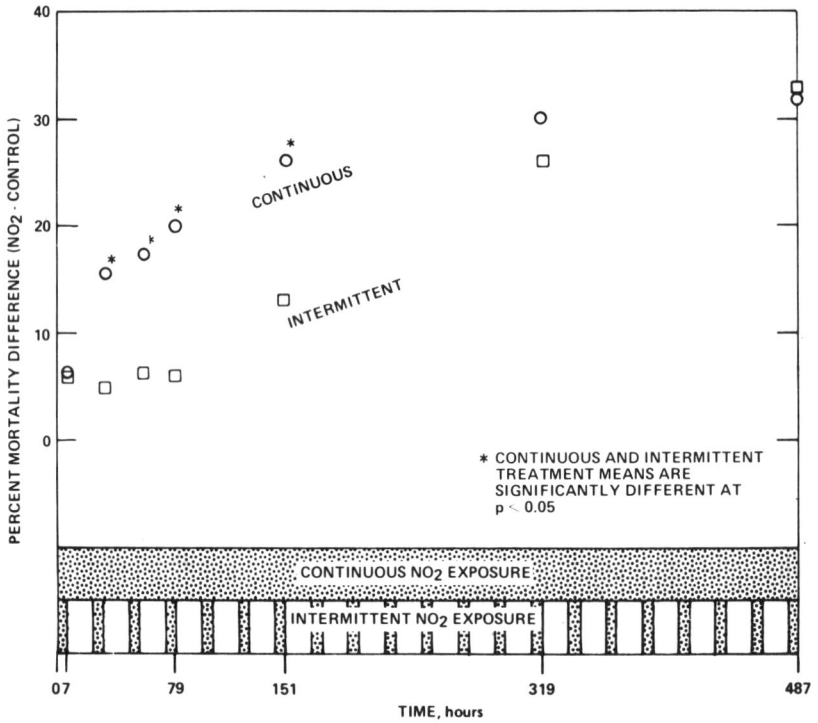

Figure 9. Percent mortality of mice vs length of either continuous or intermittent exposure to 1.5 ppm NO₂ prior to challenge with streptococci.

importance of variation in length and persistence of a peak concentration. Measurements of the persistence of the NO_2 effect were determined by maintaining the test animals (after cessation of the gas exposure) in clean air for an 18-hr postexposure recovery period prior to being exposed to the viable microorganisms.

The concentration of NO_2 used in this study was 8.1 mg/m³, (4.5 ppm), with the length of the exposure varying from 1.0 to 7.0 hr (Figure 10). In agreement with the previous studies, there was an increase in percent mortality for each experimental group with increasing length of exposure when the animals were challenged by an aerosol of microorganisms immediately after the NO_2 inhalation. However, 18 hr after the NO_2 exposure, the animals did not show any enhanced response to the microorganisms over control animals not exposed to NO_2, indicating that the animals had completely recovered from the NO_2 assault, either by direct repair of specific damages or by additional host defenses that may have been made available to protect the lung against the invading microorganisms. In support of this latter theory,

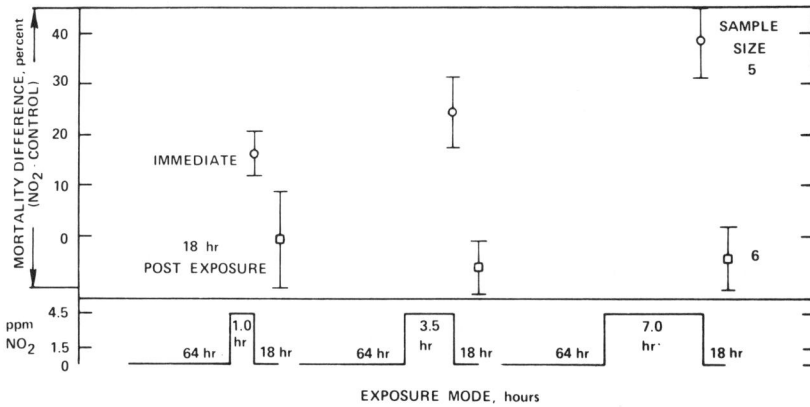

Figure 10. Persistence of NO₂ effect, as measured by the infectivity model system, for various lengths of a spiked exposure (4.5 ppm) with no background concentration of NO₂.

investigators have shown that an influx of alveolar macrophages and polymorphonuclear leukocytes can occur after cessation of the NO₂ exposure [13,21].

Objective Five

To achieve persistence of effect for various lengths of a spiked exposure superimposed on a background concentration of NO₂ was the fifth objective.

Since concentration and duration of exposure are subject to many chemical and physical interactions in the urban atmosphere, the NO₂ concentration profile usually observed contains the irregular occurrence of exposure spikes superimposed on a lower basal background level. The effects of this background concentration on the peak exposure are shown in Figure 11. Although the concentrations used in these studies are above those found in the urban atmosphere, the 3:1 ratio of peak to basal concentration is very realistic, based on available air quality data [1].

The concentration of the peaks and the length of the peak exposure were the same as in the previous study, but in addition, the test animals were exposed to 2.8 mg/m³ (1.5 ppm) of NO₂ for 64 hr prior to the peak exposure and again for 18 hr after the peak. As before, the microorganism aerosol was given either immediately after the peak exposure or following the 18-hr postpeak-exposure. These experiments indicate that if the bacterial challenge was given immediately after the peak exposures, the 64-hr preexposure to 2.8 mg/m³ (1.5 ppm) added no significant enhancement to the effect observed from the peak exposure only, regardless of the length of the peak exposure.

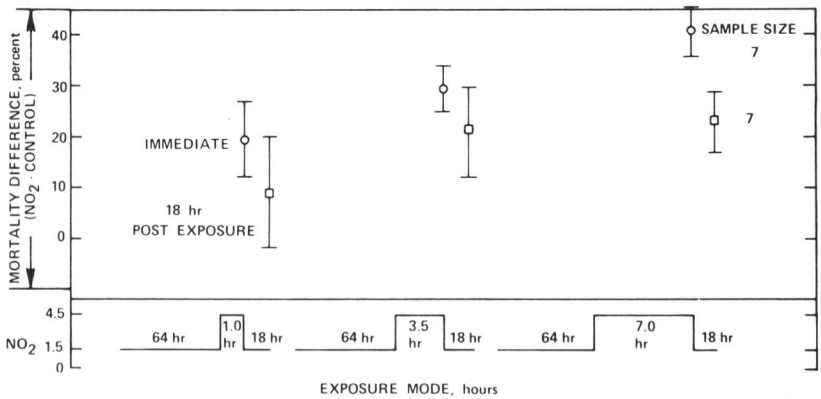

Figure 11. Persistence of NO_2 effect, as measured by the infectivity model system, for various lengths of a spiked exposure (4.5 ppm) with a background exposure to 1.5 ppm before and after the spiked exposure.

Of significant interest was the comparison of the response observed in the previous study, where the animals were allowed to recover in clean air for 18 hr after the peak expsoure, with the response observed when animals were maintained for 18 hr at a reduced NO_2 level. In the latter group, the additional 18-hr baseline exposure influenced the response of the test animals, as indicated by an increased mortality, which was significant for those animals receiving peaks for 3.5 and 7.0 hr. Based on the hypothesis stated in the previous study, these results allow one to further expand these theories and include the results of this study. In this case, the continual exposure to the lower NO_2 level after the peak exposure significantly increased the mortality rate. This may be attributed to either: (1) under continual exposure to the gas the animals could no longer recover from or adapt to the assault; or (2) the influx of new respiratory defense mechanisms (i.e., alveolar macrophages, polymorphonuclear leukocytes) cannot defend the lung because they are immediately affected by the inhaled NO_2. Further research is necessary to determine whether these or other factors are responsible for the effects observed in these two studies.

Objective Six

Effects of multiple spikes on a continuous background concentration of NO_2 was the sixth objective.

Although these are preliminary studies, they are presented here to illustrate the complexity of the problem and to indicate the present research activities of the laboratory.

The background and peak NO_2 concentrations were the same as in the

last two investigations. The spikes 8.1 mg/m³ (4.5 ppm) were of 1-hr duration, given twice a day, at 8:30–9:30 AM and 3:00–4:00 PM. The background level was 2.8 mg/m³ (1.5 ppm) for the total length of the 15-day
study, including weekends. On weekends the animals received only the background level, since air quality data indicate that peak levels are unusual on
weekends (presumably due to low automobile traffic).

In this first study, the laboratory-induced infection was given immediately
after the morning peak exposure. The animal response to multiple spikes of
NO_2 superimposed on a background level of 2.8 mg/m³ (1.5 ppm) is seen in
Figure 12. For comparison, the response resulting from continuous exposure
to 2.8 mg/m³ (1.5 ppm) without any spikes is also shown in this figure.

The first measured animal response to the 8.1 mg/m³ (4.5 ppm) peak of
NO_2 is similar to what has been shown previously, that is, approximately
20% enhancement over controls regardless of the preexposure background.
However, subsequent peak exposures did not maintain this level of effect,
and a reduction in percent mortality was observed until the animals had been
exposed to this regimen for 12 days. At that time, and for the following test
days, the level of response remained elevated and not significantly different
from the predicted response from continuous exposure to 2.8 mg/m³ (1.5
ppm) of NO_2.

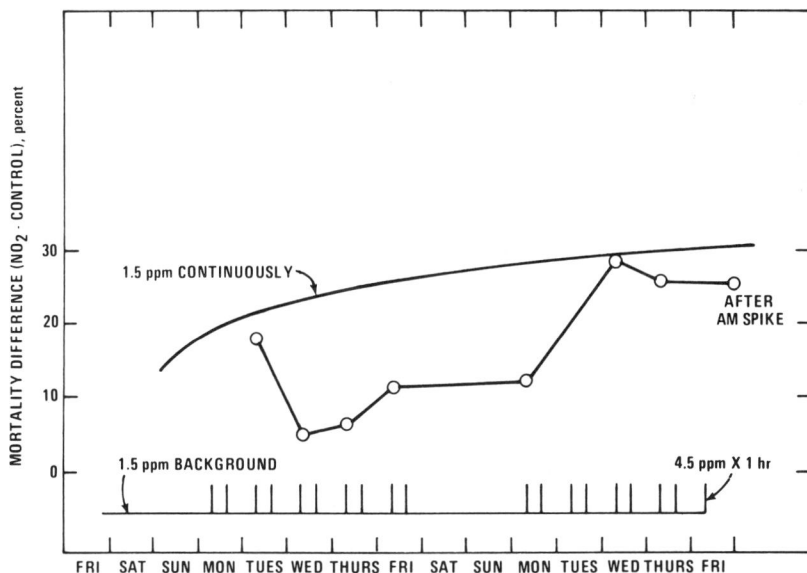

Figure 12. Enhancement of mortality from laboratory-induced infection resulting from
exposure to multiple spikes (4.5 ppm × 1 hr) on a continuous background concentration of 1.5 ppm. This is compared with continuous exposure to 1.5 ppm for the
same time period.

In the second experiment, the exposure pattern to NO_2 was identical to the above study, except that the bacterial challenge was administered immediately before the morning spike exposure. Although the mortality observed in this group was always significantly greater than the control mortality, there was no steady increase in percent mortality with increased length of exposure.

These types of studies indicate the importance for future research to be designed around an exposure pattern that mimics urban diurnal concentration patterns. The lack of such studies critically hampers the environmentalist in making useful scientific decisions in the assessment of NO_2 toxicity.

Objective Seven

To determine the effects on the dose–response relationship through the interaction of NO_2 with a photochemical oxidant was the seventh objective.

Efforts to determine the effects of NO_2 exposure on the health of the population at risk are confounded because NO_2 comprises only a portion of the complex conglomeration of pollutants in the ambient air. Adverse effects may result from exposure to individual or multiple compounds or to the product of atmospheric interaction between specific compounds. Such interactions of pollutants will result in an additive, synergistic or antagonistic response in the host.

Since NO_2 and O_3 are present simultaneously in the urban atmosphere, it was of interest to determine whether an acute exposure to those combined gases significantly alters the dose required to elicit a significant effect from exposure to either pollutant alone. In this study [22], mice were exposed for 3 hr to selected concentrations of NO_2, O_3, $NO_2 + O_3$ or to clean air. Immediately after this exposure, the animals received the aerosol of *Streptococci*. The data in Table IV indicate that the effect of NO_2 and O_3 mixtures are additive in these 3-hr exposure studies, since in most cases the differences in mortality rates were equivalent to the sum of those resulting from inhalation of each individual pollutant.

Further studies using the infectivity model have been conducted to determine the effects of longer exposure to these combined gases. In these studies, the test animals were continuously exposed for 14 days to a background of either NO_2 (0.05, 0.5, 1.2 ppm); O_3 (0.05, 0.1 ppm); or to a combination of these two pollutants. Also, during these exposure periods the concentrations of these gases were raised daily for 1 hr to simulate spikes found in the environment. The spike concentrations for NO_2 were approximately twice that of the background level. The O_3 concentrations were spiked to either 196 $\mu g/m^3$ (0.1 ppm) or 588 $\mu g/m^3$ (0.3 ppm). At the end of the 14-day exposure period, the animals were then combined and challenged with an aerosol of microorganisms. Table V gives the mortality for each of the test groups as well as the number of animals employed in each experiment. At the concentrations used in both experiments 1 and 2, the mortality rate in the mixture of gases greatly exceeded the effects seen with either gas alone. In both cases,

Table IV. Excess Mortalities in Mice Exposed for 3 Hours to Pollutants and Challenged with *Streptococcus* Over Those in Corresponding Infected Controls

Concentrations		Excess Mortality (%)			
NO_2 [mg/m^3 (ppm)]	O_3 [mg/m^3 (ppm)]	0 (0)	0.1 (0.05)	0.2 (0.1)	0.98 (0.5)
0		0	5.4	7.2	28.6[a]
2.82 (1.5)		−1.7	4.6	4.2	23.9[a]
3.76 (2.0)		14.3[a]	22.0[a]	–	56.2[a]
6.58 (3.5)		28.2[a]	–	38.5[a]	68.7[a]
9.40 (5.0)		35.7[a]	–	–	65.3[a]

[a]Significant change in mortality from corresponding infected controls (p ≤ 0.05).

Table V. A Comparison of the Effects of NO_2, O_3 and NO_2-O_3 Mixtures on Susceptibility to Pulmonary Infection

Experiment No.	Background	Spikes (2 × 1 hr)	Mortality[a] (exposed-control)	No. of Animals (dead/total no.)
1	NO_2–1.2 ppm	2.5 ppm	13.7[a]	135/315
	O_3–0.1 ppm	0.3 ppm	14.3[a]	139/319
	Combination	Combination	49.0[a]	249/319
	Control		0	93/319
2	NO_2–0.5 ppm	1.0 ppm	12.4[a]	85/240
	O_3–0.05 ppm	0.1 ppm	7.2	72/238
	Combination	Combination	30.0[a]	125/236
	Control		0	55/239
3	NO_2–0.05 ppm	0.1 ppm	6.5	185/476
	O_3–0.05 ppm	0.1 ppm	1.8	163/477
	Combination	Combination	6.1	183/477
	Control		0	155/479

[a]Significant change in mortality from corresponding infected controls (p ≤ 0.05).

the mortality rate of the mixtures was greater than the sum of the individual pollutants; however, the concentrations tested in the third experiment failed to produce any significant enhancement in mortality rate, regardless of the treatment group.

It is of special interest to compare the response of the single 3-hr exposure to the NO_2 + O_3 mixture with the continuous 14-day exposure. In the former study, the response to the combined gases was additive; however, by increasing the length of the exposure, the response became more than additive, that is synergistic.

CHRONIC RESPIRATORY DISEASE MODEL

Given the propensity for NO_2 to affect host defense mechanisms against acute bacterial respiratory diseases, a new model system is presently being developed to assess pollutant effects on a laboratory-induced chronic pulmonary infection using the microorganism *Mycoplasma pneumoniae*. Details describing this animal model system have been presented elsewhere [23].

Mycoplasma pneumoniae is one of the most common human respiratory tract pathogens. It is associated with the exacerbation of chronic bronchitis in adults; acute pneumonitis of individuals living in close quarters (i.e., college campuses, military barracks); and is a common cause of respiratory disease in children [24,25].

As a result of this prevalence, it is of interest to study the effects of environmental chemicals on the pathogenesis of this chronic disease. Initial studies indicate that a similar chronic respiratory infection can be induced in the laboratory via inhalation. Our preliminary studies addressed the technical reproducibility of this model system. To date, we have been examining the deposition and clearance rate of this organism in Golden Syrian hamsters. It was important first to establish in the normal animal the natural transport or physical clearance, as well as the decline of mycoplasma viability within the lung. Now that these data are available, a number of other biological endpoints can be added to define any adverse effects of NO_2 on this form of chronic pneumonia.

Thus far, two experiments have been completed to compare the rate of deposition and clearance of labeled *M. pneumoniae* in normal animals with a similar group exposed for 3 hr to either 0.95 or 1.8 mg/m^3 (0.5–1.0 ppm) of NO_2. In these studies, NO_2 did not alter the deposition or clearance rate of the inhaled mycoplasma. This study will be continued to further investigate the interaction of toxic chemicals with this common pathogen.

CONCLUSIONS

Occupational and environmental toxicologists are charged with the responsibility of providing a sound scientific data base to ensure that man can conduct his daily activities without undergoing any undue risk that might potentiate the development of disease. There are substantial data on the biological effects associated with exposure to NO_2; however, many of the data involve exposure at levels higher than those normally observed in ambient air and unrealistic patterns of exposure. The evidence presented here indicates that the biological mechanism of response can differ depending on concentration, interaction with other pollutants and the exposure mode.

Laboratory studies are frequently used to support and predict the possible health consequences to the humans when exposed to environmental chemicals. The problems associated with such extrapolations are numerous. Therefore, it is imperative in designing such environmental studies that every effort should be devoted to mimicking or simulating the exposure patterns, concentrations and route of exposure. In addition, more research is needed to evaluate the possibility of interaction with other environmental factors, such as diet, temperature, humidity and stresses such as exercise. Such studies must attempt to simulate the entire milieu in which man lives. Although scientists are presently unaware of the precise manner in which host variations, diet, genetic constitution or prior disease states indirectly influence the degree of toxicity experienced by man, they agree that these factors can modify the host response and increase their susceptibility and vulnerability to disease. The data presented here indicate the importance of some of these variables in determining the toxic response to NO_2.

DISCLAIMER

This report has been reviewed by the Health Effects Research Laboratory, U.S. Environmental Protection Agency and approved for publication. Mention of trade names or commercial products does not constitute endorsement or recommendation for use.

REFERENCES

1. "Air Quality—1975 Annual Statistics Including Summaries with Reference to Standards," EPA-450/2-77-002 (1977).
2. "1SAROAD, 1975," Data reported were abstracted from the 1975 SAROAD Raw Data File maintained at the National Air Data Branch of the U.S. Environmental Protection Agency, Durham, NC (1975).
3. Green, G. M. "Lung Defense Mechanisms," *Med. Clin. N.A.* 57:547–562 (1973).
4. Kass, E. H., G. M. Green and E. Goldstein. "Mechanisms of Antibacterial Action in the Respiratory System," *Bact. Rev.* 30:488–497 (1966).
5. Pecora, D. V., and D. Yegian. "Bacteriology of the Lower Respiratory Tract in Healthy and Chronic Diseases," *New England J. Med.* 258:71–74 (1978).
6. Gardner, D. E., F. J. Miller, J. W. Illing and J. M. Kirtz. "Alterations in Bacterial Defense Mechanisms of the Lung Induced by Inhalation of Cadmium," *Bull. Eur. Physio. Resp.* 13:157–174 (1977).
7. Coffin, D. L., and D. E. Gardner. "Interaction of Biological Agents and Chemical Air Pollutants," *Ann. Occup. Hyg.* 15:219–234 (1972).
8. Gardner, D. E., and J. A. Graham. "Increased Pulmonary Disease Mediated Through Altered Bacterial Defenses," in *Pulmonary Macrophages and Epithelial Cells*, Proceedings 16th Annual Hanford Biology Symposium (September 27–29, 1976), pp. 1–21.
9. Motomiya, T., K. Ito, G. Yoshida, H. Otsu and Y. Nakishima. "The Effects of Exposure to NO_2 Gas in the Infection of Influenza Virus of Mouse: Long-term Experiments in Low Concentrations," *Rep. Environ. Res. Organ. Chiba Univ.* 1:27–33 (1973).

10. Sherwin, R. P., and D. A. Carlson. "Protein Content of Lung Lavage Fluid of Guinea Pigs Exposed to 0.4 ppm Nitrogen Dioxide," *Arch. Environ. Health* 27:90–93 (1973).

11. Goldstein, E., M. C. Eagle and P. D. Hoeprich. "Effects of Nitrogen Dioxide on Pulmonary Bacterial Defense Mechanisms," *Arch. Environ. Health* 26:202–204 (1973).

12. Giordano, A. M., and P. E. Morrow. "Chronic Low Level Nitrogen Dioxide Exposure and Mucociliary Clearance," *Arch. Environ. Health* 25:443–449 (1972).

13. Gardner, D. E., R. S. Holzman and D. L. Coffin. "Effect of Nitrogen Dioxide on Pulmonary Cell Population," *J. Bacteriol.* 98:1041–1043 (1969).

14. Ehrlich, R., E. Silverstein, R. Maigetter, J. D. Fenters and D. E. Gardner. "Immunologic Response in Vaccinated Mice During Long-term Exposure to Nitrogen Dioxide," *Environ. Res.* 10:217–223 (1975).

15. Gardner, D. E. "Alteration in Host-Bacteria Interaction by Environmental Chemicals," *Proc. Am. Chem. Soc. Meeting*, Chicago, IL, 1978 (In press).

16. Miller, F. J., J. W. Illing and D. E. Gardner. "Effect of Urban Ozone Levels on Laboratory Induced Respiratory Infections," *Toxicol. Lett.* 2:163–169 (1978).

17. *Federal Register* 41(53):11260 (1976).

18. Gardner, D. E., F. J. Miller, E. J. Blommer and D. L. Coffin. "Relationship Between Nitrogen Dioxide Concentration, Time, and Level of Effect Using the Animal Infectivity Model," in *Proc. Int. Conf. Photochemical Oxidant Poll. and its Control*, Vol. 1, EPA-600/3-77-001a, U.S. Environmental Protection Agency (1977), pp. 513–525.

19. Ehrlich, R., and M. C. Henry. "Chronic Toxicity of Nitrogen Dioxide: I. Effect on Resistance to Bacterial Pneumonia," *Arch. Environ. Health* 17(12):860 (1968).

20. Gardner, D. E., D. L. Coffin, M. A. Pinigin and G. I. Sidorenko. "Role of Time as a Factor in the Toxicity of Chemical Compounds in Intermittent and Continuous Exposure," *J. Toxicol. Environ. Health* 3:811–823 (1977).

21. Freeman, G., S. C. Crane, R. J. Stephen and N. J. Furiosi. "Pathogenesis of the Nitrogen Dioxide-Induced Lesion in the Rat Lung. A Review and Presentation of New Observations," *Am. Rev. Resp. Disease* 98:429–443 (1968).

22. Ehrlich, R., J. C. Findlay, J. D. Fenters and D. E. Gardner. "Health Effects of Short-term Exposure to Inhalation of NO_2-O_3 Mixtures," *Environ. Res.* 14:223–231 (1977).

23. Hu, P. C., J. M. Kirtz, D. E. Gardner and D. A. Powell. "Experimentation Infection of the Respiratory Tract with *Mycoplasma pneumoniae*," *Proc. Symp. Exp. Models for Pulmonary Res.* (In press).

24. Foy, H. M., H. Ochs and S. D. Davis. "*Mycoplasma pneumoniae* Infection in Patients with Immunodeficiency Syndromes: Report of Four Cases," *J. Infect. Dis.* 127:388–393 (1973).

25. Dajani, A. S., W. A. Clyde and F. W. Denny. "Experimental Infection with *Mycoplasma pneumoniae* (Eaton's Agent)," *J. Exp. Med.* 121:1071–1086 (1965).

SECTION III

Effects of Nitrogen Oxides
on Humans

CHAPTER 18

RECENT EVIDENCE ON THE HUMAN HEALTH EFFECTS OF NITROGEN DIOXIDE

Carl M. Shy
Professor

Gory J. Love
Research Associate Professor
Department of Epidemiology
School of Public Health
University of North Carolina at Chapel Hill
Chapel Hill, North Carolina 27514

INTRODUCTION

This chapter reviews the literature published since 1971 on the direct human health effects of nitrogen dioxide (NO_2) at ambient concentrations and presents new epidemiological data obtained in a study of community exposure to NO_2.

Three types of human response to NO_2 have been demonstrated: (1) increased airway resistance; (2) increased sensitivity to bronchoconstrictors; and (3) enhanced susceptibility to respiratory infections. We will treat each of these response categories in succession.

INCREASED AIRWAY RESISTANCE

The evidence here is largely derived from short-duration (15 min–2 hr) experimental exposures of healthy volunteer subjects to pure NO_2 gas. Table I summarizes the most informative studies. In general, these investigations are in agreement in finding no alteration of flow resistance or related ventilatory parameters in healthy or bronchitic male volunteer subjects acutely exposed

Table I. Experimental Exposures of Volunteer Subjects to NO_2

	NO_2 Concentration ($\mu g/m^3$)	(ppm)	No. of Subjects	Exposure Time (min)	Reported Effects	Reference
1.	3000–3700	1.6–2.0	88 chronic bronchitis subjects	15	Significant increase in airway resistance.	7
	940–2800	0.5–1.5	Same as above	15	No increase in airway resistance.	7
2.	1150	0.62	15 healthy subjects	120	No significant change in cardio-vascular or pulmonary function after 15, 30 or 60 minutes of exercise.	1
3.	540	0.29	7 subjects, some "hypersensitive"	120	No change in pulmonary function, even when combined with 0.5 ppm ozone.	2–4
4.	1860 4700	1.0 2.5	8 healthy subjects	120	No increase in airway resistance at 1 ppm. Increase at 2–5 ppm.	6
5.	1860	1.0	16 healthy subjects	120	No changes in pulmonary function.	5
6.	940	0.5	20 subjects with asthma or bronchitis, 10 normal subjects	120	No significant changes in pulmonary function. Chest tightness, nasal discharge or slight headache reported by some, but these did not show functional changes.	8

to NO_2 concentrations of $\leqslant 2800$ $\mu g/m^3$ (1.5 ppm). Thus, a number of investigators, including Horvath and Folinsbee [1], Hackney et al. [2-5], Beil and Ulmer [6], von Nieding and Krekeler [7], and Kerr et al. [8] reported no statistically significant effect of NO_2 on flow resistance, with or without exercise in healthy subjects exposed for short periods over a concentration range of 560-2800 $\mu g/m^3$ (0.3-1.5 ppm). Ventilatory function in bronchitics appeared to be affected only at NO_2 concentrations above 3000 $\mu g/m^3$ (1.6 ppm). Kerr and co-workers [8] reported the lowest concentration at which subjective symptoms have been observed in exposed subjects. These investigators noted that 7 of 13 asthmatics, 1 of 7 bronchitics and 1 of 10 normal subjects reported chest tightness, burning of the eyes, headache or dyspnea with exercise at 940 $\mu g/m^3$ (0.5 ppm) NO_2; the symptoms were said to be mild and were unaccompanied by objective evidence of decreased lung function. The pathophysiological significance of these induced symptoms in the absence of functional changes remains to be evaluated.

There is no evidence from these experimental results that bronchitics are more sensitive to low concentrations of NO_2 in terms of changes in flow resistance or in other ventilatory parameters. Even when an exercise regimen is included in the study protocol, none of the standard ventilatory function measurements were affected by 1-2 hr NO_2 exposures of less than 3000 $\mu g/m^3$ (1.6 ppm).

INCREASED SENSITIVITY TO BRONCHOCONSTRICTORS

The experimental investigations of Orehek et al. [9] are well known and have been the subject of intense discussion. In 13 of 20 asthmatics, 1-hr exposure to 210 $\mu g/m^3$ NO_2 (0.11 ppm) enhanced the bronchoconstrictive effect of carbachol, as demonstrated by a significant decrease in the dose of this bronchoconstricting agent required to induce a 100% increase in airway resistance. Only three of the asthmatics had definitely increased airway resistance on exposure to NO_2 alone, and it is not possible to determine whether these increases were random responses or reproducible effects of NO_2. The study as reported has at least one important possible bias. The 20 asthmatics were divided into 13 responders and 7 nonresponders, after results were analyzed. Nonresponders were much more sensitive to carbachol under control .(non-NO_2) conditions, requiring about half the dose of carbachol to induce the same percent increase in airway resistance as for responders. Significant effects of NO_2 on reactivity to carbachol were found only among the responders, who were initially less sensitive to carbachol. This "after the fact" separation of study subjects and subsequent hypothesis testing on each separate group is statistically invalid. It is axiomatic that one cannot generate *and* test a hypothesis from the same data set. The results would be more acceptable if a subsequent experiment were performed in which responders and nonresponders were first identified on the basis of their baseline response to carbachol alone, and it could then be shown that the "responders" were made significantly more sensitive to carbachol by exposure to NO_2.

Reviewers of the Orehek et al. study have speculated whether the carbachol-NO_2 combination is relevant, especially in quantitative terms, to the way an asthmatic would respond to the combination of ambient NO_2 and naturally occurring bronchoconstrictors. It is possible that much higher ambient NO_2 concentrations are required to augment the bronchoconstricting effect of natural substances such as pollens, cold air, dust or smoke. Carbachol may artifically biomagnify the response of the asthmatic to otherwise imperceptible stimuli. These issues need to be addressed experimentally before we can interpret the health significance of the Orehek findings. In any event, the basic findings of Orehek should at least be replicated before we draw firm conclusions from them.

Von Nieding et al. [10] reported a similar enhancement of the bronchoconstricting effect of acetylcholine when healthy volunteers were exposed to a combination of 0.025 ppm ozone, 0.05 ppm NO_2 (100 $\mu g/m^3$) and 0.1 ppm SO_2. However, the study was not designed to assess the enhanced bronchoconstrictive effect of NO_2 alone, and it is not possible to determine whether NO_2, ozone (O_3) or SO_2 alone or in combination were responsible for the findings.

ENHANCED SUSCEPTIBILITY TO RESPIRATORY INFECTIONS

Subsequent to the original Chattanooga studies, Melia et al. [11] reported more cough, "colds going to the chest" and bronchitis in 6- to 11-year-old children in British homes in which gas was used for cooking, compared with children from homes where electricity was used. The "cooking effect" was not explained by distribution patterns of age, social class, latitude, population density, family size, overcrowding, outdoor concentrations of smoke or sulfur dioxide (SO_2), or type of fuel used for heating. The authors concluded that elevated indoor levels of NO_2 arising from gas stoves "might be the cause of the increased respiratory illness." Indoor NO_2 levels were not measured in this study. One potential confounding factor here is parental smoking habits. Several investigators (Love et al. [12] and Finklea et al. [13]) reported excess rates of acute respiratory disease among children of smoking parents vs children of nonsmokers. No data were obtained by Melia et al. on parental smoking habits, although it is unlikely that a parental smoking effect would remain after the adjustments were made for social class, family size and population density, which are often correlated in their distribution with smoking habits. However, by contrast to these results, Mitchell et al. [14] failed to find an association between acute respiratory disease and use of gas stoves in middle class homes of a suburb of Columbus, Ohio. The divergent results of these two studies may be explained by three differences:

1. In the Melia et al. study, information on respiratory disease was obtained retrospectively for the previous 12 months, while Mitchell et al. obtained this information prospectively at 2-week intervals for 12 months. Retrospective recall is considerably more subject to measurement error, especially when common acute, short-lived illness episodes are the subject of query.
2. Only natural gas was used in the Ohio gas stoves, while in Britain a changeover was made from coal gas to natural gas during 1973 (as reported by Melia et al.), the very year of the British study. The combustion products of coal gas and natural gas may well differ in some important characteristics.
3. Melia et al. studied acute illness in children; Mitchell et al. in housewives.

This chapter reports for the first time a followup study to the original Chattanooga school children studies of Shy et al. [15,16] and Pearlman et al. [17]. The complete description of methods and results is given by Love et al. [18]. Briefly, in 1972 and 1973 we repeated the longitudinal surveys of acute respiratory disease incidence among residentially stable families (who had not moved within the three years prior to the study) residing in the same study areas as originally described: (1) a high NO_2 exposure community surrounding a TNT plant; (2) an intermediate exposure community seven miles southwest of the plant; and (3) a low NO_2 exposure community more distant from the plant. Socioeconomic and demographic characteristics were determined by direct questionnaire and were found to be essentially the same for the three communities. NO_2 concentrations in each community were measured in 1972 and 1973 by the Saltzman and chemiluminescent methods. Total particulates, sulfates and nitrates were also monitored by the usual methods. SO_2 concentrations in all areas were less than 40 $\mu g/m^3$, so were not monitored regularly. Trained interviewers telephoned each study family biweekly and asked about any new acute respiratory illness experienced during the two-week period. A standardized questionnaire was used to record information. Respiratory illness data were obtained for 12 weeks in 1972, from February through April, and for the same 12 weeks in 1973. Information on potential confounders was evaluated, particularly that of smoking within the home.

Yearly average NO_2 concentrations for the three communities, shown in Figure 1 for the years 1968–1973, reflect the steady decline in production and improved emission controls at the TNT plant. A shutdown of the plant due to a labor strike from January–March of 1973 accounts for the low NO_2 levels in the high NO_2 area that year. Community differences in short-term peak NO_2 concentrations were larger in magnitude than were differences in average levels. Peak NO_2 concentrations are given in Table II for each community in terms of 90th percentiles for 24-hour concentrations and for maximum daily hourly concentrations. One monitoring station in the high NO_2 community recorded 24-hr levels of 200 $\mu g/m^3$ (0.12 ppm) or more on 10% of the days in 1972, with maximum hourly concentrations of 815 $\mu g/m^3$ (0.44 ppm) or more, whereas in the low NO_2 community, 24-hr levels of 62 $\mu g/m^3$ (0.03 ppm) and maximum hourly concentrations of 113 $\mu g/m^3$ (0.06 ppm) were exceeded on 10% of the days in 1972. These differences were not reflected in 1972 annual average values for NO_2, which

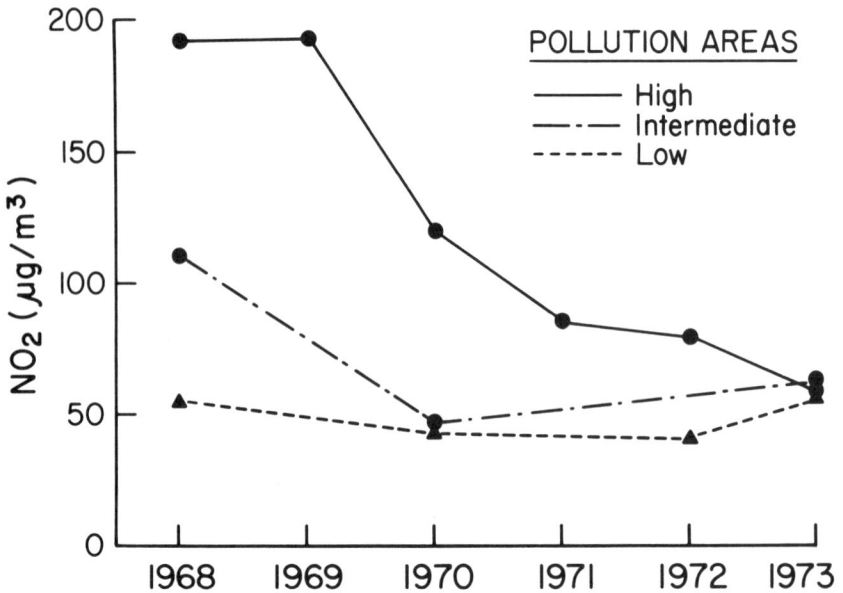

Figure 1. Annual mean NO$_2$ concentration by study areas, 1968–1973.

were 94 μg/m^3 (0.06 ppm) in the high and 43 μg/m^3 (0.02 ppm) in the low NO$_2$ communities.

Total suspended particulate and sulfate levels were very similar in the three communities from 1970 through 1973 (Table III). Suspended nitrates generally corresponded with community differences in NO$_2$ levels, as expected.

Figures 2 and 3 show respiratory incidence rates for upper and lower respiratory symptoms respectively, by year, community and family segment. In 1972, upper respiratory incidence rates (Figure 2) generally followed the NO$_2$ exposure gradient. During the 1973 illness surveillance period, half of which occurred simultaneously with the strike at the TNT plant, upper respiratory illness rates were considerably lower than in 1972 and community differences were less; however, highest incidence rates persisted in the high NO$_2$ area. These findings are suggestive of a close temporal relationship between peak NO$_2$ levels and upper respiratory illness incidence, although other explanations for rate differences between 1972 and 1973 are possible, particularly normal yearly variations in respiratory epidemics. 1972 and 1973 were influenza epidemic years (U.S. Public Health Service) [19,20], although these epidemics were apparently more severe in other parts of the country than in Tennessee. It is possible that in the spring of 1972 the epidemic was manifested by a higher incidence of upper respiratory tract disease, but we have no direct evidence for this.

Table II. 90th Percentiles of 24-hr and of Daily Hourly Maximum NO_2
Concentrations by Study Area, 1972 and 1973

| Community Exposure Ranking | 90th Percentile Concentrations ($\mu g/m_3$) | | | |
| | 24-hr | | Daily Hourly Maximum | |
	1972[a]	1973	1972[a]	1973
Low (1 station)	62	84	113	150
Intermediate (2 stations)	64–78	62–89	116–143	124–164
High (4 stations)	75–220	55–118	228–815	157–522

[a]September through December, 1972 only.

Table III. Suspended Particulate Concentrations by Study Area, 1968–1972

| Pollutant | Community Exposure Ranking[a] | Average Annual Concentrations ($\mu g/m^3$) | | | | |
		1968–1969[b]	1970[c]	1971	1972	1973
Total Suspended Particulates[d]	Low	62	64	64	68	60
	Intermediate	72	63	62	62	56
	High	81	50	52	50	41
Suspended Sulfates	Low	10	13	10	10	11
	Intermediate	10	13	10	11	11
	High	12	13	10	11	11
Suspended Nitrates	Low	2	1	2	1	1
	Intermediate	3	2	3	2	1
	High	6	3	6	3	1

[a]Averages are for two sites in low and intermediate areas and five sites in high exposure area.
[b]Based on daily monitoring in November 1968 and March 1969 and monitoring one in four days from December 1968 to February 1969 and April 1969.
[c]Based on daily measurements January–December 1970.
[d]TSP values are geometric means, except 1968–1969, which are arithmetic means. Other values are arithmetic means.

As shown in Figure 3, temporal patterns for lower respiratory disease incidence rates were generally opposite to those of upper tract illness, but differences between years were lesser in magnitude. In 1972, lower tract incidence rates again generally followed the NO_2 exposure gradient. These community differences persisted among children in 1973, but not among adults, who overall experienced a 50% lesser incidence of lower respiratory tract disease. Several alternative explanations may account for these

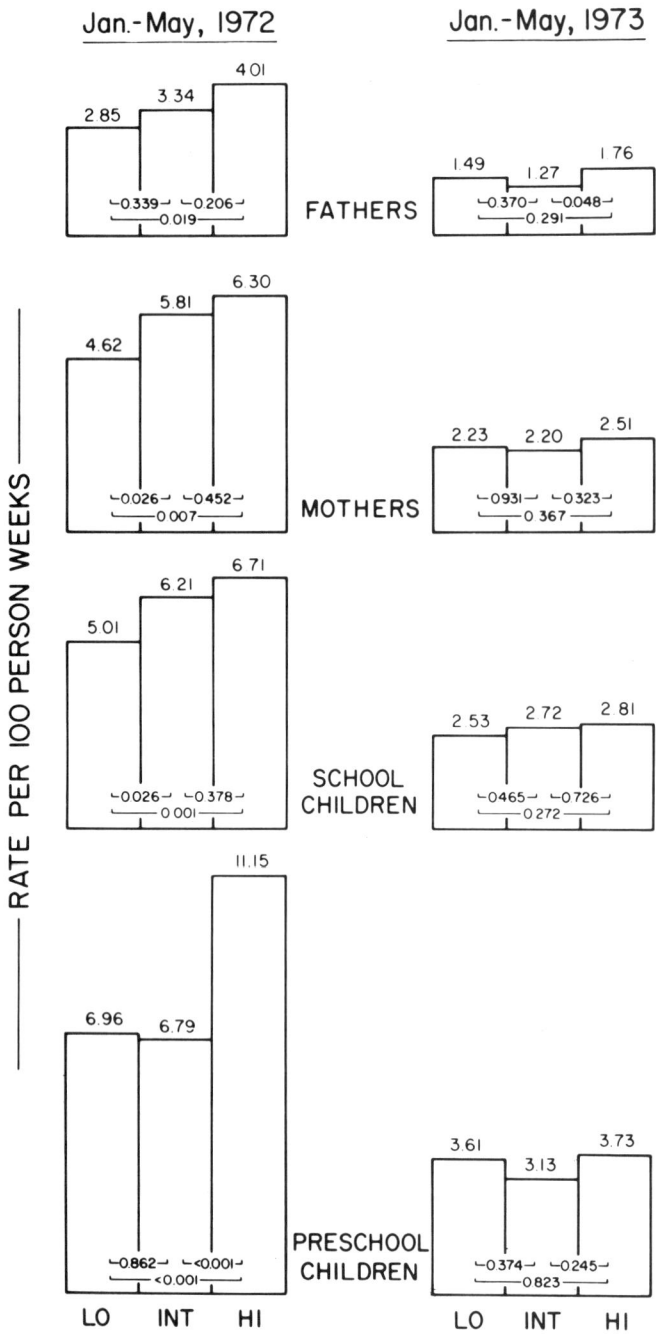

Figure 2. Upper respiratory tract incidence rates by year, community and family segment (p values within bars).

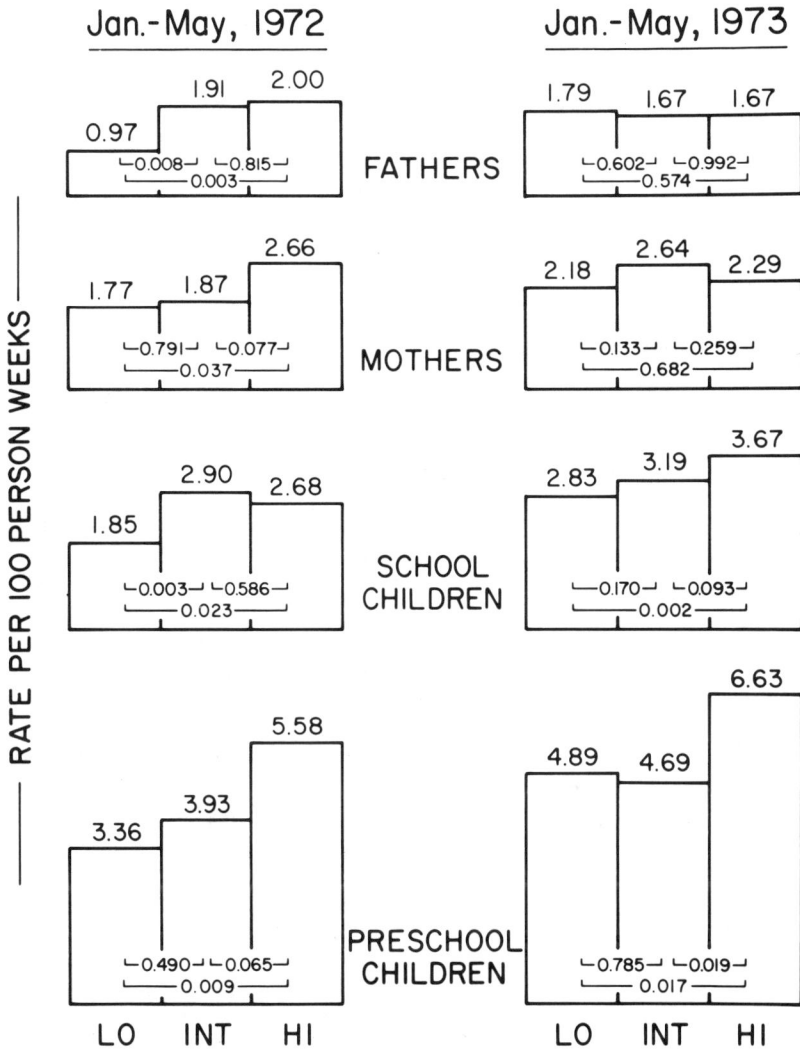

Figure 3. Lower respiratory tract incidence rates by year, community and family segment (p values within bars).

differences by year and family segment. The 1973 influenza epidemic may have induced more lower tract illness among children than in 1972. The lower ambient NO_2 levels associated with the 1973 strike may have selectively benefited adult parents in the high NO_2 community, but not the children whose risk may have been offset by epidemic conditions in 1973. Lower respiratory tract incidence rates generally followed the NO_2 exposure gradient among all family segments in 1972 and among children in 1973. The complicating factors of the strike in 1973 and the presence of influenza epidemics in different parts of the U.S. in 1972 and 1973 make it exceedingly difficult to interpret the change in community rankings for adult lower respiratory tract disease in 1973.

Several questions were included in the biweekly telephone survey to assess complications and severity of respiratory illness episodes. Thus, queries were made concerning the presence of fever, restricted activity, visit to a physician, and physician's diagnosis of otitis media. Figure 4 shows data on the percentage of reported respiratory illnesses requiring a visit to a physician. Overall the pattern for children in 1972 is somewhat similar to that of lower respiratory tract incidence rates given in Figure 3. In the low pollution area, visits to the physician were more frequent in 1973 than in 1972, perhaps reflecting that lower respiratory disease more frequently requires evaluation by a physician. In 1973 there was no consistent pattern of illness severity relating to the NO_2 community exposure gradient. Other indices of illness severity, including fever, restricted activity and otitis media percentages followed the same pattern as physician visits. These results also argue against overreporting of illness by residents of the high NO_2 community. Minor symptoms and illness could be exaggerated or more readily perceived in one of the communities, but if this were so, we would anticipate finding in that community a lower percentage of total illnesses requiring a visit to the physician. Such does not appear to be the case in the high NO_2 community.

Respiratory tract illness rates (combined upper and lower) were stratified by community for the presence or absence of smoking parents (Figure 5). Within smoking categories, children in the high NO_2 area usually were found to have higher illness rates, while within the same communities, children of smoking parents did not have a consistent pattern of higher or lower rates than children of nonsmokers. The combined effect of each factor, smoking and pollution, was less than additive, suggesting independent action of each risk factor.

These results of the followup studies in Chattanooga suggest a continuing effect of moderately elevated ambient NO_2 concentrations on risk of acute respiratory disease. Pollution effects appeared to be more consistent among children than adults. The fact that preschool children (0-4 yr) in the high NO_2 area still manifested excess risk of disease in 1973 suggests that the high ambient NO_2 concentrations associated with the peak TNT production years of 1966-1969 could not account for all of the reported disease excess. Since yearly average NO_2 concentrations for 1971-1973 were not strikingly different in these communities, we postulate an effect of repeated short-term peak NO_2 concentrations on respiratory disease risk.

Among adults and school children, it is possible that excess disease can be attributed to residual effects of previous higher levels of NO_2. However, the sharp decline and the disappearance of community differences

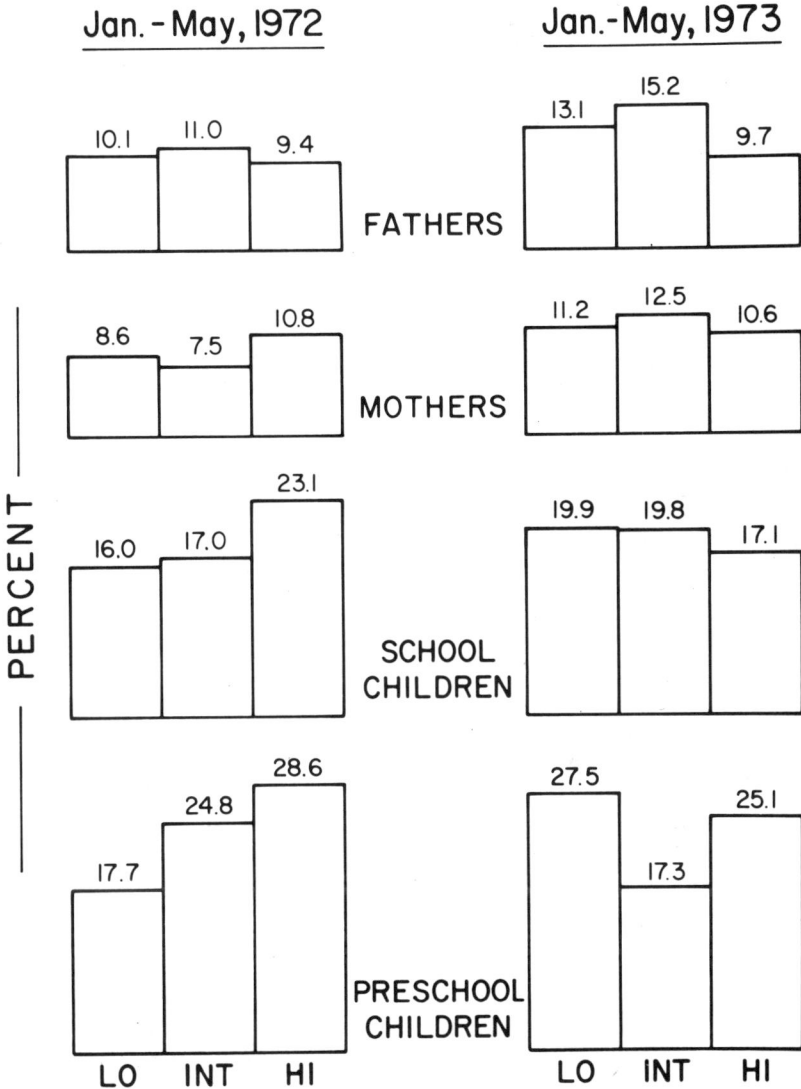

Figure 4. Percentage of all respiratory diseases requiring a physician's visit by year, community and family segment.

in adult respiratory illness rates in 1973, when NO_2 levels fell because of the strike, suggests a close temporal relationship with NO_2 levels. This pattern of decline also was revealed in upper respiratory tract illness rates of children, but less so for lower tract disease rates, possibly due to the effects of an epidemic that affected the lower respiratory tract more in children than in

SCHOOL CHILDREN

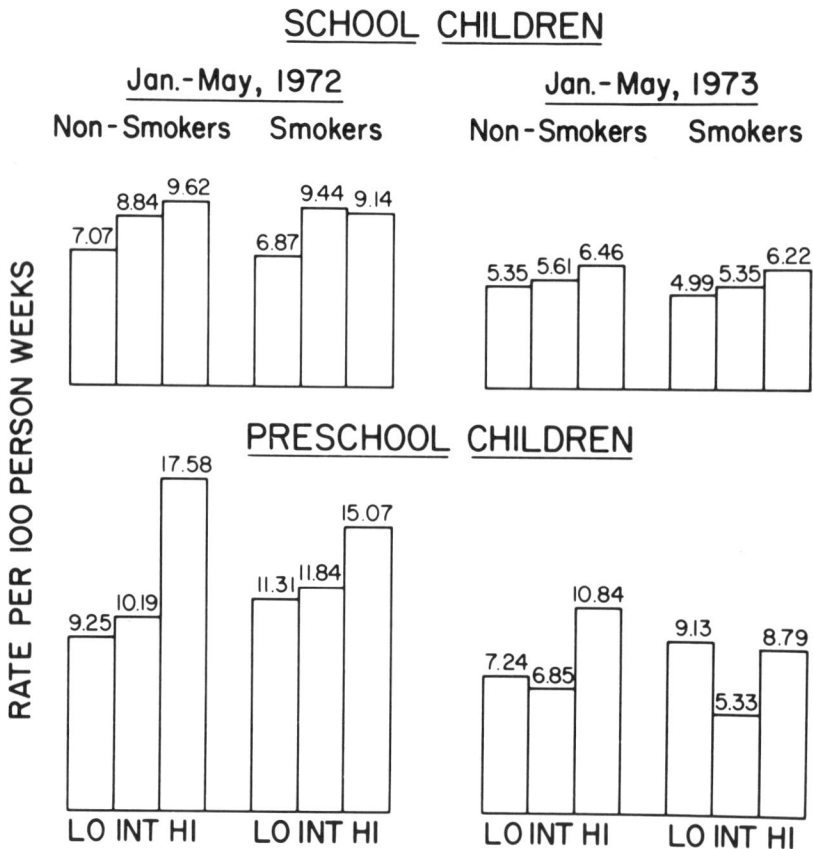

Figure 5. All respiratory illness in children by year, community and smoking habits of parents.

adults. None of the community differences in illness rates could be accounted for by smoking habits of parents or by other socioeconomic or demographic factors.

Other commonly monitored pollutants—total suspended particulates, sulfates or SO_2—were unrelated to community differences in illness rates. Nitrate levels were, as expected, closely related to NO_2 concentrations. Measurements of nitric acid or sulfuric acid aerosols were not feasible at that time, and it is possible that these by-products of the TNT manufacturing process may have played some role in the induction of susceptibility to illness.

As in any epidemiological study of populations exposed to complex mixtures of pollutants, we cannot state unequivocally that atmospheric levels of NO_2 were clearly responsible for the observed disease excess. Evidence from other studies, including experimental exposures of animals to NO_2 combined with infectious agents (Purvis and Ehrlich [21], Coffin et al. [22] and Goldstein et al. [23]), and epidemiological studies related to use of gas stoves in homes (Melia et al. [11]) support the conclusion that ambient NO_2 may be the responsible agent. Because of our uncertainties concerning the association of respiratory disease excess with long- vs short-term averaging times, we cannot draw definitive conclusions about the precise ambient NO_2 level associated with our observations of disease excess. In 1972, the study year prior to the labor strike, point estimates of the yearly average, 90th percentile 24-hr average and 90th percentile maximum 1-hr average in the high NO_2 area were, respectively, 94, 220 and 815 $\mu g/m^3$ (0.03, 0.12 and 0.44 ppm). Long- and short-term exposures could be expressed in other terms, e.g., medians, 95th or 99th percentiles. We conclude that the ambient environment of the high NO_2 community, characterized by any of several measures of pollution exposure related to the emission from the TNT plant, was associated with an increased risk of acute respiratory disease in the neighboring community.

REFERENCES

1. Horvath, S. M., and L. J. Folinsbee. "The Effect of NO_2 on Lung Function in Normal Subjects," Final contract report, U.S. Environmental Protection Agency Contract No. 68-02-1757 (1975).
2. Hackney, J. D., W. S. Linn, R. D. Buckley, E. E. Pedersen, S. K. Karuza, D. C. Law and D. A. Fischer. "Experimental Studies on Human Health Effects of Air Pollutants. I. Design Considerations," *Arch. Environ. Health* 30:373–378 (1975).
3. Hackney, J. D., W. S. Linn, J. G. Mohler, E. E. Pedersen, P. Breisacher and A. Russo. "Experimental Studies on Human Health Effects of Air Pollutants. II. Four-Hour Exposure to Ozone Alone and in Combination with Other Pollutant Gases," *Arch. Environ. Health* 30:379–384 (1975).
4. Hackney, J. D., W. S. Linn, D. C. Law, S. K. Karuza, H. Greenberg, R. D. Buckley and E. E. Pedersen. "Experimental Studies on Human Health Effects of Air Pollutants. III. Two-Hour Exposure to Ozone Alone and in Combination with Other Pollutant Gases," *Arch. Environ. Health* 30:385–390 (1975).
5. Hackney, J. D. "Effects of Atmospheric Pollutants on Human Physiologic Function," Final report, U.S. Environmental Protection Agency, Contract No. R-801396 (1976).
6. Beil, M., and W. T. Ulmer. "Wirkung von NO_2 in MAK-Bereish auf Atemmechanik und Acelytcholinempfindlichkeit bei Normalpersonen," *Intern. Arch. Occup. Environ. Health* 38:31–44 (1976).
7. von Nieding, G., and H. Krekeler. "Pharmakologische Beeinflussung der akuten NO_2-Wirkung auf die Lungenfunktion von Gesunden und Kranken mit einer chronischen Bronchitis," *Intern. Arch. Arbeitsmed.* 29:55–63 (1971).

8. Kerr, H. D., T. J. Kulle, M. L. McIlhany and P. Swedersky. "Effects of Nitrogen Dioxide on Pulmonary Function in Human Subjects. An Environmental Chamber Study," U.S. EPA Publication EPA-600/1-78-025. U.S. Environmental Protection Agency, Research Triangle Park, NC (1978).

9. Orehek, J., J. P. Massari, P. Gayrard, C. Grimaud and J. Charpin. "Effect of Short-Term, Low-Level Nitrogen Dioxide Exposure on Bronchial Sensitivity of Asthmatic Patients," *J. Clin. Invest.* 57:301–307 (1976).

10. von Nieding, G., H. M. Wagner, H. Lollgen and H. Krekeler. "Acute Effects Function in Normal Subjects," Final contract report, U.S. Environmental Protection Agency Contract No. 68-02-1757 (1977).

11. Melia, R. J. W., C. du. V. Florey, D. S. Altman and A. V. Swan. "Association Between Gas Cooking and Respiratory Disease in Children," *Brit. Med. J.* 2:149–152 (1977).

12. Love, G. J., A. A. Cohen, J. F. Finklea, J. G. French, G. R. Lawrimore, W. C. Nelson and P. B. Ramsey. "Prospective Surveys of Acute Respiratory Disease in Volunteer Families: 1970–1971 New York Studies," in *Health Consequences of Sulfur Oxides: A Report from CHESS, 1970–1971*, EPA-650/1-74-004, U.S. Environmental Protection Agency, Research Triangle Park, NC (1974).

13. Finklea, J. F., J. G. French, G. R. Lawrimore, J. Goldberg, C. M. Shy and W. C. Nelson. "Prospective Surveys of Acute Respiratory Disease in Volunteer Families: Chicago Nursery School Study, 1969–1970," in *Health Consequences of Sulfur Oxides: A Report from CHESS, 1970–1971*, EPA-650/1-74-004, U.S. Environmental Protection Agency, Research Triangle Park, NC (1974).

14. Mitchell, R. I., R. Williams, R. W. Cote, R. R. Lanese and M. D. Keller. "Household Survey of the Incidence of Respiratory Disease in Relation to Environmental Pollutants," WHO International Symposium Proceedings: Recent Advance in the Assessment of the Health Effects of Environmental Pollutants, Paris, June 24–28 (1974).

15. Shy, C. M., J. P. Creason, M. E. Pearlman, K. E. McClain, F. B. Benson and M. M. Young. "The Chattanooga School Children Study: Effects of Community Exposure to Nitrogen Dioxide. I. Methods, Description of Pollutant Exposure and Results of Ventilatory Function Testing," *J. Air Poll. Control Assoc.* 20(8):539–545 (1970).

16. Shy, C. M., J. P. Creason, M. E. Pearlman, K. E. McClain, F. B. Benson and M. M. Young. "The Chattanooga School Study: Effects of Community Exposure to Nitrogen Dioxide. II. Incidence of Acute Respiratory Illness," *J. Air Poll. Control Assoc.* 20(9):582–588 (1970).

17. Pearlman, M. E., J. E. Finklea, J. P. Creason, C. M. Shy, M. M. Young and R. J. M. Horton. "Nitrogen Dioxide and Lower Respiratory Illness," *Pediatrics* 47(2):391–398 (1971).

18. Love, G. J., et al. "Acute Respiratory Illness in Families Exposed to Nitrogen Dioxide Ambient Air Pollution," Unpublished manuscript. Department of Epidemiology, University of North Carolina at Chapel Hill (1979).

19. U.S. Public Health Service, National Center for Disease Control. "Weekly Mortality and Morbidity Reports," Annual Supplement (1972).

20. U.S. Public Health Service, National Center for Disease Control. "Weekly Mortality and Morbidity Reports," Annual Supplement (1973).
21. Purvis, M. R., and R. Ehrlich. "Effect of Atmospheric Pollutants on Susceptibility to Respiratory Infection. II. Effect of NO_2," *J. Infect. Dis.* 113:72–76 (1963).
22. Coffin, D. L., D. E. Gardner and E. J. Blommer. "Time-Dose Response for Nitrogen Dioxide Exposure in an Infectivity Model System," *Environ. Health Persp.* 13:11–15 (1976).
23. Goldstein, E., M. C. Eagle and P. D. Hoeprich. "Effect of Nitrogen Dioxide on Pulmonary Bacterial Defense Mechanisms," *Arch. Environ. Health* 26:202–204 (1973).

CHAPTER 19

HUMAN HEALTH EFFECTS OF NITROGENOUS AIR POLLUTANTS: RECENT FINDINGS FROM CONTROLLED-ENVIRONMENT CLINICAL STUDIES

Jack D. Hackney, William S. Linn, Edward L. Avol, Michael P. Jones, Michael T. Kleinman and Ronald M. Bailey

Rancho Los Amigos Hospital Campus
University of Southern California
School of Medicine
Downey, California 90242

INTRODUCTION

Human clinical studies of air pollution effects involve deliberate exposure of volunteers to air pollutants in a laboratory setting and observation of their health status to determine whether adverse health effects occur as a result. If meaningful results are to be obtained, the experimental exposure conditions must simulate real-world pollution exposure conditions, and the exposure atmosphere must be controlled and characterized well. Furthermore, the most sensitive health-effect measures available must be used, and the volunteer subjects must be reasonably good representatives of the population whose risk from air pollution is to be evaluated. Human experimentation is ethically limited to the investigation of relatively mild, readily reversible, short-term health effects. Within this limit, an initial investigation is often designed to simulate the most severe real-world exposure considered likely to occur (worst-case condition), to document the presence or absence of detectable adverse effects under this condition. If effects are found under the worst-case condition, further studies are performed under less severe exposure conditions to obtain dose–response information, which can then be used to estimate an apparent "threshold"—a dose below which no effect is detectable. Of course, the "threshold" level is meaningful only in relation to the specific health effects measured; introduction of a more sensitive measurement technique may result in a lowering of the "threshold." In practical circumstances, dose–response curves usually have rather broad confidence limits, limiting the precision with which a "threshold" can be estimated. Not only are responses highly variable among different individuals, but effective doses of pollutants may be difficult to estimate, since they have a complex dependence on physical properties, chemical properties, concentration, exposure time and characteristics of breathing. Even when relatively precise dose–response information is available, based on a sensitive and reliable health measurement, one still faces the problem of generalizing the results from the relatively small number of subjects studied in the laboratory to much larger and more diverse populations at risk from pollution exposure. For these reasons, considerable caution must be exercised in interpreting findings concerning "thresholds" and translating them into air quality criteria.

Nitric oxide (NO), the predominant nitrogenous pollutant in combustion effluents, appears to have no important health effects at concentrations attainable in ambient air. In the atmosphere, NO is converted to nitrogen dioxide (NO_2) and nitrate aerosols; this process also contributes to the photochemical production of ozone (O_3) under appropriate conditions. Ozone appears to be the most toxic of common air pollutants, and thus perhaps the strongest motivation for controlling nitrogen oxide emissions is to minimize O_3 pollution episodes. Biological effects of O_3 have been reviewed recently elsewhere [1,2]. Nitrogen dioxide is an irritating, oxidizing gas like O_3. While NO_2 is a less powerful oxidizing agent and shows lower toxicity than O_3 in animal studies, it is considered a potentially hazardous air pollutant. Serious illness or death results from accidental exposures to high concentrations (100 or more times ambient concentrations), and pathological changes can be documented in animals exposed to concentrations above maximum ambient levels. Epidemiological studies of exposed human populations, while invariably confounded by the presence of other air pollutants, have shown statistical associations between increases in ambient concentrations of NO_2 and/or nitrate-containing particulates, and increases in certain indices of respiratory ill health. These various findings have been reviewed elsewhere [3–6]. This chapter deals only with recent controlled human studies, some of which are reviewed briefly in relation to the problem of defining "threshold" levels (as needed to help formulate reasonable air quality criteria) for NO_2 and for nitrate aerosols. The findings discussed are subject to various uncertainties, as previously mentioned, and relate only to short-term health effects. Possible long-term effects of repeated exposures present a different and more difficult scientific problem.

NITROGEN DIOXIDE STUDIES

Early controlled-exposure studies in humans generally dealt with industrial exposure concentrations (1–5 ppm). The lowest concentrations at which even mild decrements in lung function were found more or less consistently were

1.5-2 ppm, about twice the worst-case ambient concentration. More recently, exposure chamber studies have been conducted at concentrations more comparable to ambient levels (0.1-1 ppm). We recently reported findings [7] in a group of 16 healthy male volunteers exposed to 1.0 ppm NO_2 2 hr/day on 2 successive days, with intermittent light exercise during exposure. The exposure conditions were intended to simulate a worst-case ambient exposure. There were no meaningful changes in conventional measures of lung function—volume and flowrates of a maximal forced expiration, resistance of airways or intrapulmonary distribution of ventilation (evaluated in terms of the pattern in which residual air is washed out of the lungs in a single maximal expiration following a single maximal inspiration of oxygen). Some subjects reported minor increases in respiratory symptoms during exposure, but the overall change in symptoms (expressed as a semiquantitative score) was not statistically significant. Blood biochemical studies were performed on a subgroup of these subjects [8]. They showed small, statistically significant decreases in hemoglobin concentration and in activity of red-cell membrane acetylcholinesterase, possibly suggesting an adverse effect of NO_2 exposure on red cells. However, similar (though even smaller) changes were found in control studies (exposures to purified air). The health significance of these biochemical changes and the degree to which NO_2 is specifically responsible for them are not yet understood.

Folinsbee et al. [9] measured cardiac and metabolic functions, as well as lung function, in 15 healthy male volunteers exposed to 0.6 ppm NO_2 for 2-hr periods, including moderate exercise for 15, 30 or 60 min. None of the physiological measurements showed significant changes attributable to the exposure, regardless of the amount of exercise.

Kerr et al. [10] studied male and female volunteers, including asthmatics and chronic bronchitics, exposed for 2 hr to 0.5 ppm NO_2 with a single 15-min period of moderate exercise. Health effects were evaluated in terms of pulmonary function and symptoms. Minimal symptoms were reported by a minority of individual subjects, mostly asthmatics. A few statistically significant group mean pulmonary function changes were found, but these did not correlate with reported symptoms. The apparent individual and group responses might have been due merely to chance, given that a relatively large number of measurements and statistical analyses were performed. The authors considered the results to be generally negative.

From the abovementioned findings, one might conclude that the threshold concentration for healthy people exposed to NO_2 for no more than 2 hr is near 1.0 ppm, if the few symptoms reported at that concentration were considered the earliest manifestation of an adverse health effect. Similarly, the threshold for asthmatics might be 0.5 ppm—a concentration attainable in ambient pollution episodes. On the other hand, one might reasonably attribute the few observed symptom responses to chance, and thus conclude that "threshold" levels are substantially higher. In either case, the results relate only to symptoms and commonly used measures of lung function, which perhaps are not the most sensitive means of detecting adverse health effects. Orehek et al. [11] have employed a possibly more sensitive measurement technique and have reported positive findings at an NO_2 concentration much lower than those used elsewhere—0.1 ppm. These investigators exposed asthmatic subjects for 1 hr at rest. They evaluated responses in terms of airway resistance per se, and also in terms of the increases in airway resistance brought about by inhalation of aerosols containing carbachol, a powerful bronchoconstricting drug. The response to carbachol is considered to be an indicator of the reactivity of the subject's airways to inhaled irritants or allergens. Asthmatics, as expected, typically react more markedly and at lower doses of carbachol than do healthy subjects. A carbachol dose—response curve was generated for each subject on two separate occasions, once after breathing NO_2 and once after a similar exposure to room air with no NO_2 added. For the entire group of 20 subjects studied, the mean response to carbachol apparently did not differ statistically between NO_2 exposure and control conditions. However, 13 of the 20 subjects, termed NO_2 responders, appeared to have a greater increase in carbachol response with NO_2 than would be expected in the absence of deliberate pollutant exposure. (The expected variability of the carbachol response was determined in a different group of 12 subjects.) In the NO_2 responders subgroup, there was a significant mean increase in carbachol response with NO_2 as compared to control, as well as a small increase in airway resistance measured before carbachol challenge.

The findings of Orehek et al. raise the question whether quite low levels of NO_2, frequently exceeded in polluted environments, may render exposed individuals more susceptible to asthmatic attacks caused by other substances in the environment, even when the NO_2 produces little or no response by itself. Unfortunately, the interpretation of these results is subject to even more uncertainty than is the case for the studies previously mentioned. First, it is not known whether the bronchoconstrictive response to deliberate inhalations of a drug is a good predictor of responses to the many possible inhalable irritants in the real-world environment. Second, determination of carbachol dose—response curves is technically difficult, since it is hard to tell how much aerosol is present initially, how much is inhaled, how much is deposited in the respiratory tract, and where it is deposited. Unexpected variations in any one of these factors, as well as in others, such as small changes in baseline airway resistance, might affect the test results, the reproducibility of which has been documented only to a limited extent. Finally, the positive statistical results of Orehek et al. depend on a somewhat arbitrary designation of a reactive subgroup of subjects. From a biostatistical standpoint, this may raise some doubts as to the interpretation of the overall findings, although from a clinical standpoint it is often considered appropriate to identify and direct primary attention to a reactive subgroup. At present, then, it is not clear whether increased bronchoconstrictor response is, in fact, produced by NO_2 exposure specifically and, if so, whether this represents a genuine health hazard. Considerable further investigation is needed of this and other possible health responses to NO_2.

NITRATE STUDIES

Biological effects of nitrates have been studied much less extensively than effects of NO_2. Particulate nitrates are relatively difficult to monitor in the atmosphere, so reliable monitoring data from which to estimate typical exposure concentrations are scarce. Controlled generation of particulate pollutants in the exposure laboratory is also much more difficult than generation of common gases. Apparatus to generate and monitor large quantities of aerosols as required for exposure chambers is available only in a few laboratories. Studies often must be done with small quantities of aerosols delivered to the subjects by mouthpiece or mask, a relatively unrealistic simulation of ambient exposure conditions. The dose of any particulate substance inhaled and deposited in the respiratory tract is strongly dependent on particle size, as well as on concentration of particles and breathing characteristics. Thus, there are more variables to be considered than in gas exposures.

Utell et al. [12] exposed 10 normal and 11 mildly asthmatic volunteers for 16-min periods to sodium nitrate ($NaNO_3$) aerosol at a concentration of 7000 $\mu g/m^3$, much higher than ambient nitrate concentrations. The aerosol, delivered by mouthpiece, had a mass median aerodynamic diameter (MMAD) of 0.46 μm, with a geometric standard deviation (σg) of 1.7. In control experiments, subjects breathed similar aerosols of sodium chloride, considered innocuous. Health effects were measured in terms of lung function test performance, respiratory symptoms and response to carbachol inhalation. No statistically significant changes relatable to nitrate exposure were found, although two asthmatic subjects showed minimally greater responses to carbachol after $NaNO_3$.

We recently studied 20 normal subjects exposed to ammonium nitrate (NH_4NO_3) in an environmental chamber for 2 hr with intermittent light exercise, a protocol similar to that used for the NO_2 exposures described previously. The nominal exposure concentration was 200 $\mu g/m^3$; the aerosol MMAD was 0.8 μm with σg of 2.2. These conditions were intended to simulate the worst case, 2-hr average nitrate concentration monitored in the Los Angeles area [13]. Control studies (exposures to purified background air only) were performed under otherwise similar conditions on the day preceding the actual exposures. The first subjects to be studied (the investigators) were accidentally exposed to about twice the nominal concentration because a malfunctioning diluter produced a negative error in the aerosol monitor readout. Neither these nor any of the other subjects experienced any obvious increase in respiratory symptoms or decline in lung function test performance during exposure. Selected pulmonary function data for all 20 subjects are shown in Table I. The function tests include forced vital capacity (FVC), the volume of a maximal breath forced out; one-second forced expiratory volume (FEV_1), the maximum amount that can be expired in the first second of expiration; total lung capacity (TLC); delta nitrogen, a measure of the rate of change of expired nitrogen concentration during expiration after a breath of

Table I. Selected Mean Lung Function Measures for 20 Normal Men
Exposed to NH_4NO_3 Aerosol

Test	Measurement	Control	Exposure	p^a
FVC (liters)	Preexposure	4.91	4.88	0.19
	Change	+0.04	+0.04	0.80
FEV_1 (liters)	Preexposure	4.07	4.05	0.35
	Change	+0.07	+0.08	0.40
TLC (liters)	Preexposure	6.26	6.37	0.01
	Change	+0.01	−0.04	0.13
Delta Nitrogen ($\%N_2/l$)	Preexposure	0.773	0.750	0.54
	Change	−0.004	+0.032	0.35
R_t (cm H_2O)/(liter/sec)	Preexposure	3.48	3.49	0.97
	Change	+0.09	−0.15	0.15

[a]Probability that difference between control and exposure means is due to chance; $P < 0.05$ considered statistically significant.

oxygen (a measure intended to detect abnormal regional distribution of ventilation within the lungs); and total resistance of the respiratory tract (R_t) as measured by a method of forced oscillation. The mean value measured "preexposure," i.e., during the first few minutes in the exposure chamber, is given for each test day; below each of these values is the corresponding mean change observed during the time spent in the chamber. None of the tests showed more than slight changes, either on control days or on aerosol exposure days. In no case was the change during exposure significantly different from the corresponding control value, although preexposure TLC showed a small difference between the two days. The statistical test employed was repeated-measures analysis of variance, which, in these particular cases, gives the same results as the familiar Student paired test, since there are only two groups to compare. Respiratory symptoms reported by the subjects were evaluated by assigning scores based on the number and severity of symptoms, ranking the scores and testing the distribution of ranks by the Friedman rank sums test. The results showed essentially no difference between control and exposure days in the group's overall symptom experience.

The foregoing results, while admittedly very limited, suggest that in normal subjects, the threshold concentration for short-term effects of nitrate-containing aerosols is probably above the ambient concentration range.

DISCUSSION AND CONCLUSIONS

Although some experimental findings suggest that NO_2 may have adverse effects, controlled human studies so far have provided no conclusive evidence that exposure to nitrogenous air pollutants at ambient concentrations is harmful to health. Thus, in general, they have failed to support the apparently positive findings from epidemiological studies. At present then, it would seem that the most important short-term health consequence of nitrogen oxide emissions relates to their role in the photochemical formation of O_3. Epidemiology, controlled studies of humans and animal toxicology all provide reason to suspect that ambient O_3 concentrations are sufficient to affect respiratory health adversely, at least under severe pollution conditions.

For situations in which ozone formation is not important but nitrogenous pollutant levels are high, the question of adverse health consequences remains open. Controlled human studies can never entirely rule out the possibility of harmful short-term effects, since only small numbers of people can be studied. Even when a deliberate effort is made to recruit subjects likely to be unusually sensitive to pollutant exposure, segments of the population most at risk may not be represented. Furthermore, controlled studies limited to one pollutant, or even to mixtures of several pollutants, may underestimate the health effects of actual ambient mixtures, in which many other coexisting species may interact additively and synergistically. To help resolve the remaining uncertainties, provide more basis for estimates as to how many people are likely to experience adverse effects, and help arrive at reasonable cost/benefit decisions regarding control strategies, continuing health-effect studies are needed—human-exposure studies as well as those with more traditional approaches.

ACKNOWLEDGMENTS

This work has been supported by Project CAPM-31-78 of the Coordinating Research Council, Inc. and by Project RP 1225-1 of the Electric Power Research Institute.

REFERENCES

1. Committee on Medical and Biologic Effects of Environmental Pollutants. *Ozone and Other Photochemical Oxidants* (Washington, D.C.: National Academy of Sciences, 1977).
2. U.S. Environmental Protection Agency. *Air Quality Criteria for Ozone and Other Photochemical Oxidants*, Report No. EPA-600/8-78-004 (Washington, D.C.: U.S. Government Printing Office, 1978).
3. Committee on Medical and Biologic Effects of Environmental Pollutants. *Nitrogen Oxides* (Washington, D.C.: National Academy of Sciences, 1977).
4. Morrow, P. E. "An Evaluation of Recent NO_x Toxicity Data and an Attempt to Derive an Ambient Air Standard for NO_x by Established Toxicological Procedures," *Environ. Res.* 10:92 (1975).
5. Knelson, J. H., and R. E. Lee. "Oxides of Nitrogen in the Atmosphere: Origin, Fate, and Public Health Implications," *Ambio* 6:126 (1977).

6. "Nitrogen Oxides: Current Status of Knowledge," Electric Power Research Institute, Report No. EA-668, Palo Alto, CA (1978).

7. Hackney, J. D., F. C. Thiede, W. S. Linn, E. E. Pedersen, C. E. Spier, D. C. Law and D. A. Fischer. "Experimental Studies on Human Health Effects of Air Pollutants. IV. Short-Term Physiological and Clinical Effects of Nitrogen Dioxide Exposure," *Arch. Environ. Health* 33:176 (1978).

8. Buckley, R. D., C. Posin, K. Clark, J. D. Hackney, M. P. Jones and J. V. Patterson. "Nitrogen Dioxide Inhalation and Human Blood Biochemistry," *Arch. Environ. Health* 33:318 (1978).

9. Folinsbee, L. J., S. M. Horvath, J. F. Bedi and J. C. Delehunt. "Effect of 0.62 ppm NO_2 on Cardiopulmonary Function in Young Male Non-smokers," *Environ. Res.* 15:199 (1978).

10. Kerr, H. D., T. J. Kulle, M. L. McIlhany and P. Swidersky. "Effects of Nitrogen Dioxide on Pulmonary Function in Human Subjects," Environmental Protection Agency, Report No. EPA-600/1-78-025, Health Effects Research Laboratory, Research Triangle Park, NC (1975).

11. Orehek, J., J. P. Massari, P. Gayrard, C. Grimaud and J. Charpin. "Effect of Short-Term, Low Level Nitrogen Dioxide Exposure on Bronchial Sensitivity of Asthmatic Patients," *J. Clin. Invest.* 57:301 (1976).

12. Utell, M. J., A. J. Swinburne, D. M. Speers, J. W. Shigeoka, F. R. Gibb, P. E. Morrow and R. W. Hyde. "Airway Reactivity to Nitrate Aerosols in Normals and Asthmatics," *Am. Rev. Resp. Dis.* 115:248 (1977).

13. Hidy, G. M. *California Aerosol Characterization Study, 1971–1974*, Final Report, California Air Resources Board, Contract No. 358, Sacramento, CA (1974).

CHAPTER 20

EFFECT OF EXPERIMENTAL AND OCCUPATIONAL EXPOSURE TO NO$_2$ IN SENSITIVE AND NORMAL SUBJECTS

G. von Nieding, H. M. Wagner, H. Casper,
A. Beuthan and U. Smidt

Hospital Bethanien, D-413 Moers
Federal Public Health Office
D-1000 Berlin
Federal Republic of Germany

INTRODUCTION

A number of authors have described the acute effect of NO$_2$ on human lung function in healthy subjects and individuals with chronic bronchitis [1-4]. These investigations have shown that NO$_2$ in the MAK (MAK ≈ TLV [5]) concentration range (5 ppm) and below may have negative effects on human lung function, especially on the mechanics of breathing. The thresholds for the reaction of the bronchial system on acute NO$_2$ inhalation were between 1.5 and 2.5 ppm [1,2].

Many questions on NO$_2$ effects remain unanswered, especially that of the individual reactivity and reproducibility of the reactions caused by NO$_2$ inhalation and that of the extrapolation of effects observed in acute experiments to the effects seen in long-term animal experiments and indicating emphysema-like changes of the lung [6].

One of the aims of this study was to investigate the variation of the reaction of the respiratory tract of different individuals, i.e., to see whether among randomly selected groups of healthy subjects and those with chronic bronchitis there are individuals showing a stronger reaction to NO$_2$ inhalation than others. In the second part of the study the influence of chronic exposure to air pollutants under occupational conditions on the development of chronic respiratory disease was investigated. It is well known that correlations between prevalence and incidence of chronic lung disease and air pollution from epidemiological studies must be interpreted with great care. In this case the situation was even more complicated by the fact that numerous substances with similar effects had to be taken into account.

When assessing the influence of air pollutants on human health data from occupational medicine should be included. In this field, exposure to higher concentrations and more defined compositions of pollutants can be expected. The conditions also may permit a better differentiation between the effect of single or defined combinations of pollutants.

A longitudinal epidemiological study of the Deutsche Forschungsgemeinschaft (DFG) (German Research Association) in which the correlation between chronic dust exposure in the working environment and chronic bronchitis was examined [7,8] was extended by including the investigation of SO_2, NO_2 and O_3 exposure. Such studies have been recommended in the WHO document "Health Criteria for Oxides of Nitrogen" [9] and the CEC document "Preparatory Study for Establishing Criteria for Nitrogen Dioxide" [10].

METHODS

Acute NO_2 exposure experiments were performed with NO_2 concentrations in the MAK range under controlled laboratory conditions. Chronic NO_2 effects were investigated in occupationally exposed subjects.

Clinical Experiments Under Controlled Laboratory Conditions

The exposure experiments were performed on 49 healthy male subjects and 40 patients with chronic bronchitis (age range from 20–72 years). The patients with chronic bronchitis were in-patients who had been admitted to the hospital due to an exacerbation of their illness; part of these individuals (12 subjects) were affected additionally by coalminer's silicosis. At the time of the investigation their condition had improved significantly and was considered stable. Most of them did not suffer from severe obstruction or hypoxemia at the time of the investigation and they were not under the influence of bronchodilating drugs.

Our main interest focused on the effect on air resistance, measured as total airway resistance [11] and the thoracic gas volume (TGV) (volume constant, humidity and temperature-compensated body plethysmograph [12]). Measurement of R_{aw} has been a sensitive indicator of changes in lung function associated with gaseous pollutants at low concentrations [1–3]. In some of the subjects additional parameters of respiratory gas exchange such as the alveolar–arterial oxygen partial pressure difference ($AaDO_2$) and the arterial to alveolar carbon dioxide partial pressure difference ($aADCO_2$) were determined. (The reproducibility of the methods used and their variability in intraindividual measurements have been the subject of earlier papers [13–15].) The following experiments were conducted:

Exposure to 5–8 ppm NO$_2$ up to a Maximum of 5 min

Group 1: *14 healthy subjects and 14 patients with chronic bronchitis.* Examination of the reactivity of the bronchi and the reproducibility of airway resistance (judged by ΔR_{aw} from the values before and after exposure) on four different days of a week at the same time of the day and approximately equal concentration of the pollutant.

Exposure to 5 ppm NO$_2$ up to 5 min

Group 2: *30 healthy subjects and 40 subjects with chronic bronchitis including those from group 1.* The reactivity of the bronchi (R_{aw} and ΔR_{aw}) were tested as a function of the initial R_{aw} value.

Exposure to 5 ppm NO$_2$, 5 ppm SO$_2$ and 0.1 ppm O$_3$
for 2 hr Alone and in Combination

Group 3 consisted of 11 healthy subjects including 2 atopics with a pollen allergy.

Application of Pollutants

In the 5-min inhalation experiments the subjects were exposed to pollutants applied via mouthpiece from gastight plastic bags with prepared mixtures. The 2-hr experiments were conducted in an exposure chamber as described previously [16].

Cross-Sectional Epidemiological Study

The epidemiological study conducted to test the prevalence of chronic bronchitis among persons chronically exposed to air pollutants encompassed a total of 925 male workers and employees of a steel works. This number was divided into several groups depending on the degree and type of air pollutants in the working environment (Figure 1):

- Group 1: exposed to fine dust $\leqslant 1.0$ mg/m^3
- Group 2: exposed to fine dust at a concentration above 1.0 mg/m^3
- Group 3: exposed to SO$_2$ and dust
- Group 4: exposed to NO$_2$ and dust
- Group 5: exposed to NO$_2$ peak concentrations ($\geqslant 0.5$ ppm); simultaneous exposure to SO$_2$ and dust
- Group 6: controls not exposed to the air pollutants at work

In addition to thorough lung function test including the measurement of airway resistance and respiratory gas exchange with the methods described above, the workers were interviewed by a physician following a standardized questionnaire of the European Community for Coal and Steel (ECCS) [17]. Additionally, the data on absentees were extracted from the medical files of the factory for the past six years. Total number of days of absence and all episodes due to disease of the bronchopulmonary system except tuberculosis, were assessed.

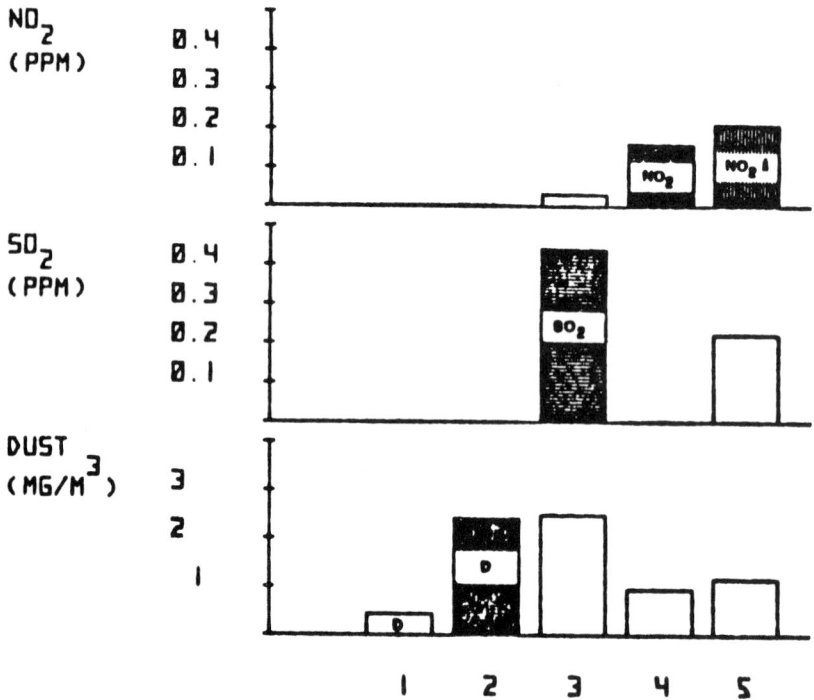

Figure 1. Mean concentrations of dust, SO$_2$ and NO$_2$ at the working areas of the different groups: group 1 = fine dust $\leqslant$ 1.0 mg/m^3; group 2 = fine dust > 1.0 mg/m^3; group 3 = SO$_2$ + fine dust; group 4 = NO$_2$ + fine dust; group 5 = characterized by NO$_2$ peak concentrations.

Pollutant Analysis

NO$_2$ analyses were made using the manual Saltzman method [18]. Simultaneously, NO$_2$ concentrations were monitored continuously by an automatic procedure developed by Breuer [19] and manufactured by Hartmann & Braun under the name "PICOS." The same methods were used to measure the concentrations in the working environment.

SO$_2$ concentrations in the clinical experiments and in working areas were measured with a "Total Sulfur Analyzer" (Bendix) in combination with an H$_2$S scrubber. To check the data gained with this instrument, SO$_2$ was also measured discontinuously and manually by the standard method of West and Gaeke according to the VDI Guideline 2451 [20].

O$_3$ concentrations were monitored by the highly specific and technically very reliable photoluminescence method (Bendix). The instrument was calibrated by the neutral-buffered potassium iodide method according to the VDI Guideline 2468 [21].

Suspended particulate matter was collected by the dust collector VC 25 and subsequently analyzed by a method employing the quenching of β-radiation [22].

Statistical Evaluation

Statistical evaluation of the data from the clinical and epidemiological investigations was performed using standard statistical procedures by means of a laboratory computer (PDP 12) and a special program for evaluation [23]. The calculation of the statistical error and p-values was made with Wilcoxon's [24] ranking test and the chi-square test, respectively. Also, tables for statistical evaluation, published by Koller [25] were used.

RESULTS

The results of the clinical experiments are demonstrated in Figures 2, 3 and 4. The epidemiological findings are summarized in Figure 5a–d.

Clinical Studies

Group 1 (Healthy Subjects)

The upper part of Figure 2a shows the behavior of the airway resistance (R_{aw}) before and after inhalation of NO_2 concentrations between 5 and 8 ppm up to 5 min in 14 healthy subjects. Resistance of every subject was measured on four separate days at the same time of day. The exposure concentrations were the same on every day. The initial value is marked by a point.

The variability of the initial values was very small in some individuals (e.g., subject 4) and greater in others (e.g., subject 8). The R_{aw} increase (ΔR_{aw}) after NO_2 inhalation was different for the individual subjects: the mean ΔR_{aw} varied between the 0.35 ± 0.09 and 1.03 ± 0.15 cm H_2O/liter/sec.

It became evident that at the concentrations chosen for the experiment, there had been an increase in R_{aw} in every case. In one case (subject 8), however, NO_2 exposure did result in pathological values ($\geqslant$3.5 cm H_2O/liter/sec).

The lower part of Figure 2a shows the absolute ΔR_{aw} values for each individual: this form of presentation clearly indicates that all subjects showed a significant mean R_{aw} increase of 0.5 cm H_2O/liter/sec or more after exposure to NO_2.

In most of the 14 subjects the reaction on NO_2 inhalation on the four different days was relatively uniform. In general, the ΔR_{aw} values did not vary more than 50% in each individual. In some subjects, major differences between the results as measured on four different days could be noticed. However, in these cases R_{aw} also increased. Furthermore, there was no essential difference in reactivity between smokers and nonsmokers.

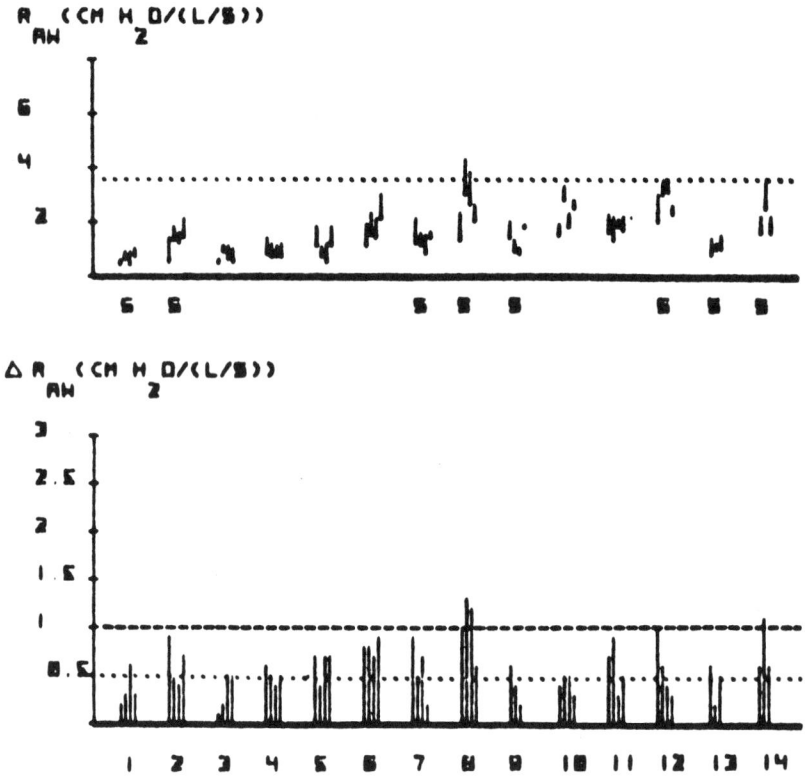

Figure 2a. Behavior of airway resistance (R_{aw}) before and after inhalation of 5–8 ppm NO_2 for up to 5 min on four different days of a week (14 healthy subjects). Dots in the upper part = initial R_{aw} values; the lower part shows the absolute ΔR_{aw} values for each individual.

Group 1 (Patients with Chronic Bronchitis)

In contrast to the healthy group, the patients showed a considerable variation of the initial R_{aw} values as can be seen in Figure 2b. On any of the four days most of these individuals showed initial values above the normal range ($\geqslant 3.5$ cm H_2O/liter/sec).

The individuals of this group also reacted with a stronger increase in R_{aw}. In one case an R_{aw} decrease was seen on one of the four days of the experiment. The individual R_{aw} increase on the four different days was quite similar in most of the patients; the mean increase after NO_2 exposure in this group was 1.07 ± 0.1 cm H_2O/liter/sec.

An R_{aw} increase over 2.0 cm H_2O/liter/sec occurred in three patients

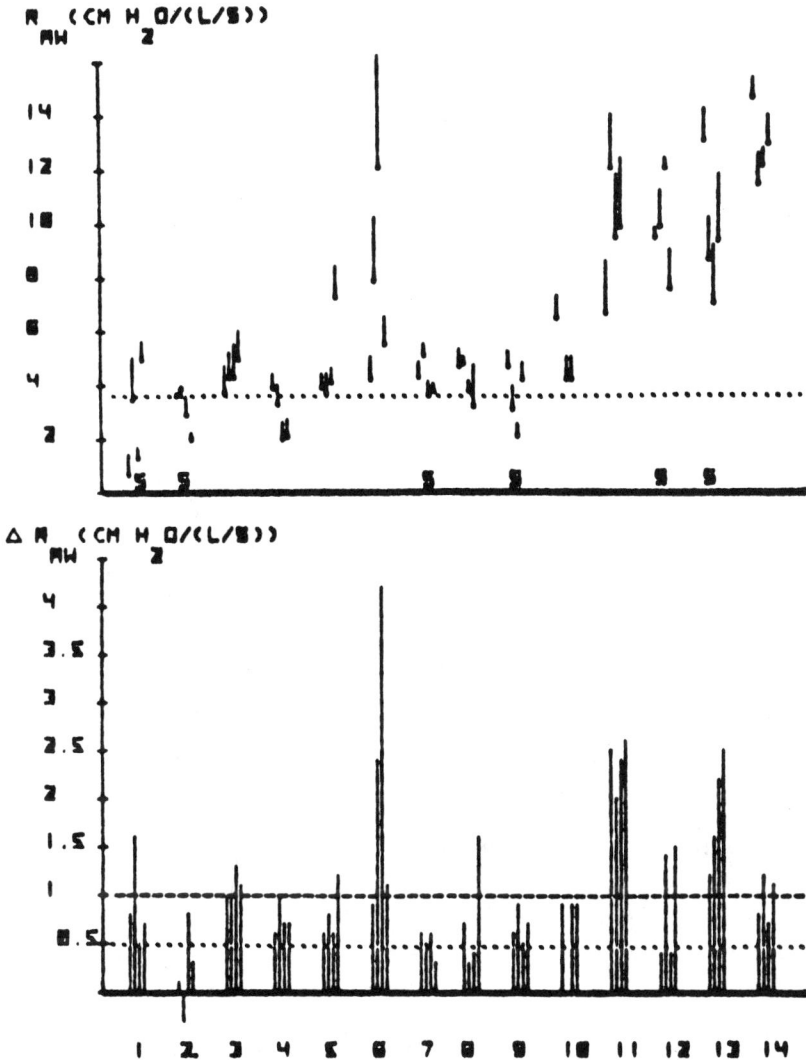

Figure 2b. Same as in Figure 2a in 14 patients with chronic bronchitis.

(patients 6, 11 and 13): If the mean R_{aw} value was calculated excluding these three patients, a mean value of 0.76 ± 0.12 cm H_2O/liter/sec resulted for the rest. This mean R_{aw} increase differed only slightly from the mean R_{aw} increase of the healthy individuals in group 1 (0.58 ± 0.27 cm H_2O/liter/sec). These increases of R_{aw} in all cases can hardly be termed dramatic.

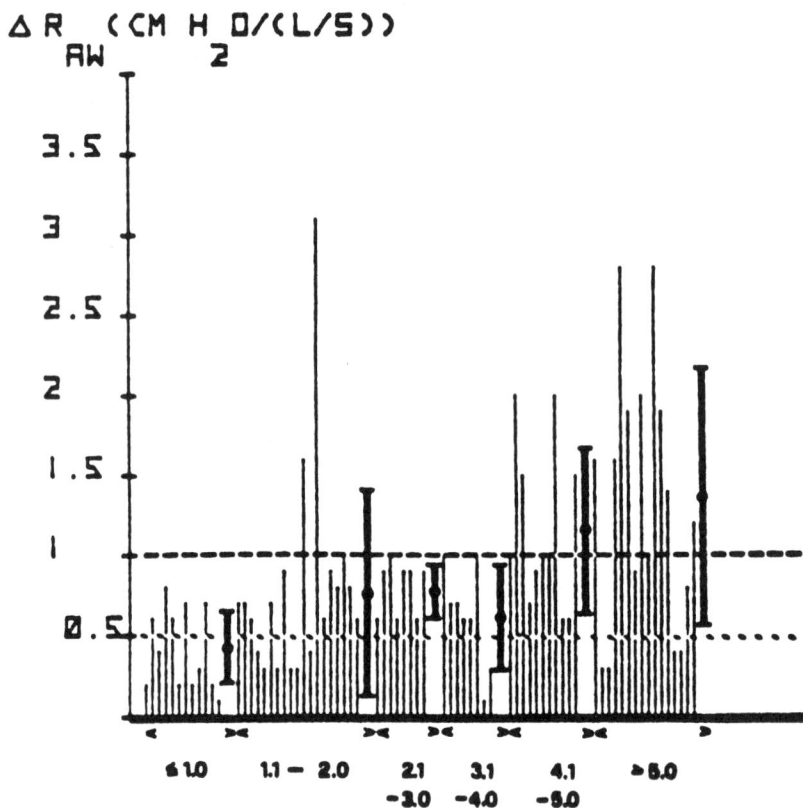

Figure 3. ΔR_{aw} in healthy subjects and patients with chronic bronchitis resulting from NO_2 inhalation at TLV levels for 5 min as plotted against the initial values. The latter (x-axis) are subdivided into corresponding groups of different R_{aw} ranges $\leqslant 1.0$, 1.1–2.0, 2.1–3.0, 3.1–4.0, 4.1–5.0 and $\geqslant 5.1$ cm H_2O/liter/sec.

There is a clear indication, however, that some individuals among the bronchitics reacted more severely upon NO_2 exposure than did the rest of the group.

Group 2 (Healthy Subjects and Chronic Bronchitics)

Figure 3 demonstrates the ΔR_{aw} values in healthy subjects and chronic bronchitics as plotted against initial R_{aw} values. The latter are subdivided into groups corresponding to different R_{aw} ranges. This figure includes besides the data for day 1 of group 1 also the values for 46 healthy subjects and chronic bronchitics that had been exposed to 5 ppm NO_2. The <4.0-cm H_2O/liter/sec group included three bronchitics, the group with an initial

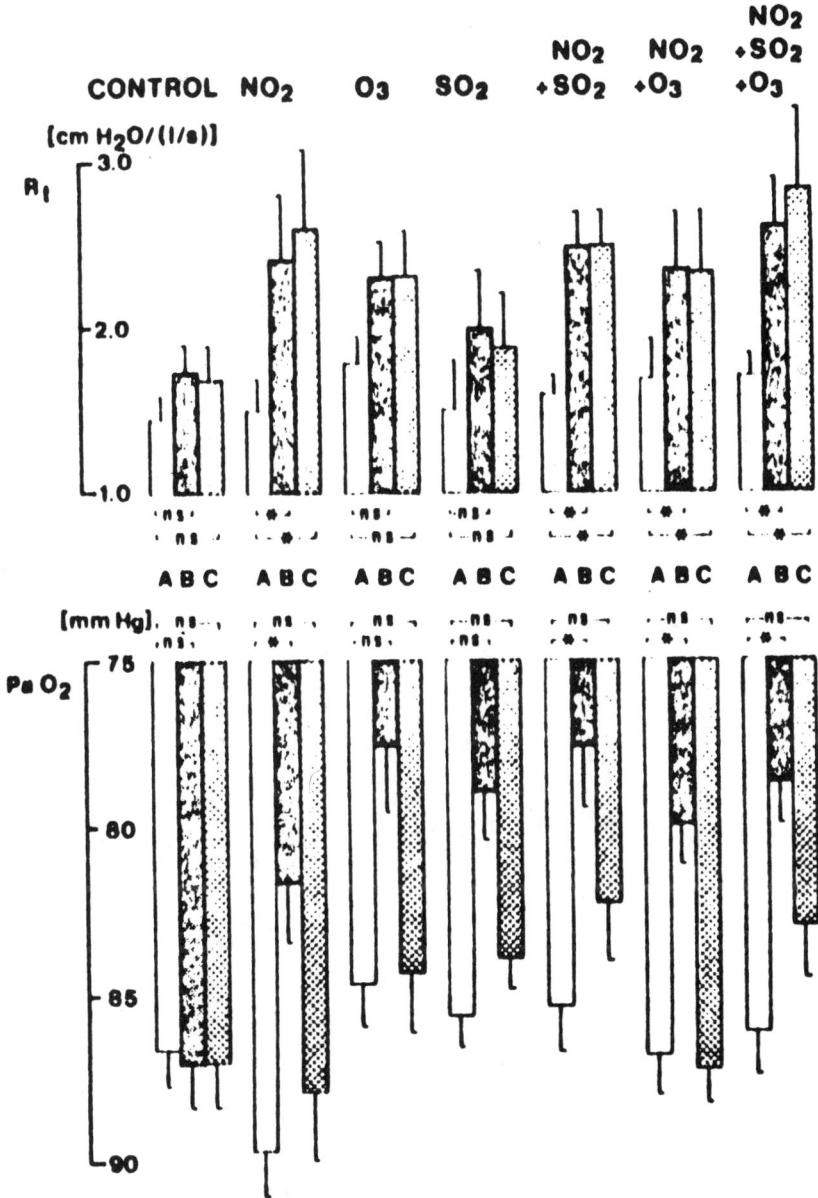

Figure 4. Influence of the single and combined action of NO_2, O_3 and SO_2 at TLV concentration on airway resistance R_t and arterial PO_2 PaO_2; exposure phase (B = 2 hr) as compared to the pre-exposure phase (A = 1 hr), the postexposure phase (C = 1 hr), and to the control (mean values ± S.D.M.). P ⩽ 0.02.

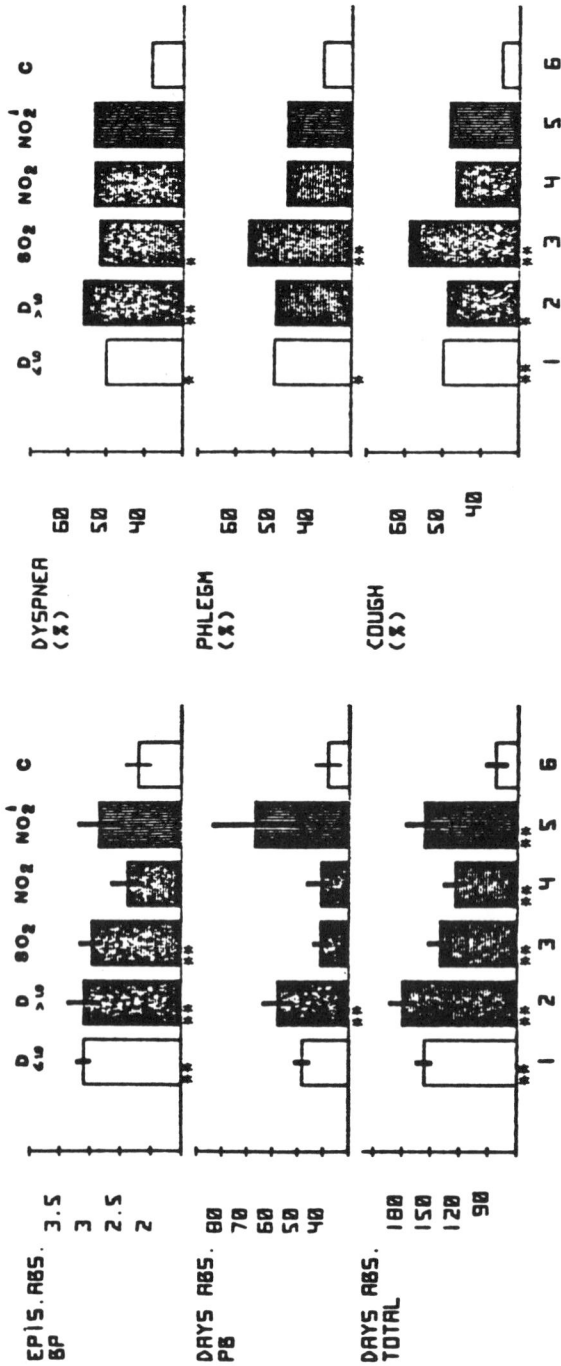

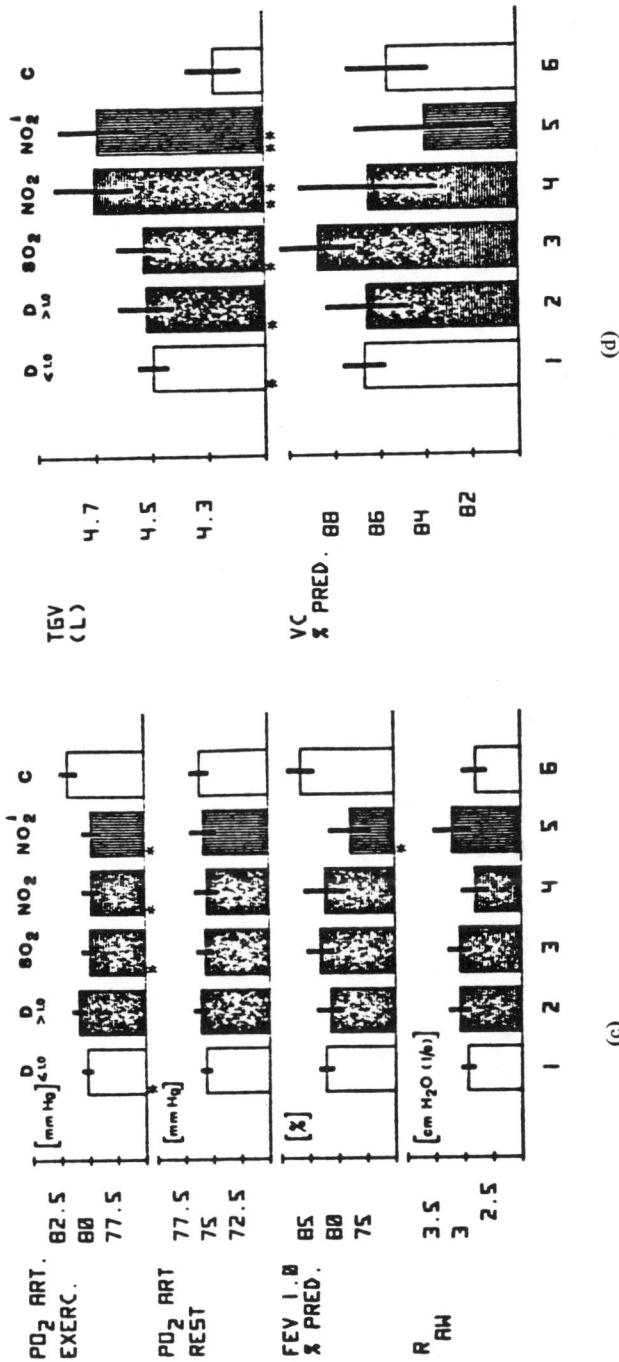

Figure 5. (a) Days of absence from work (days abs. total), days absent from work and episodes absent from work because of broncho-pulmonary disease (days abs. BP and epis. abs. BP, resp.) in the polluted groups as compared to the control group; (b) cough, phlegm and dyspnea at exercise in the polluted groups as compared to the control group; (c and d) differences in functional parameters of groups 1–5 as compared to the control group 6 (* = 1% < P ≤ 5%; ** = P ≤ 1%).

value of $\geqslant 4.1$ cm H_2O/liter/sec had only bronchitics. It was in this group that the two subjects with an initial value of over 10 cm H_2O/liter/sec were found.

Similar to the data demonstrated in Figures 2a and b in which a significant R_{aw} increase was seen after NO_2 inhalation, the ΔR_{aw} values of this group were not greater than 1.0 cm H_2O/liter/sec for initial R_{aw} values below 4.0 cm H_2O/liter/sec. With initial values above 4.0 cm H_2O/liter/sec, an increasing number of ΔR_{aw} values above 1.0 cm H_2O/liter/sec (r = 0.54) was found. Similar to Figure 2b it is evident that certain individuals may react much more strongly upon NO_2 inhalation than others. Although such persons could also be found in the group of the healthy subjects, they occurred far more frequently among the group of bronchitics.

Group 3 (Healthy Subjects)

The results and methods of evaluation for group 3 (Figure 4) have been described elsewhere [15]; they have shown that exposure to 5 ppm NO_2 for 2 hr resulted in a significant increase of R_{aw} in the range between 0.5 and 1.5 cm H_2O/liter/sec and a decrease of PaO_2 of about 7 mm of mercury. This phenomenon appeared also after 2 hr of exposure to SO_2 and O_3 in MAK concentrations ($SO_2 \approx 5$ ppm, $O_3 \approx 0.1$ ppm). However, the differences between the initial values and control were not significant.

The effect of NO_2 inhalation upon resistance values or PaO_2 was not enhanced by combining NO_2 with O_3, with SO_2 or with both SO_2 and O_3. In contrast to the exposure experiments with single pollutants, exposure to a combination of NO_2, SO_2 and O_3 led to a persistent increase of R_{aw} after the termination of exposure. This increase after a 1-hr postexposure phase, however, was not significant as compared to the 2-hr exposure value.

Epidemiological Studies

Results of the Pollutant Measurements

Figure 1 shows that groups 1–5 differed significantly regarding the composition of pollutants occurring in the various working areas.

- *Group 1* was exposed to respirable dust in a concentration $\leqslant 1.0$ mg/m³; the mean concentration was 0.43 mg/m³.
- *Group 2* was exposed to respirable dust concentrations > 1 mg/m³ with a mean value of 2.42 mg/m³.
- *Group 3* was exposed simultaneously to mean values of 0.42 ppm SO_2 and 2.48 mg/m³ dust; in these working areas only traces of NO_2 (mean value 0.03 ppm) were present.
- *Group 4* was exposed to mean NO_2 and dust concentrations of 0.16 ppm and 0.94 mg/m³, respectively.
- *Group 5.* The exposure pattern in this group was characterized by peak concentrations of NO_2 (i.e., $\geqslant 0.5$ ppm) which occurred at least five times during one working shift. The mean NO_2 values were also higher than in the previous groups. At the same time mean concentrations for SO_2 were 0.22 ppm and for respirable dust 1.16 mg/m³.
- *Group 6.* No continuous or discontinuous measurements of pollutants were made at the workplaces of the office employees. Spot checks with manual procedures indicated extremely low background concentrations of the pollutants.

Clinical and Functional Data

Regarding age, size and weight, the values for the experimental groups resembled those of the control group with a mean age of 47.6 ± 0.81 years, a mean size of 174.0 ± 0.62 cm and a mean weight of 80.4 ± 1.08 kg. The slight differences found were not significant, however. Smoking habits and number of smokers were relatively uniform with the highest numbers in group 5 with 93% and in the control group with 83% of the total numbers being smokers.

Looking at the days of absence from work (Figure 5a) it is obvious that all groups exposed to air pollutants showed a greater incidence of sick-leave days than the nonexposed control group. These differences were highly significant for all groups. If one examines the absenteeism due to bronchopulmonary diseases there were greater differences for group 2 (high dust exposure) and group 5 (exposure to NO_2 at peak concentrations) but only the difference between control group and group 2 was significant. Episodes due to illnesses of the bronchopulmonary system occurred far more frequently among groups 1–3.

Symptoms like cough, phlegm and dyspnea (Figure 5b) were more frequent in all groups compared with the controls. The difference was, however, significant only for groups 1–3.

In virtually all groups, the same functional parameters (Figure 5c and d) were found to be significantly different on the 1 and 5% level, respectively. These parameters were PO_2 during exercise and thoracic gas volume (TGV). In group 5 (with NO_2 peak exposure), FEV 1.0 was significantly lower than in all other groups.

PO_2 at rest and also VC and R_{aw} did not show a significant change. In group 5, however, there seemed to be a tendency towards higher R_{aw} values. The R_{aw} values for groups 2, 3 and 5 were between 3.0 and 3.5 cm H_2O/liter/sec, corresponding to the range of the threshold to pathological values.

Table I gives a synopsis of the statistically significant differences (on the 1 and 5% level, respectively) between the control group and the exposure groups.

DISCUSSION

The results of the clinical studies have shown that NO_2 inhalation caused an increase of airway resistance in both healthy subjects and patients with chronic bronchitis: the increase of R_{aw} was within a range of 0.5 and a maximum value of 1.5 cm H_2O/liter/sec in healthy subjects, while these values were slightly higher in the group of chronic bronchitics. However, in both groups individuals with a markedly stronger reaction were found.

In none of the cases the increase of R_{aw} values reached levels like those recently observed by Ulmer [26] after inhalation of 7 ppm SO_2 for 5 min in clinically treated chronic bronchitics: After SO_2 inhalation, R_{aw} increased up to values of 30 cm H_2O/liter/sec in certain patients, requiring acute therapeutic measures. Healthy subjects did not show such excessive reactions.

Table I. Parameters Investigated in Six Groups of Employees of a Steel Works Differently Exposed to Dust, SO_2 and NO_2; Levels of Significant Differences Between Groups 105 and the Control Group [26]

Group	Dust $\leqslant 1$ mg/m³	Dust > 1 mg/m³	Mean SO_2[a]	Mean NO_2[a]	Peak NO_2[b]	Control
	1	2	3	4	5	6
Episodes–bronchopulmonary	xx	xx	xx	–	–	
Days–bronchopulmonary	–	xx	–	–	–	
Days total	xx	xx	xx	xx	xx	
Dyspnea	x	xx	x	–	–	
Phlegm	x	–	xx	–	–	
Cough	xx	x	xx	–	–	
PO₂ exercise	x	–	x	x	x	
PO₂ rest	–	–	–	–	–	
FEV 1.0	–	–	–	–	x	
R_{aw}	–	–	–	–	–	
TGV	x	x	–	xx	xx	
VC	–	–	–	–	–	
n	414	125	139	83	58	110

x	$\hat{=} 1\%$	P	$\leqslant$	5%
xx	$\hat{=}$ P	$\leqslant$	1%	

[a]Mean concentration for one work shift.
[b]Peak concentrations $\geqslant 0.5$ ppm.

As both investigations were performed under similar conditions and the groups exposed to NO_2 and SO_2 were clinically very similar, different mechanisms for the effects of NO_2 and SO_2 inhalation must be postulated: from published data there is evidence [27–29] that SO_2 acts by reflex bronchoconstriction and NO_2 probably by release of mediators like histamine [4].

As can be seen from Figure 2a and b the individual reaction upon NO_2 inhalation was relatively uniform on different days in the individual subjects, i.e., the day-to-day variation of the R_{aw} values in the individual subjects corresponded to the statistical variation on the method used. The variation of the individual initial values on different days in healthy subjects was very small, generally, whereas in chronic bronchitics the variation became greater, especially with higher initial values.

It is obvious that in the concentration range used practically all subjects reacted with an R_{aw} increase. Inferring that the method shows a certain variability to both positive and negative values, this means that the effects were so strong that R_{aw} practically always increased. At lower concentration ranges this is not the case [3].

As shown in Figure 3 there seems to be a dependency between the extent of the R_{aw} increase and the initial R_{aw} value after NO_2 exposure. The reason for this could be that changes of the bronchial diameter influence airway resistance at a ratio of four powers of the radius. This would mean relatively small changes at low initial R_{aw} and stronger changes at higher initial R_{aw}.

Both Figures 2b and 3 include individual subjects who showed stronger reactions on NO_2 inhalation than did the rest of the subjects. This indicates that some subjects were especially sensitive to NO_2 inhalation.

From the clinical point of view, laboratory experiments are in principle a quite simplified model when trying to assess the effects of air pollutants under ambient or realistic conditions. These experiments may, however, furnish precise data regarding dose/effect relationships and the intra- and interindividual variability of individuals.

However, an extrapolation from the effects seen in these acute exposure experiments to long-term impairments caused by chronic exposure usually is not possible or only with very strong limitations. These long-term effects can only be assessed by epidemiological methods.

From the data of this epidemiological study it becomes evident that the dust-exposed group, the NO_2-exposed groups and the SO_2- and dust-exposed groups clearly showed more symptoms of bronchopulmonary disease, more days of absence from work and also lung function impairments than a control group of office employees.

As can be seen from Figure 5b the frequency of the occurrence of symptoms like cough, phlegm and dyspnea was quite high also in the control group of the employees: this might be due to influences like smoking or due to the relatively high ambient air pollution level in the areas of residence of these individuals.

A direct, toxicological comparison of the single groups exposed to different pollutants is possible only to a limited degree because of the complexity of the conditions in the working environment within a steel works. Besides the inhalative pollutants themselves, a series of factors might be responsible in this case, e.g., work in great heat, working outdoors, in open drafty halls or the different pathogenicity of the inhaled dust. All these factors might even act synergistically. This might be an explanation for the relatively similar changes in all exposure groups.

When comparing the individual exposure groups of Table I with the control group the least frequency of significant changes is found in the NO_2-exposed groups. Although there is a clear trend as can be seen in Figure 5 of a higher frequency of cough, phlegm and dyspnea at exertion and higher values for R_{aw} in group 5, the group with the NO_2 peak exposure, are relatively high, these differences are not statistically significant because of the limited number of cases. Furthermore, it must be taken into account that in group 4 which had been exposed mainly to NO_2, the mean NO_2 concentration was relatively low (0.16 ppm). This value was even below that of the ambient air standard for short-term exposure (TA-Luft [30]). The most frequent changes of lung function parameters were found in group 5, the group with the NO_2 peak concentrations and the highest NO_2 mean exposure values. In this group, the possibility of a combined effect of SO_2 also has to be taken into account, as can be seen from Figure 1. A combined effect of SO_2 and NO_2 is also indicated in the experimental studies of the combined effects of NO_2, SO_2 and O_3 which resulted in a prolongation of the recovery time (Figure 4).

In this NO_2 peak group besides an increased TGV, there was also a decrease of FEV 1.0 in combination with a decrease of arterial PO_2 during exercise. This pattern might be interpreted as an indication of emphysematous changes. However, a clear differentiation between the groups exposed to NO_2 and those exposed to dust was not possible because all groups had been exposed to dust and the group exposed to the smallest dust concentrations also showed similar changes.

The presence of dust in the working place even in low concentrations seems to be a decisive factor in the development of chronic lung disease. The results seem to indicate, however, that there is also an additive effect of gaseous pollutants and particulates. It will be indispensable, besides reducing the concentration of gaseous pollutants in the working environment, also to reduce occupational exposure to suspended particulates.

SUMMARY

The effects of NO_2 on healthy subjects and chronic bronchitics were investigated under controlled laboratory conditions. The results showed that amongst a group of bronchitics there were individuals that react stronger to exposure than the rest. However, airway resistance increase never reached the values seen upon exposure to SO_2 under similar conditions.

Epidemiological studies of workers exposed to different concentrations of pollutants in working areas, such as dust, NO_2 and SO_2, indicated that in all the exposed groups there were clearly more symptoms of bronchopulmonary disease, more days of absence from work and lung function impairments than in the control group of unexposed office employees. The presence of dust in the working environment, even in low concentrations, seems to be a decisive factor in the development of chronic lung disease. The results seem to indicate, however, that there is also an additive effect of gaseous pollutants and particulates. It is considered as a necessity, in addition to reducing the concentration of gaseous pollutants in the working environment, also to reduce the occupational exposure to suspended particulates.

ACKNOWLEDGMENTS

Supported by grant No. 145-77-1 ENV D (European Community) and grant No. 104 01 023 (German Federal Government).

REFERENCES

1. Beil, M., and W. T. Ulmer. *Int. Arch. Occup. Environ. Health* 38:31 (1976).
2. von Nieding, G., et al. *Int Arch. Arbeitsmed.* 27:234 (1970).
3. von Nieding, G., et al. *Int. Arch. Arbeitsmed.* 27:338 (1971).
4. von Nieding, G., et al. *Int. Arch. Occup. Environ. Health* (In press).
5. "Maximale Arbeitsplatz Konzentrationen–1977," Deutsche Forschungsgemeinschaft, Bonn-Bad Godesberg, FRG.
6. Freeman, G., et al. *Arch. Environ. Health* 17:181 (1968).
7. Valentin, H., et al. *Arb. Med., Soz. Med., Präv. Med.* 14:25 (1979).
8. "Chronic Bronchitis," *Deutsche Forschungsgemeinschaft* (Boppard: Boldt-Verlag, 1978).
9. "Oxides of Nitrogen," *Environmental Health Criteria 4*, World Health Organization, Geneva, Switzerland (1977).
10. Wagner, H. M. Commission of the European Communities Report No. EUR 5436e (1976).
11. Nolte, D., E. Reiff and W. T. Ulmer. *Respiration* 25:14 (1968).
12. Muysers, K., U. Smidt and F. W. Buchheim. *Pflügers Archiv.* 307:211 (1969).
13. Krekeler, H., et al. *Pneumonologie* 146:34 (1971).
14. Löllgen, H., and G. von Nieding. *Verhandl. Dtsch. Gesellsch. Inn. Med.* 84:389 (1978).
15. von Nieding, G., et al. *Prax. Pneumonologie* 146:858 (1977).
16. von Nieding, G., and H. Krekeler. *Int. Arch. Arbeitsmed.* 29:55 (1971).
17. Bolt, W., et al. "Communaute européenne du charbon et de l'acier," *Collection d'hygiene et de medecine du travail No. 11. Aide-memoire pour la pratique de l'examen de la fonction ventilatoire par la spirographie.* Luxembourg (1971), 127 pp.
18. Saltzman, B. E. *Anal. Chem.* 26:1949 (1954).
19. Breuer, W. *VDI Report 149* (Düsseldorf: VDI, 1970).
20. *VDI Guideline 2451* (Düsseldorf: VDI, 1968).
21. *VDI Guideline 2468* (Düsseldorf: VDI, 1974).
22. Coehnen, W. *Staub-Reinhaltung Luft* 35:452 (1975).
23. Smidt, U. *Meth. Inform. Med.* 16:96 (1977).
24. Wilcoxon, F. *Biometrics* 1:30 (1945).
25. Koller, S. *Neue graphische Tafeln zur Beurteilung statistischer Daten* (Darmstadt: Steinkopff-Verlag, 1969).
26. Ulmer, W. T., and M. S. Islam. *VDI Report 135* (Düsseldorf: VDI, 1978), p. 135.
27. Islam, M. S., E. Vastag and W. T. Ulmer. *Int. Arch. Arbeitsmed.* 29:-21 (1972).
28. Nadel, J. A., et al. *Arch. Environ. Health* 10:175 (1965).
29. Tomono, Y. *Japan J. Ind. Health* 3:77 (1961).
30. *Technische Anleitung Luft* (Köln: Heymanns-Verlag, 1974).

CHAPTER 21

LUNG FUNCTION STUDIES ON INTERMITTENTLY EXERCISING HIGH SCHOOL STUDENTS EXPOSED TO AIR POLLUTION

Jun Kagawa and Kieko Tsuru

Department of Environmental Medicine and Occupational Health
Tokai University School of Medicine
Isehara-shi, Japan

Tohru Doi

Department of Epidemiology
College of Health Sciences
University of the Ryukyu
Naha, Japan

Tohru Tsunoda, Toshio Toyama and Masahiro Nakaza

Department of Preventive Medicine and Public Health
Keio University School of Medicine
Tokyo, Japan

INTRODUCTION

There are few studies of the acute effects of air pollution on the lung function of exercising students. Wayne et al. [1] noted high correlations of decreased athletic performance and increased oxidant levels. Lebowitz et al. [2] found a postexercise decrease of forced vital capacity (FVC) and one-second forced expiratory volume in children and adolescents produced by exposures to pollutants of either oxidant or *reducing type* and high temperature. McMillan et al. [3] measured resting lung function of Los Angeles elementary school children in relation to pollution levels and found no significant changes in peak expiratory flowrate, which was correlated with acute changes in air pollution. However, Kagawa and Toyama [4] found

333

significant correlations among changes in the parameters of specific airway conductance (G_{aw}/V_{tg}) and flow volume curves of resting elementary school children with temperature, nitric oxide (NO), ozone (O_3), hydrocarbons, sulfur dioxide (SO_2), nitrogen dioxide (NO_2), relative humidity, suspended particulate matter (SPM) and oxidant, in that order, and some children seemed to react quite sensitively to environmental factors. Among the environmental factors, temperature affected various lung function parameters more than the others.

To investigate the acute effects of air pollution, we studied lung function in exercising students exposed to ambient atmospheres. More specifically, we attempted to delineate whether lung function would be significantly influenced by specific environmental factors.

METHODS

Selection of a Place for Investigation and Study Subjects

Ebara high school was chosen for the study because of its willingness to cooperate. We studied 9 18-year-old boys who were considered healthy on the basis of their physical status. None was a smoker or exsmoker.

Environmental Factors

The air pollutant monitoring system installed at the school provided the concentrations of 5 different air pollutants and the temperature continuously (Table I).

Table I. Measurement of Environmental Factors

Environmental Factors	Measurement Methods	Measurement Interval
Oxidant (pphm)	Neutral buffered potassium iodide method	Continuous and 1-hr avg.
	Coulometric method	
Ozone (pphm)	Ethylene-chemiluminescence method	Continuous and 1-hr avg.
Nitrogen Dioxide (pphm)	Colorimeter with Saltzman method	1-hr avg.
Sulfur Dioxide (pphm)	Conductimetric method	1-hr avg.
Suspended Particulate Matter (mg/m³)	Light scattering method	1-hr avg.
Temperature (°C)	Platinum resister	Continuous

Lung Function Tests

Measurements of airway resistance (R_{aw}) and volume of thoracic gas (V_{tg}) were made by the pressure-type body plethysmographic method [5,6]. R_{aw} values were converted to the reciprocal, G_{aw}. All data were calculated as G_{aw}/V_{tg}, where V_{tg} was the lung volume at which R_{aw} was measured. Six measurements of G_{aw}/V_{tg} were averaged. As a next step, a flow volume curve was recorded at least three times on an X-Y recorder (Hewlett Packard, Model 7041M) using a Wedge-type spirometer (Med Science, Model 570). The flow volume curves were obtained, breathing air, then after 5 inspirations (3 vital capacity and 2 tidal inspirations) of a 80% He-20% O_2 gas mixture. For FVC, $\dot{V}_{max}$ at 50% or 25% FVC, the averages of the two best and most nearly identical flow volume curves were selected for analysis.

As the previous study [4] showed that temperature affected various lung function tests significantly, lung function was measured every 30 minutes between 10:45 A.M. and 14:45 P.M. for 2 consecutive 5-day periods in July, 1977 to minimize the effect of temperature variation on lung function. We asked the subjects to exercise as the controlled human exposure studies [7,8] had shown that the effects of air pollutants on lung function could be enhanced by exercise.

After the measurement of 3 baselines (for example, subject 1 performed lung function tests at 10:45, 11:15 and 11:45 A.M. to get a baseline value), the subject exercised intermittently on an outdoor playing field riding a bicycle ergometer at an approximately 50-W, 50-rpm workload for 15-minute periods, alternating with a period of rest for 2 hours from 11:45 A.M. to 13:45 P.M. Lung function measurement was made before each exercise, so four measurements were obtained during each exercise period. After exercise, two measurements were followed by two 30-minute recovery periods.

Data Analysis

The specific aim of this study was to see whether lung function would be significantly influenced by acute exposure to specific environmental factors by evaluating alteration in lung function due to various levels of environmental factors. Effects of environmental factors on lung function were evaluated as follows:

1. The effect of exercise was calculated as the percent change from baseline to exercise period each day for each subject separately to exclude the influence of individual variations.
2. Environmental factors corresponding to the measurement period of lung function were also calculated as the percent change.

Simple correlation coefficients were calculated between the percent changes of lung function and environmental factors during the exercise period to the baseline period. A multiple linear regression analysis [9] was performed using various combinations of the weather and pollution measurements as the independent variables and the lung function measurements as the dependent variable. The square of the multiple correlation coefficient (R^2) was calculated for each subject. R^2 could then be compared among subjects and used as a relative measure to explain variance of lung function.

RESULTS

Variations of Environmental Factors

During the experiment, the temperature varied between 22.5 and 32.1°C; SO_2 between 0.3 and 1.3 pphm; NO_2 between 1.1 and 11.2 pphm; O_3 between 0.3 and 18.0 pphm; oxidant between 1.9 and 16.2 pphm; and SPM ranged between 0.02 and 0.102 mg/m³. The differences in environmental factors during intermittent exercise from those during baseline measurement of lung function were −0.1~2.6°C for temperature; −0.3~0.2 pphm for SO_2; −6.2~1.6 pphm for NO_2; −7.0~0.6 pphm for O_3; −5.9~4.8 pphm for oxidant; and −0.025~0.008 mg/m³ for SPM.

Correlation Coefficients Between Lung Function Measurements and Environmental Factors

Table II shows the correlation coefficients among environmental factors during the measurement of lung function. Highly significant positive correlation coefficients were seen between SO_2 and NO_2, temperature and oxidant, oxidant and O_3, SPM and NO_2, SPM and O_3, SPM and oxidant.

Selected measurements of lung function (with means and standard deviations) and the number of study days are listed in Table III. Although subject 9 showed the low values of lung function tests, which suggested the existence of airway obstruction, he had neither a history of lung disease nor any respiratory symptoms.

Table IV shows the significance of correlation coefficients between the percent changes of lung function and environmental factors for each subject.

Table II. Correlation Coefficients Between Environmental Factors[a]

	Temperature	SO_2	NO_2	O_3	Oxidant	SPM
Temperature		−0.001	−0.373[b]	0.325[b]	0.362[c]	−0.094
SO_2			0.609[c]	−0.407[b]	−0.221	0.134
NO_2				0.001	−0.047	0.619[c]
O_3					0.978[c]	0.665[c]
Oxidant						0.546[c]
SPM						

[a]Number of measurements: 38–50.
[b]Level of significance: $P < 0.05$.
[c]Level of significance: $P < 0.01$.

Table III. Values for Selected Lung Function Measurements
with Mean and Standard Deviation

Subject No.	Standing Height (cm)	Study Days	G_{aw}/V_{tg} (1/cmH$_2$O·sec)	$\dot{V}_{max}$ (50% FVC) (liter/sec)	$\dot{V}_{max}$ (25% FVC) (liter/sec)
1	167.0	5	0.207 ± 0.010	6.95 ± 0.41	3.92 ± 0.32
2	171.5	6	0.214 ± 0.010	5.94 ± 0.28	3.54 ± 0.22
3	167.5	10	0.193 ± 0.012	10.13 ± 0.80	5.39 ± 0.53
4	166.0	10	0.187 ± 0.014	5.28 ± 0.25	2.74 ± 0.23
5	178.5	9	0.210 ± 0.009	11.31 ± 0.70	6.14 ± 0.74
6	161.0	5	0.214 ± 0.011	7.25 ± 0.51	3.94 ± 0.33
7	174.5	10	0.195 ± 0.009	7.11 ± 0.41	3.90 ± 0.24
8	162.5	4	0.206 ± 0.006	5.66 ± 0.41	2.80 ± 0.28
9	166.0	10	0.113 ± 0.012	3.24 ± 0.22	1.65 ± 0.12

The lung function parameters, which were significantly influenced by environmental factors, appeared to be G_{aw}/V_{tg}. Among environmental factors, oxidant and SPM significantly influenced various lung function parameters. If one considers the correlation coefficients at the level of $p < 0.01$ only, it becomes clear that oxidant and SPM were significantly correlated with lung function.

Square of the Multiple Correlation Coefficient (Coefficient of Multiple Determination) Between Lung Function Measurements and Environmental Factors

Multiple linear regression analyses were performed for each subject, utilizing each of lung function tests (G_{aw}/V_{tg}, $\dot{V}_{max}$ at 50% and 25% FVC) as the dependent variable and each of environmental factors (temperature, SO$_2$, NO$_2$, oxidant and SPM) as the independent variable. The percentage increase of $\dot{V}_{max}$ at 50% FVC while breathing He-O$_2$ as compared to air was not used, since this index showed little significant correlation with environmental factors.

R^2 between each lung function test and each environmental factor for every subject is shown in Figure 1. For example, 62.5% of the total variance of G_{aw}/V_{tg} of subject 1 was attributable to the combined effects of temperature, SO$_2$, NO$_2$, oxidant and SPM, and the increment of R^2 attributable to SO$_2$ was 17.1% in the presence of the other four variables (temperature, NO$_2$, oxidant and SPM). The significance of the increment of R^2 attributable to SO$_2$ and others was evaluated from the significance of the standardized partial correlation coefficient of SO$_2$ and others. Although it is impossible to evaluate separately the degree of each environmental factor's contribution because significant correlations between environmental factors are seen in

Table IV. Subjects Who Showed the Significant Correlation Between Lung Function Measurement and Environmental Factors

Subject No.	1				2				3				4				5			
	G/V[a]	V50	V25	H/A	G/V	V50	V25	H/A	G/V	V50	V25	H/A	G/V	V50	V25	H/A	G/V	V50	V25	H/A
Temperature	⊛[b]		⊛	*[c]		⊛	⊛		*											
SO₂	*																			
NO₂			*										⊛⊛							
O₃									*	**[d]										⊛
Oxidant	*					*			**	**					*					
SPM			**				**		*	**	**	**				*		**		

Subject No.	6				7				8				9			
	G/V	V50	V25	H/A	G/V	V50	V25	H/A	G/V	V50	V25	H/A	G/V	V50	V25	H/A
Temperature										*						
SO₂							*				*					
NO₂	*															
O₃											*					
Oxidant		*							*	**					*	
SPM					*		*	*								

[a]G/V: G_{aw}/V_{tg}, V50 and V25: V̇max at 50% and 25% FVC, H/A: Percent increase of V50 while breathing He-O₂ as compared to air.
[b]Positive correlation.
[c]Level of significance: $P < 0.05$.
[d]Level of significance: $P < 0.01$.

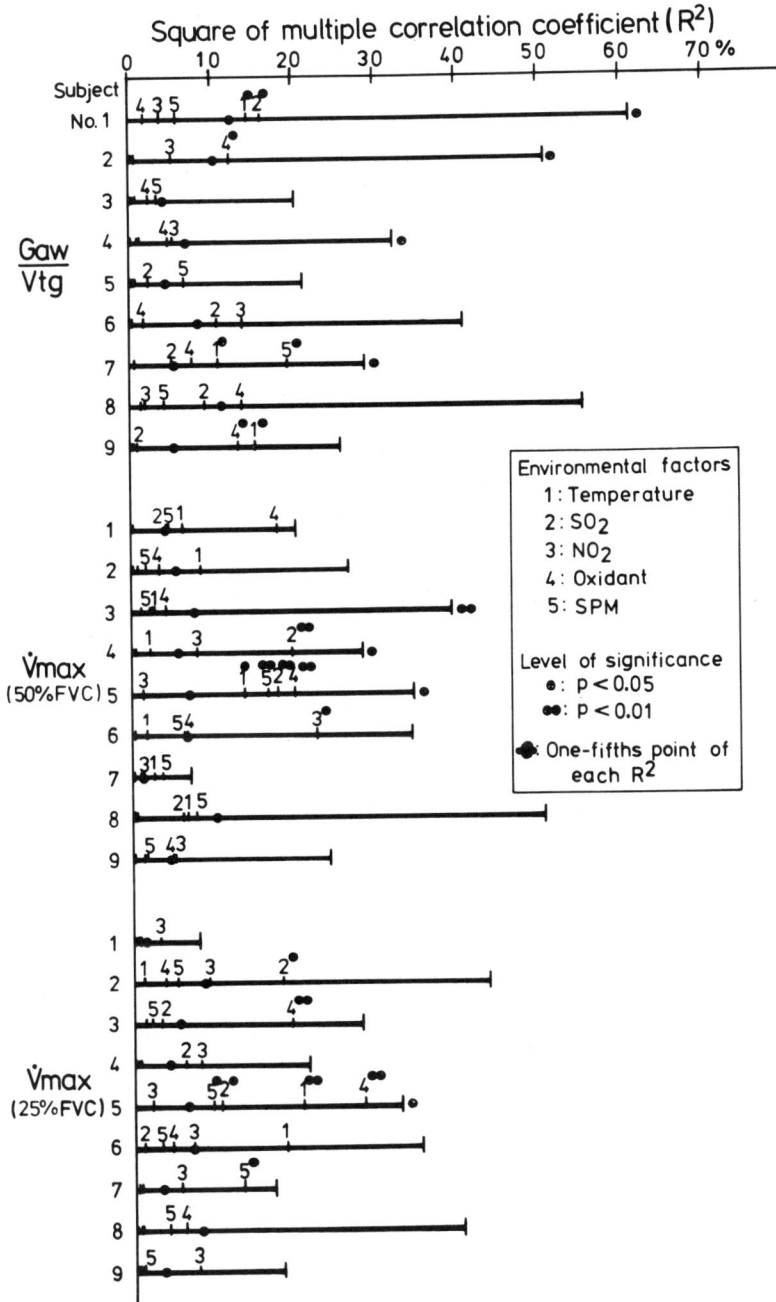

Figure 1. The square of multiple correlation coefficient between lung function tests and environmental factors.

this study, one may estimate the independent contribution of one factor from the degree of the increment of R^2 attributable to one factor in the presence of the other four factors. It may be considered that the factor will contribute considerably to the total variance of lung function if the increment of R^2 attributable to this factor is large. One-fifths point of R^2 may be considered a critical point for estimating the contribution of each environmental factor to the total variance of lung function from the degree of the increment of R^2. For example, it was considered that the contribution of SO_2 or temperature to the total variance of G_{aw}/V_{tg} of subject 1 was significant among environmental factors because their increments of R^2 were greater than one-fifths point of R^2.

As for G_{aw}/V_{tg}, the significant R^2 was shown in subjects 1, 2, 4 and 7. The significant standardized partial correlation coefficient (SPCC) was shown for SO_2 and temperature in subject 1, oxidant in subject 2, SPM and temperature in subject 7, temperature and oxidant in subject 9.

For $\dot{V}_{max}$ at 50% FVC, the significant R^2 was shown in subjects 3, 4 and 5. The significant SPCC was shown for SO_2 in subject 4; oxidant, SO_2, SPM and temperature in subject 5; NO_2 in subject 6.

As for $\dot{V}_{max}$ at 25% FVC, the significant R^2 was shown only in subject 5. The significant SPCC was shown for SO_2 in subject 2; oxidant in subject 3; oxidant, temperature, SO_2 and SPM in subject 5; SPM in subject 7. The increments of R^2 attributable to these factors showing the significant SPCC are over one-fifths point of R^2. Among these factors, oxidant seems to have a considerable effect on total variance of lung function from the viewpoint of increment of R^2.

DISCUSSION

Lung function studies have been used in a variety of epidemiological respiratory studies to correlate the severity of the airway obstruction with environmental quality. The airway obstruction is usually evaluated by measuring FVC and $FEV_{1.0}$, which is an integral part of all epidemiological respiratory studies. However, percent $FEV_{1.0}$ is considered less sensitive for detecting mild airway obstruction. Lung function tests should be sufficiently sensitive to detect the changes of airway to environmental changes in ambient atmospheric conditions. The most sensitive methods selected for this epidemiological study were the measurements of G_{aw}/V_{tg} to examine the change in the upper airway, and flow volume curves obtained by breathing air and after washing with 80% He–20% O_2 gas mixture to study the change in the lower airway.

The results indicated several significant negative correlation coefficients between some lung function tests and some environmental pollutants in each subject, except subjects 4 and 5, who showed a positive correlation with SO_2. Subjects 1 and 3 showed many significant correlations between their lung function tests and environmental factors, and were considered sensitive to the environmental factors. Among lung function parameters used in this study, G_{aw}/V_{tg} appeared to be most appropriate for examining the acute changes of lung function associated with environmental factors. Of those environmental factors considered, oxidant was shown to be highly correlated with lung function changes. The correlation between lung function tests and NO_2 exposure was less significant than oxidant.

Do these results suggest that some environmental factors contribute more significantly toward combined change in lung function measurements? Multiple regression analysis was performed to evaluate these problems. When we considered the environmental conditions (consisting of five different environmental factors—temperature, SO_2, NO_2, oxidant and SPM), 4 subjects among 9 showed significant R^2 for G_{aw}/V_{tg}; 3 subjects for $\dot{V}_{max}$ at 50% FVC; and only 1 for $\dot{V}_{max}$ at 25% FVC. This suggests the combined contribution of temperature, SO_2, NO_2, oxidant and SPM to the total variance of lung function in about half the subjects. However, there is considerable individual variability of response to the environmental factors. Generally speaking, the total variance of lung function was attributable to the combined effects of environmental factors rather than to one specific environmental factor, as shown in Figure 1. However, oxidant seems to play a significant role in total variance of lung function, and NO_2 seems to contribute in combination with other environmental factors, but not independently. From the finding that R^2 for G_{aw}/V_{tg} in each subject had a tendency to show a greater value than $\dot{V}_{max}$, the upper airway might be more influenced than the lower airway by specific environmental factors. This finding is supported by the results of the controlled human exposure studies [7,8], showing that O_3 may have the greater effect on lung function in comparison with SO_2 and NO_2 in the same concentration from the 2-hour exposure studies with intermittent exercise. And G_{aw}/V_{tg} shows the significant change even if $\dot{V}_{max}$ and gas distribution index (ΔN_2) do not show any.

Thus, the statistically significant correlations observed in this study indicate that the normal variations of lung function may be significantly influenced by specific environmental factor(s). Approximately 30–40% of the total variance of lung function may be attributable to the combined effects of temperature, SO_2, NO_2, oxidant and SPM, which are considered to have a great influence on lung function among environmental factors. The remaining 60–70% may be attributable to other factors, such as host. It might be possible to observe greater influence of environmental factors on the change in lung function if a more sophisticated analysis were used, such as a multiple regression analysis including a quadratic equation.

CONCLUSIONS

The acute effects of air pollution on the lung function of 9 high school students with two hours intermittent exercise were investigated during 2 consecutive 5-day periods. Multiple regression analysis was performed for each subject, utilizing each lung function test as the dependent variable and each environmental factor as the independent variable. The evaluation of the square of the multiple correlation coefficient suggests that the significant contributions of the environmental factors to the total variance of lung function were seen in about half the subjects.

ACKNOWLEDGMENTS

This study was supported by a Tokyo Photochemical Air Pollution Research Grant from the Department of Hygiene, Tokyo Metropolis. M. Nagasaki, M.D. and I. Tominaga aided in this study.

REFERENCES

1. Wayne, W. S., P. F. Wehrle and R. E. Carroll. "Oxidant Air Pollution and Athletic Performance," *J. Am. Med. Assoc.* 199(12):901–904 (1967).
2. Lebowitz, M. D., P. Bendheim, G. Cristea, D. Markovitz, J. Misiaszek, M. Staniec and D. V. Wyck. "The Effect of Air Pollution and Weather on Lung Function in Exercising Children and Adolescents," *Am. Rev. Resp. Dis.* 109:262–273 (1974).
3. McMillan, R. S., D. H. Wiseman, B. Hanes and P. F. Wehrle. "Effects of Oxidant Air Pollution on Peak Expiratory Flow Rates in Los Angeles School Children," *Arch. Environ. Health* 18:941–949 (1969).
4. Kagawa, J., and T. Toyama. "Photochemical Air Pollution: Its Effects on Respiratory Function of Elementary School Children," *Arch. Environ. Health* 30:117–122 (1975).
5. DuBois, A. B., S. Y. Botelho and J. H. Comroe, Jr. "A New Method for Measuring Airway Resistance in Man using A Body Plethysmograph: Values in Normal Subjects and in Patients with Respiratory Diseases," *J. Clin. Invest.* 35:327–335 (1956).
6. DuBois, A. B., S. Y. Botelho, G. N. Bedell, R. Marshall and J. H. Comroe, Jr. "A Rapid Plethysmographic Method for Measuring Thoracic Gas Volume," *J. Clin. Invest.* 35:322–326 (1956).
7. Kagawa, J., and K. Tsuru. "Respiratory Effects of 2-hour Exposure to Ozone and Nitrogen Dioxide Alone and in Combination in Normal Subjects Performing Intermittent Exercise," *Jap. J. Thor. Dis.* 17:(1979) (in press).
8. Kagawa, J., and K. Tsuru. "Respiratory Effects of 2-hour Exposure with Intermittent Exercise to Ozone and Sulfur Dioxide Alone and in Combination in Normal Subjects," *Jap. J. Hyg.* 34:(1980) (in press).
9. Draper, N. R., and H. Smith. *Applied Regression Analysis* (New York: John Wiley & Sons, Inc., 1966).

CHAPTER 22

HEALTH EFFECTS OF INDOOR NO$_2$ EXPOSURE PRELIMINARY RESULTS

Frank E. Speizer

Channing Laboratory
Department of Medicine
Harvard Medical School and
Peter Bent Brigham Hospital
Boston, Massachusetts 02115

Benjamin Ferris, Jr.

Department of Physiology

Yvonne M. M. Bishop

Department of Biostatistics

John Spengler

Department of Environmental Health Sciences
Harvard School of Public Health
Boston, Massachusetts 02115

INTRODUCTION

There is little doubt that nitrogen dioxide (NO$_2$) at high concentrations can be associated with acute pulmonary edema and death. Early descriptions of Silo Filler's disease adequately described the indoor exposure of farmers to concentrations of NO$_2$ in excess of 200 ppm, with a resultant acute occurrence of pulmonary disease and occasionally death [1]. Those subjects surviving such exposures might develop pulmonary fibrosis or recover in a few days or weeks.

Concern over the effects of chronic exposure indoors to lesser concentrations of NO$_2$ has led to studies of children [2] and housewives [3] that have given inconsistent results. Melia [2] reported excess lower respiratory disease rates in British school children living in households where gas stoves were used for cooking compared with children from households using electric stoves. These differences could not be explained by social class gradients or differences in household size; however, this study did not account for the smoking habits of the parents. Subsequently, Melia et al. [4] documented that households with gas stoves had seven times the concentration of NO_2 in the kitchen than households with electric cooking devices. Similar studies [5] in the United States have indicated about fourfold higher levels of NO_2 in kitchens of households with gas stoves. The source of the NO_2 appears to be the rapid oxidation of nitrous oxide (NO), which is generated when natural gas as a cooking fuel is burned in the atmosphere. The conversion is rapid, and levels of NO_2 disperse rapidly throughout the house. In contrast to Melia's study, a study of adult women does not suggest an effect in terms of respiratory disease rates from living (and working) in households with gas rather than electric stoves [3].

The results reported here were obtained as part of a long-range prospective study on health effects of exposure to ambient levels of pollutants resulting from the burning of fossil fuels. In each of six communities in the eastern United States, random samples of adults aged 25–74 are seen every three years. School children (initially seen in grades 1 and 2) are seen annually. In the children, information was collected from a questionnaire on home heating and cooking sources, air conditioning smoking status of parents and a number of health-related parameters. The children are seen in school, where height and weight are measured and spirometric pulmonary function testing is conducted. We measured the indoor home exposure to a number of pollutants, including NO_2. This chapter discusses the initial measurements of pulmonary function and information on respiratory diseases obtained in children in the six cities and relates these health measures to their potential indoor exposure [6].

METHODS

Study Design

A total of 9280 children participated in the initial surveys. They represented 12 separate cohorts from six cities. Two cities were surveyed for three years, and a new group of first-grade school children were added each year; thus, these cities provided six cohorts. Two cities were surveyed for 2 years giving four more cohorts, and 2 cities were surveyed once (Table I). More than 95% of the eligible children enrolled at the schools were studied.

Information about the children's exposure was obtained from the parental response to a questionnaire on the nature of the home cooking device, the home heating fuel, and the presence or absence of both air conditioning and adult smokers in the household. The effects of these conditions on the air

Table I. Study Design for School Children: Times at Which
New Cohorts of Children Were First Seen

City	Year 1		Year 2		Year 3	
	Fall 1974	Spring 1975	Fall 1975	Spring 1976	Fall 1976	Spring 1977
Watertown, Massachusetts	√		√		√	
Kingston-Harriman, Tennessee		√		√		√
St. Louis, Missouri			√		√	
Steubenville, Ohio				√		√
Portage, Wisconsin					√	
Topeka, Kansas						√

quality were also measured in selected households in each community. Measurement was carried out by a home sampling unit, which was placed in an "activity room," specifically defined as *not* being the kitchen or bedroom. The monitors were run for 24 hours every 6 days during at least one spring and one fall season in each city. Mass respirable particulates (mass median diameter of 3.5) were collected on Millipore filters, and NO_2 was collected by a bubbler technique and measured by the U.S. Environmental Protection Agency (EPA) reference method for NO_2, which is a modified sodium arsenite method [7].

Forced expiratory measurements were performed using a water-filled, low-inertia recording spirometer. The children did not wear nose-clips and performed the task standing. Each child had a minimum of five and a maximum of eight attempts in an effort to obtain at least three acceptable tracings. Forced vital capacity (FVC) and forced expiratory volume in one second (FEV_1) were read from each tracing. Values were corrected to body temperature and pressure saturated with water vapor (BTPS) and summarized as the mean of the three best efforts within 170 ml. Standing height and weight in stocking feet were recorded for each child.

There were 8866 children (95.5% of the total seen) who were between 72 and 119 months at the time of their initial survey. The sample was further reduced to 8120 children by limiting the analyses to the white population. This restriction was necessary because for a given height, black children have smaller lungs than white children [8]. Thus, 746 nonwhite children (8.4%) were removed from the sample (Table II).

Statistical Methodology

To establish the appropriate adjustment of the pulmonary function for body size in these children, a variety of models using age, height, weight and

Table II. Distribution by Sex of White and Nonwhite Children Aged 72–119 Months in Each City

		Portage	Topeka	Watertown	Kingston-Harriman	St. Louis	Steubenville	Total
Males	Total	363	816	763	709	1107	798	4556
	White	362	700	752	654	1078	659	4205
	%	99.7	85.8	98.6	92.2	97.4	82.6	92.3
	Nonwhite	1	116	11	55	29	139	351
	%	0.3	14.2	1.4	7.8	2.6	17.4	7.7
Females	Total	346	859	688	665	943	809	4310
	White	344	709	676	614	917	655	3915
	%	99.4	82.5	98.3	92.3	97.2	81.0	90.7
	Nonwhite	2	150	12	51	26	154	395
	%	0.6	17.5	1.7	7.7	2.8	19.0	9.3

percentile height-for-age and percentile weight-for-height variables were tested [9]. In addition, because of the multicolinearity of the independent variables, a number of models using ordinary least squares, principal components and ridge analysis were undertaken [10]. These assessments were needed before the data can be used in a prospective fashion, which requires extension beyond the present narrow prepubertal age range. In the restricted age range of 72–119 months, adjusting the data for height alone appears to be as effective as any procedure, but the development of the standardization equations required that we only consider children of normal size for their age.

To develop standardization equations, the children (aged 72–119 months) seen in Watertown and Kingston-Harriman in the third year of the study were selected. They included the three intake cohorts for each city, so some of the children had performed spirometry testing at least once. Reasons for selecting these cities were as follows

1. They were the first cities for whom intake was completed and thus had the largest sample available.
2. By choosing a single year seen by the same interview team cohort effect need not be taken into account.
3. By the third year the instructions to the field team had become standardized [11].

Various regression procedures were tried relating lung function to age and height. We found that a well-fitting model was obtained by selecting those white children whose height fell within the 5–95 normal percentile range for U.S. children of that age [12]. The equations were computed for each sex and for FVC and FEV_1 separately (Table III). Inspection of residuals did not show any trend with age.

The restriction to the 5–95 percentile range of expected height was a device for stabilizing the standardization equation. When analyzing the data, all children within the 1–99 percentile range were included. Exclusion of the upper and lower 1% ensured that the analyses were not distorted by children whose heights were incorrectly recorded. Lung function predicted for each

Table III. Equation for Standardized Lung Function in Children Aged 72–119 Months (whites only)[a]

		Constant	Ht (cm)	R^2	N
Males	FEV_1	−2.4066	0.0310	0.602	915
	FVC	−2.9790	0.0374	0.625	915
Females	FEV_1	−2.1567	0.0286	0.595	904
	FVC	−2.6340	0.0338	0.628	904

[a]Based on children whose height for age was between 5th and 95th percentile of expected value from National Center for Health Statistics growth chart [12].

child's height was computed from the standard equation, and the difference between their observed lung function and this predicted value was obtained. These residuals were analyzed using standard analysis of variance techniques.

The reported disease rates were analyzed using log-linear models to determine the significant interactions between the variables of disease, age, sex, cohort, city and other influencing home variables. Adjusted rates were computed based on models that included the significant interactions [13].

The data on air pollution levels were first adjusted to account for missing values using a linear model for day of observation and site. The influence of home variables was determined by analysis of variance, with appropriate adjustment of the residual degrees of freedom.

RESULTS

Assessment of Exposure to NO_2

Although gas and electric cooking stoves were each found to account for about half the homes, there were considerable differences between cities. The distribution of the children by home cooking device ranged from a high of 82.2% gas-cooking homes in St. Louis to a low of 4.6% in Kingston-Harriman (Table IV). The remaining 6% of the homes either used some other form of cooking device, alone or in conjunction with gas and/or electricity (1.9%), or the nature of the cooking device was not reported (4.1%).

Our information regarding the differences in air quality associated with different cooking devices was obtained by setting up indoor–outdoor monitors in samples of homes. These homes were not necessarily the homes of children in the study but were selected to represent the various living patterns in each community. The homes were sampled every sixth day for 24 hours from May 1977 through April 1978.

Although the number of homes studied in each community was not large (between 5 and 11), the number of 24-hour periods for which matched indoor and outdoor data were available was several hundred (Table V). The homes were divided between gas and electric cooking devices, except for Kingston-Harriman, in which no homes with gas stoves were studied. The results showed a gradient of NO_2 levels in homes with electric stoves, which reflects outdoor sources of NO_2, and a substantial increase in NO_2 levels in homes with gas stoves, which reflects the addition of indoor sources to the outdoor level of NO_2, as we were dealing with 24-hr integrated averages collected in "an activity" room not the kitchen. In some cities, the daily levels encountered in some gas stove households exceeded the federal standard for the annual average of the 24-hour NO_2 levels ($100 \ \mu g/m^3$) (Table VI). Such levels for integrated 24-hour values indicate that peak exposures must be substantially higher.

The NO_2 levels at each site within a city were analyzed to determine the significance of the observed variability between days, gas homes, electric homes, and gas and electric homes (Table V). Missing values were filled to

Table IV. Distribution of Children Aged 72–119 Months by City and Nature of Home Cooking Stoves

Type of Stove	Portage	Topeka	Watertown	Kingston-Harriman	St. Louis	Steubenville	Total
Gas	245	406	946	58	1639	482	3776
(%)	(34.7)	(28.8)	(66.2)	(4.6)	(82.2)	(36.7)	(46.5)
Electric	446	982	296	1020	323	792	3859
(%)	(63.2)	(69.7)	(20.7)	(80.4)	(16.2)	(60.3)	(47.5)
Other	15	21	65	8	20	27	156
(%)	(2.1)	(1.5)	(4.6)	(0.6)	(1.0)	(2.1)	(1.9)
Unknown	0	0	121[a]	182[a]	13	13	329
(%)	–	–	(8.5)	(14.4)	(0.7)	(1.0)	(4.1)
Total	706	1409	1428	1268	1995	1314	8120

[a]Nature of cooking device not asked at initial survey. In the two cities indicated, 121 and 182 children, respectively, were not available for followup.

Table V. Comparisons of Variability in Six Cities of NO$_2$ Levels Between Days, Electric Homes and Gas Homes, and Between Gas and Electric Homes

	No. of Days	No. of Homes		Between Days		Between Electric Homes		Between Gas Homes		Gas vs Electric	
		Electric	Gas	MS[a]	F[b]	MS	F	MS	F	MS	F
Outdoor											
Portage	50	8	3	11.935	31.22[c]	9.621	25.17[c]	0.475	1.24	4.064	10.63[c]
Topeka	57	6	1	1.404	5.96[c]	2.809	11.92[c]	–	–	0.329	1.40
Kingston-Harriman	56	8	0	1.330	16.29[c]	2.849	34.89[c]	–	–	–	–
St. Louis	58	3	6	1.504	7.17[c]	1.419	6.77[c]	0.981	4.68[c]	1.760	8.35[c]
Steubenville	61	2	3	0.775	2.19[c]	0.000	0.00	5.441	15.40[c]	0.371	1.05
Watertown	59	2	5	1.320	11.02[c]	7.404	61.80[c]	0.589	4.92[d]	0.001	0.01
Indoor											
Portage	50	8	3	7.628	12.59[c]	28.381	46.69[c]	0.027	0.04	211.508	349.22[c]
Topeka	57	6	1	0.781	2.63[c]	3.148	10.62[c]	–	–	11.649	39.28[c]
Kingston-Harriman	56	8	0	0.720	4.17[c]	7.234	41.90[c]	–	–	–	–
St. Louis	58	3	6	1.215	3.08[c]	28.18	71.52[c]	7.02	17.82[c]	87.64	222.42[c]
Steubenville	61	2	3	1.027	1.35[e]	55.090	72.37[c]	39.915	52.44[c]	3.770	4.95[e]
Watertown	59	2	5	0.702	4.90[c]	1.050	7.32[d]	2.120	14.78[c]	6.168	43.01[c]

[a] mean square.
[b] F ratio.
[c] $p < 0.001$.
[d] $p < 0.01$.
[e] $p < 0.05$.

Table VI. Indoor and Outdoor Levels of NO_2 Each City (May, 1977–April, 1978)

| City | Median Level of NO_2 ($\mu g/m^3$) | | | | 95th Percentile Level of NO_2 ($\mu g/m^3$) | | | |
| | Outdoor | | Indoor | | Outdoor | | Indoor | |
	Electric	Gas	Electric	Gas	Electric	Gas	Electric	Gas
Portage[a]	9.8	8.4	5.3	21.7	31.8	25.4	17.6	39.3
Topeka	19.2	17.0	20.8	36.9	42.4	40.7	41.6	73.6
Kingston-Harriman	18.1	–	10.7	–	38.4	–	29.8	–
St. Louis	36.9	40.0	19.3	46.5	64.3	70.9	63.3	79.3
Steubenville	37.7	36.0	30.8	29.7	82.9	87.8	74.5	103.9
Watertown	50.0	50.6	39.5	54.3	101.6	106.3	95.2	116.3

[a]Based on a 10-month sample.

conform with the linear model, and the degrees of freedom associated with the error mean square were appropriately adjusted [14,15]. This means that differences between sites could not be attributed to missing days.

There was significant daily variation in NO_2 levels reflecting seasonal effects, but this was generally less indoors than outdoors. Similarly, for there were significant differences between all homes, but the variability indoors was greater than that outdoors. There were, however, two exceptions: In Portage gas homes, the variability both indoors and outdoors was small; in Watertown electric homes, the variability was greater outdoors. The latter is accounted for by the proximity of traffic to the Watertown homes.

The most striking aspect of these analyses was the difference between gas and electric homes. When we look at the outdoor values, the variance is of the same order of magnitude as the variance between homes with the same cooking device. By contrast, the indoor values showed very large differences between gas and electric homes in every city. The indoor difference between gas and electric homes was greater than the differences between homes with the same device and, except for Steubenville, ranged from 3–7 times larger.

Health Measures

Two specific measures of children's health effects were available: (1) historic data on illness reported on questionnaires completed by parents; and (2) the measurement of current pulmonary function status. Responses to three health-related questions were analyzed: (1) whether the child had ever had bronchitis (diasnosed by a physician); (2) whether the child had had a serious respiratory disease before age 2; and (3) whether the child had had a respiratory illness in the last year. Both the history of illness responses and the pulmonary function measures were tested for relationship to several household variables. Nature of cooking device, nature of fuel used for heating, presence of adult smokers, presence of air conditioning and socioeconomic status of the family were evaluated. The measure of socioeconomic level used gave equal weight to a parent's occupation and educational attainment.

The three reported disease rates were analyzed by fitting log-linear models [13]. The effect of each of the household variables on each disease rate was investigated. Two of the variables—home heating and air conditioning—were not related to the disease rates. Social class, parental smoking and type of cooking stove had variable effects on the three diseases when tested alone (Table VII). As the effects were likely to be interrelated, we then evaluated each disease by including these three home variables simultaneously. For respiratory disease before age 2, we found that the effect of cooking stove now became significant, as well as the effect of parental smoking, sex of the child and city–cohort. However, for this disease, age at time of reporting was not significant. When we corrected rates for smoking and city–cohort, we found a difference of 35/1000 in males and 30/1000 in females. Lower rates were found in children of households with electric stoves (Table VIII).

The effects of parental smoking and city–cohort on respiratory disease before age 2 was not independent, but the effect of cooking stove was unrelated to the other home variables. Thus, when we corrected for smoking for each cohort, we found that the rates varied between cohort, but the direction of the cooking stove effect did not vary.

To assess the effect of home factors on pulmonary function in these children, we calculated for each child the difference between the expected and observed FVC and FEV_1. The effect of cohort (year of study and city) and the same home variables on the residual pulmonary function were

Table VII. Relation Between Home Variables and Reported Diseases

	Social Class			Parental Smoking			Home Cooking		
	G^2 [a]	df	P	G^2 [a]	df	P	G^2 [a]	df	P
History of Bronchitis	0.29	1	0.59	1.32	1	0.25	18.31	1	<0.001
Serious Respiratory Illness Before Age 2	4.77	1	0.029	10.29	1	0.001	3.85	1	0.050
Respiratory Illness in Last Year	9.78	1	0.002	4.99	1	0.026	0.03	1	0.86

[a]G^2 is the likelihood statistic and is distributed as a chi square, with the appropriate degrees of freedom (df).

Table VIII. Respiratory Illness Before Age 2 by Cohort and Nature of
Home Cooking Device (standardized for parental smoking)

City[a]	Sex	Year 2 Gas	Year 2 Electric	Year 3 Gas	Year 3 Electric
Watertown	M	236	201	284	246
	F	193	164	237	203
Kingston-Harriman	M	276	237	248	212
	F	229	195	204	173
St. Louis	M	149	126	313	271
	F	120	101	262	225
Steubenville	M	241	206	279	241
	F	198	168	232	198
Portage	M			173	146
	F			140	118
Topeka	M			212	189
	F			181	153
Adjusted by City-Cohort	M	249	214		
	F	205	175		

[a]No home cooking stove data were collected in year 1.

assessed by analysis of variance. Preliminary regression of lung function on socioeconomic status showed no relationship. For each of these variables there was a significant effect ($p < 0.01$) of cohort on both FEV_1 and FVC. This means that from city to city and from year to year there were differences in the height-adjusted pulmonary function levels in these children (Table IX). There were no significant associations between FVC or FEV_1 residuals and the presence of air conditioning in the home. Although the association between parental smoking and FVC was significant at the 5% level, the effect was opposite to that anticipated and was regarded as a random phenomenon. Home heating and FEV_1 residuals were also significantly associated at the 5% level. The overall means covered a 30-ml range, with the lower level in children from homes using oil, the highest for those from electrically heated homes.

Although FEV_1 residuals were affected by home heating fuels, the most consistent and significant finding was the lower levels of both FVC and FEV_1 in children whose homes had gas cooking stoves, compared with those whose homes had electric stoves. The overall effect of home cooking, after correcting for cohort effect, was 20 ml for both FEV_1 and FVC. This effect was apparent in almost all cohorts. For FVC, only one cohort (St.

Table IX. Analysis of Variance of Children's Lung Function
for Various Home Variables[a]

| | | | F-Ratios | | |
Home Variable	Lung Function	No. of Children	Home Variable	Cohort[b]	Interaction
Cooking:	FVC	6803	7.94[b]	10.02	1.13
Gas/Electric	$FEV_{1.0}$	6803	8.11[b]	6.25	0.82
Heating:	FVC	6734	0.76	8.94	0.73
Oil/Gas/Electric	$FEV_{1.0}$	6734	3.25[b]	5.64	1.09
Air Conditioning: None/Partial/	FVC	7126	1.22	10.90	0.73
Central	$FEV_{1.0}$	7126	0.61	6.61	0.72
Parental Smoking:	FVC	5842	6.27[c]	10.56	1.31
None/Some	$FEV_{1.0}$	5842	0.27	6.27	1.20

[a]Home variable has 1 or 2 degrees of freedom, cohort has 11 degrees of freedom, and all F-ratios have at least 5000 degrees of freedom associated with the denominator mean square.
[b]Significant at 1% level.
[c]Significant at 5% level.

Louis, year 1), did not show lower levels of pulmonary function in children living in homes with gas compared with electric stoves (Table X). For FEV_1 in 10 of 12 cohorts, the children in homes with gas stoves had lower function than children in homes with electric stoves (Table XI). An unexpected finding in these data was the low level of pulmonary function measured in Topeka, a city with generally lower levels of pollution. In an attempt to investigate this finding, we tested the effect of different interviewers; reread the spirometer tracing to test effect of readers; and compared the values obtained on each spirometer by month of study to test the possibility of a defective machine. None of these potential problems could explain the lower pulmonary function values. In addition, the distribution of height for age of the children in Topeka did not differ significantly from the other cities. It was concluded that the pulmonary function measurements in Topeka children were lower than in other cities and assumed that it was a cohort effect that is as yet unexplained. We will be interested to determine whether this persists in subsequent years.

DISCUSSION

The significant associations found in this analysis were between home cooking stove and both illness history and lung function. We also found an

Table X. FVC Redisuals in 72–119-Month-Old White Children,
by Cohort, City and Cooking Fuel

City	Cooking Fuel	Mean Residual [liters (number of children)]		
		Year 1	Year 2	Year 3
Watertown	Total	−0.031 (484)	−0.004 (304)	−0.089 (308)
	Gas	−0.045 (364)	−0.009 (236)	−0.099 (238)
	Electric	0.011 (120)	0.014 (66)	−0.057 (70)
Kingston-Harriman	Total	−0.010 (365)	0.025 (297)	0.028 (343)
	Gas	−0.008 (18)	−0.018 (17)	0.029 (21)
	Electric	0.012 (347)	0.028 (280)	0.027 (322)
St. Louis	Total		−0.013 (966)	0.023 (646)
	Gas		−0.011 (817)	0.016 (525)
	Electric		−0.026 (149)	0.053 (121)
Steubenville	Total		0.002 (739)	0.015 (422)
	Gas		−0.001 (293)	−0.020 (143)
	Electric		0.004 (446)	0.033 (279)
Portage	Total			0.016 (656)
	Gas			0.000 (233)
	Electric			0.024 (423)
Topeka	Total			−0.037 (1273)
	Gas			−0.037 (369)
	Electric			−0.036 (904)

effect of parental smoking on disease history. The importance of these findings rests with the interpretations of these changes. Sample sizes are sufficiently large to determine whether observed minor differences between groups are significantly greater than would occur by chance. The size of the differences found was consistent with the anticipated magnitude of effect of environmental agents [16], and our home measurements of air quality supported this.

That homes with gas cooking stoves have higher levels of NO_2 than those with electric stoves has been demonstrated [4,5], and peak levels over gas stoves have been reported to reach approximately 1 ppm (1880 $\mu gm/m^3$) for periods of 10–15 min [5]. Similarly, we know from both our own investigation and studies of Hinds and First [17] that the mass respirable particulate loads in households with smokers can be many times higher than in nonsmoking households. Other potential confounding factors seem to have been excluded (e.g., socioeconomic status, presence of air conditioning, nature of heating fuel).

To determine the importance of our findings it is necessary to evaluate a number of potential sources of bias. The questionnaire information on disease rates for an individual child depends on the parents' ability to recall and may

Table XI. FEV Residuals in 72–119-Month-Old White Children,
by Cohort and Cooling Fuel

		Mean Residual [liters (number of children)]		
		Year 1	Year 2	Year 3
Watertown	Total	−0.015 (484)	0.012 (304)	−0.061 (308)
	Gas	−0.025 (364)	0.009 (236)	−0.068 (238)
	Electric	0.016 (120)	0.022 (68)	−0.037 (70)
Kingston-	Total	−0.009 (365)	0.013 (297)	−0.003 (343)
Harriman	Gas	−0.031 (18)	−0.016 (17)	−0.014 (21)
	Electric	−0.008 (347)	0.014 (280)	−0.002 (322)
St. Louis	Total		−0.007 (966)	0.018 (646)
	Gas		−0.006 (817)	0.012 (525)
	Electric		−0.013 (149)	0.039 (121)
Steubenville	Total		0.012 (739)	0.014 (422)
	Gas		0.012 (293)	−0.009 (143)
	Electric		0.013 (446)	−0.026 (279)
Portage	Total			0.007 (656)
	Gas			−0.016 (233)
	Electric			0.020 (423)
Topeka	Total			−0.024 (1273)
	Gas			−0.026 (369)
	Electric			−0.023 (904)

be biased by the present status of the child. No attempt was made to have a doctor confirm diagnosed disease. It seems unlikely, however, that any such biases would be related consistently to the type of home cooking stove. The good response rate and the sampling plan, which ensures that we accept all potentially available children, means that we do not have unrepresentative samples.

The pulmonary function data are potentially subject to possible interviewer bias, malfunctioning machinery or biased reading of the spirometer tracing. However, we checked for all these sources and found them not to exist. Neither the field screeners nor the readers were aware of the individual child's home environment when the spirometry was performed or the tracings read; thus attribute association with home variables cannot be attributed to bias.

Essentially, interpretation of the pulmonary function finding relates to the sensitivity of the measure and the biological expectation of the magnitude of anticipated effect in a population of children age 72–119 months. We used FEV_1 as a measure of airflow obstruction in these children, not because we believed it to be the best measure of obstruction, but because we planned to follow these children over several years. After several years, a stable estimate of change in pulmonary function can be related to our understanding of the development of adult obstructive airways disease. In these children, many of whom can empty their entire FEV in less than two seconds, the FEV_1 does not measure obstruction as much as it measures FVC. Thus, it is reassuring to find similar changes in both measures when trying to understand the significance of any finding.

Our understanding of the biology of lung growth and the nature of the onset of obstructive lung disease in adult life led us to believe that only minor differences in the rate of functioning lung growth* in young children could lead to their not reaching their full adult lung size. We do not know whether failure to reach full adult lung size is related to the subsequent susceptibility of developing obstructive lung disease, but it is possible that those persons with minor impairment of total lung growth are at greater risk of sustaining more rapid decline in pulmonary function in adult life.

Our results differed only slightly from those reported to date. The findings of Melia et al. [2] regarding lower respiratory tract illness rates in children whose homes have gas stoves were similar. That study was criticized because it did not have data on smoking. In our study, the adjustment of rates of illness before age 2 to include smoking led to a clear association with gas-cooking devices. The study of Lutz et al. [18] of both adults and children suggests no association of gas stoves with respiratory disease rates. This study measured incidence of acute respiratory disease over one year; but the sample of children studied was quite small and did not represent a general population. Bouhuys et al. [19] did study population-based samples of children and adults, but the 7000 persons aged 7 and older were primarily adults, and only a small number (165) of children aged 7–14 were studied in the two communities investigated [20]. That he was unable to find an association with home cooking devices may be attributed to the small numbers used.

Tager et al. [21] using a different indicator of airways obstruction (mid-maxim expiratory flow), found an association between the pulmonary function levels in children and the number of smokers in the household. We did not find such an association in our sample using FEV_1. This may mean that our measure of airways obstruction was too insensitive. If we believe that FEV_1 is not a good measure of airways obstruction in these young children, but that it closely resembles FVC, we can guess at the mechanical effect of exposure to NO_2 from gas stoves, namely that it reduces the vital capacity. We cannot distinguish between two causes of loss of vital capacity, reduction in total lung capacity or increase in residual volume. Obviously, the pathological lesions could be very different, according to which process was operating to cause loss of pulmonary function.

Followup of these cohorts is underway. Initial findings reported here need replication to eliminate some subtle bias or alternative explanation. It will

---

*We are using FVC as a crude indicator of lung size recognizing that the total lung capacity (TLC) includes not only FVC but also the residual volume which is not being measured in these field studies.

be of interest to determine whether the relative position of these children's lung sizes remains the same or changes. If it changes, it will be important to assess the factors that most influence the change, whether it be: (1) changes in ambient pollution (outdoor levels) or in personal pollution (indoor exposures and cigarette smoking); (2) personal factors, such as frequency of respiratory infections; or (3) other recognized potential risk factors for developing chronic obstructive respiratory disease.

ACKNOWLEDGMENTS

This work was supported in part by grants from the National Institute of Environmental Health Service (ES0002, ES01108) and the Electric Power Research Institute, contract No. RP 1001 EPRI.

We would like to thank numerous people involved in the collection and processing of these data, particularly local school superintendents, principals and teachers who allowed us into their classrooms; and our field teams, who for these particular studies were led by C. Humble, B. Cate and S. Hancock. The indoor/outdoor monitoring was under the direction of D. Dockery, and our office coordinating staff was led by S. Puleo. Technical computing assistance was provided by J. Weener, M. Levenstein and D. Glicksburg. We also appreciate the secretarial and editorial support of H. Taplin and M. Masters.

REFERENCES

1. Lowry, T., and L. M. Schuman. "'Silo Filler's Disease': A Syndrome Caused by Nitrogen Dioxide," *J. Am. Med. Assoc.* 162:153–160 (1956).
2. Melia, R. J. W., C. Du V. Florey, D. S. Altman and A. V. Swan. "Association between Gas Cooking and Respiratory Disease in Children," *Brit. Med. J.* 2:149–152 (1977).
3. Mitchell, R. W., R. Williams, R. W. Cote, R. Lunase and M. D. Keller. "Household Survey of the Incidence of Respiratory Disease Environmental Pollutants," in *Recent Advances in the Assessment of the Health Effects of Environmental Pollutants*, WHO International Symposium Proceedings, Paris, June 24–28 (1974).
4. Melia, R. J. W., C. du V. Florey, S. C. Darby, E. D. Palms and B. D. Goldstein. "Differences in NO_2 Levels in Kitchens with Gas or Electric Cookers," *Atmos. Environ.* 12:1379–1381 (1978).
5. Wade, W. A., III, W. A. Cote and J. E. Yocum. "A Study of Indoor Air Quality," *J. Air Poll. Control Assoc.* 25:933–939 (1975).
6. Speizer, F. E., Y. Bishop and B. G. Ferris, Jr. "An Epidemiologic Approach to the Study of Health Effects of Air Pollution," in *Proc. 4th Symp. on Statistics and the Environment*, March 3–5, 1976, American Statistical Association, Washington, D.C. (1977), pp. 56–68.
7. Saltzman, B. E. "Colorimetric Microdetermination of Nitrogen Dioxide in the Atmosphere," *Anal. Chem.* 26:1949–1955 (1954).

8. Binder, R. E., C. A. Mitchell, J. B. Schoenberg and A. Bouhuys. "Lung Function among Black and White Children," *Am. Rev. Resp. Dis.* 114: 955–959 (1976).
9. Weener, J. *An Assessment of Standardizing Procedures for Test of Lung Function in Children*, Ph.D. Thesis, University of Michigan School of Public Health, Ann Arbor, MI, In preparation.
10. Levenstein, M., Y. M. M. Bishop, B. G. Ferris, Jr. and F. E. Speizer. "Six City Study Standardization of Lung Function Measurements," in *Energy and Health Science*, Sims Conference Series Number 6, SIAM, Philadelphia, PA (1979).
11. Ferris, B. G., Jr., F. E. Speizer, Y. Bishop, G. Prang and J. Weener. "Spirometry for an Epidemiologic Study: Deriving Optimum Summary Statistics for Each Subject," *Bull. Eur. Physiopath. Resp.* 14:145–166 (1978).
12. National Center for Health Statistics. "The Monthly Vital Statistics Report: National Center for Health Statistics Growth Charts, 1976," *Monthly Vital Stat. Rep.* 25(3):Suppl. (June 22, 1976).
13. Bishop, Y. M. M., S. E. Fienberg and P. W. Holland. *Discrete Multivariate Analysis: Theory and Methods* (Cambridge, MA: MIT Press, 1977).
14. Cochran, W. G., and G. M. Cox. *Experimental Designs*, 2nd ed. (New York: John Wiley & Sons, Inc., 1957), pp. 110–112.
15. Sokal, R. R., and F. J. Rohlf. *Biometry* (San Francisco, CA: W. H. Freeman and Co. Publishers, 1969), p. 320.
16. U.S. Department of Health, Education and Welfare. *Human Health and the Environment: Some Research Needs*, DHEW Publication No. NIH 77-1277 (Washington, D.C.: U.S. Government Printing Office, 1977).
17. Hinds, W. C., and M. W. First. "Concentrations of Nicotine and Tobacco in Public Places," *New England J. Med.* 292:844–845 (1975).
18. Lutz, G. A., R. I. Mitchell, R. W. Cote and M. D. Keller. "Respiratory Disease Symptom Study," American Gas Association (1977).
19. Bouhuys, A., G. L. Beck and J. B. Schoenberg. "Do Present Levels of Air Pollution Outdoors Affect Respiratory Health?" *Nature* 276:466–471 (1978).
20. Mitchell, C. A., R. S. F. Schilling and A. Bouhuys. "Community Studies of Lung Disease in Connecticut: Organization and Methods," *Am. J. Epidemiol.* 103:212–225 (1976).
21. Tager, I. B., S. T. Weiss, B. Rosner and F. E. Speizer. "Effect of Parental Cigarette Smoking on the Pulmonary Function of Children," *Am. J. Epidemiol.* 110:15–26 (1979).

CHAPTER 23

CRITERIA RELEVANT TO AN OCCUPATIONAL
HEALTH STANDARD FOR NITROGEN DIOXIDE

Trent R. Lewis

National Institute for Occupational Safety and Health
Cincinnati, Ohio 45226

INTRODUCTION

This chapter outlines the historical evolution of the federal industrial air standard for nitrogen dioxide (NO_2), with major emphasis on the criteria utilized as a basis for existing and advocated standards. A concise description of the principles basic to the establishment of federal industrial air standards is followed by a description of those factors inherent to nitrogen dioxide that determine its hazard, namely, its chemical and physical properties and its acute and chronic toxicity in humans and animals. The chapter concludes with a review of prior, existing and presently advocated standards, both nationally and internationally.

The health of American workers is protected by federally and locally mandated industrial hygiene standards. The federal standards are designed to protect the health of workers for up to a 10-hr work day, 40-hr work week, over a working lifetime (P.L. 91-596, Occupational Safety and Health Act of 1970, Section 5). Prior to the development of such standards, specific criteria are derived in a documented format, which summarizes the biological effects that can be expected from exposure to various airborne concentrations of the material under consideration. The criteria may take several forms and include carcinogenesis, teratogenesis, mutagenesis, physiological dysfunction, morphological alterations and biochemical changes. Because criteria vary in sensitivity and specificity, so do the standards on which they are based.

The basis for industrial air standards for noncarcinogenic substances is the concept of thresholds of toxicological response, i.e., although all chemical substances produce a response (toxicity, irritation, sensitization, narcosis, etc.) at some concentration, it is equally true that a concentration exists for

361

most chemical substances from which adverse effects are not expected. This concentration will not produce harmful responses on an 8- to 10-hour work day, 40-hour work week basis over a working lifetime and through retirement.

There are many factors that require consideration before scientific data can be utilized to recommend a permissible occupational exposure. One reason for this is that the human species is genetically diffuse and heterogeneous, with widely differing capacities to respond to toxic insult. The total breadth of human response is seldom known. Accordingly, a safety factor is invariably applied to a recommended occupational air standard. As the margin of protection is expanded to safeguard the more susceptible individual, the greater becomes the magnitude of the safety factor. Thus, the recommendation of a standard is based on objective judgmental evaluation of biological responses. Each value judgment must be decided independently and should not be predetermined by chemical or biological analogies. Thus, each standard and criterion on which the standard is based is subject to review and revision as necessary. Furthermore, prior to the promulgatory stage, industrial standards must be kept within the bounds of analytical and engineering practicality.

Of the eight different oxides of nitrogen identified, NO_2 poses the greatest health hazard in the workplace, both from the standpoint of toxicity and quantities present.

Occupational exposures to nitrogen dioxide are fairly common. Exposures result primarily from decomposition of nitrates (dynamite blasting, silaging); reactions of nitric acid with metals or other reducing agents (acid dipping, dye and aniline makers); from various processes in which air is heated to a high temperature (furnaces, welding, cutting torch operations); or from the exhaust of internal combustion engines (automotive, diesel, light and heavy-duty equipment operated by liquid fuels). Most occupational exposures to nitrogen dioxide exist as exposures to a mixture of airborne contaminants, and many such exposures are poorly documented as to actual atmospheric concentrations and duration of exposure. Although NO_2 has been used in industry as a nitrating or oxidizing agent and as a rocket fuel ingredient, most exposures arise from its presence as an undesirable by-product from industrial practices and/or processes.

It is difficult to estimate the number of people exposed occupationally to nitrogen dioxide because both those persons engaged in an operation that generates nitrogen dioxide and those persons in the immediate or general vicinity are exposed. Storlazzi [1] has stated that only 6000 workers were directly engaged in welding and burning operations in United States naval shipyards, but approximately 60,000 workers were indirectly exposed to effluents of nitrogen dioxide and other pollutants. The National Institute for Occupational Safety and Health (NIOSH) has estimated that at least 1.5 million workers are potentially exposed to oxides of nitrogen including nitrogen dioxide [2].

CHEMICAL AND PHYSICAL PROPERTIES

Inherent to the toxicity of any chemical are its chemical and physical properties. Nitrogen dioxide is a very pungent, reddish-brown or dark orange gas, depending on its atmospheric concentration. Nitrogen dioxide boils at 21°C; below this temperature it condenses to a yellow liquid. At ambient temperatures, NO_2 exists in equilibrium with its dimer nitrogen tetraoxide, N_2O_4, and the dimer is the form in which NO_2 is converted to a solid. The solid melts at −9.3°C. Nitrogen dioxide is a strong oxidizer, is extremely reactive and corrosive, and reacts with water to form nitric acid and either nitrous acid or nitric oxide.

One of the more significant chemical–biological attributes of nitrogen dioxide is its interconversion with nitric oxide (NO). Nitrogen dioxide is frequently produced from the oxidation of nitric oxide. Nitric oxide may be produced from nitrogen dioxide as NO_2 absorbs ultraviolet and visible radiation and dissociates to nitric oxide and atomic oxygen. Thus, workers are exposed not to either nitrogen dioxide or nitric oxide per se, but to mixtures of the two. The ratios of the two oxides of nitrogen (NO_x) depend largely on the source from which they were generated, the temperature of the effluent, the rate of dilution and cooling, and the presence or absence of moisture, sunlight and metal fumes.

A second significant chemical–biological characteristic of nitrogen dioxide is that it possesses unpaired electrons, which convey to the molecule free radical characteristics. These characteristics impart to nitrogen dioxide a similarity of biological action to ionizing radiation, or a radiomimicity. Thus, nitrogen dioxide can generate free radicals, which, in turn, may initiate a series of free radical chain reactions. The latter is one of the hypothesized mechanisms in tumorigenesis.

ACUTE EFFECTS IN MAN

Signs and symptoms of acute exposure in man include odor perception, physiological changes, nose and throat irritation, reversible pneumonia, bronchitis and bronchiolitis, to death arising from acute pulmonary edema, bronchopneumonia and/or bronchiolitis fibrosa obliterans (a disease characterized by a bronchiolitis resulting in occlusion of the lumina of bronchioles, due to the growth of fibrous connective tissue from the wall of the terminal bronchi). The acute hazard of inhaling high concentrations of NO_2 is dramatized in the report by Wade et al. [3], in which 90 deaths prior to 1930 and 47 deaths between 1930 and 1949 were attributed to exposure to high concentrations of NO_2.

The odor threshold for nitrogen dioxide has been reported to be below 0.5 ppm [4,5]. Russian investigators [6,7] have reported changes in the sensitivity of the human eye to adaptation to darkness following exposures ranging from 0.075–0.26 ppm. These effects of odor perception and darkness adaptation are both immediately reversible, and there is no evidence of evolving human pathological sequelae. Such responses are normal physiological responses to an external stimulus. Although nasal irritation and, less often, eye irritation have been reported in human volunteers exposed at concentrations of 13 ppm [8], inured workers have been reported to be able to work at average concentrations of 20–30 ppm for up to 18 months and longer [9,10].

Abe [11] exposed 5 healthy adult men to NO_2 at 4–5 ppm for 10 minutes. Measurements of pulmonary compliance, inspiratory flow resistance, maximal mid-expiratory and peak flowrates, and vital capacity were made prior to and immediately after exposure and at intervals of 10, 20 and 30 minutes after the exposure was terminated. A delayed effect of NO_2 was observed in which compliance decreased by 40% in the exposed group, and expiratory and inspiratory flow resistance increased maximally 30 minutes after cessation of exposure. Two Japanese studies are supportive of the pulmonary changes reported by Abe. Nakamura [12] reported a significant increase in airway resistance in 15 healthy individuals exposed for 5 minutes at NO_2 concentrations of 6 ppm to as high as 40 ppm. Suzuki and Ishikawa [13] measured pulmonary compliance and inspiratory and expiratory flow resistance in 10 healthy subjects exposed at 0.7–2.0 ppm of NO_2 for 10 minutes. The inspiratory and expiratory flow resistances rose to 150 and 115% of control value, respectively, and the compliance decreased to 90% of control values, 10 minutes after the exposure. One must note that although increased airway resistance resulting from NO_2 exposure has been found repeatedly, its threshold concentration is uncertain, and wide variations in individual sensitivity do occur [14].

Von Nieding and his associates [15] exposed 88 patients with chronic bronchitis (25 in the arterial pO_2 studies; 63 in the airway resistance studies) to NO_2. The latter group of 63 was exposed for 30 breaths of NO_2 at concentrations between 0.5 and 5 ppm. Airway resistance was measured by body plethysmography before and immediately after each exposure. Significant increases in airway resistance were observed at concentrations of 1.6–2.0 and above. No effects were noted on airway resistance at the lower concentrations. In the 25 chronic bronchitics, they observed a significant decrease in the arterial oxygen tension and a significant increase of the end expiratory arterial pressure differences for oxygen at 4 and 5 ppm of NO_2, but not at 2 ppm, for 15 min. These values returned to baseline within 10 min of cessation of exposure to NO_2. Von Nieding also conducted similar studies on healthy subjects exposed for 15 min at 5 ppm. The healthy subjects exhibited a diminished single-breath diffusion [16] capacity and significant increases in the alveolar-arterial pO_2 [15]. Pharmacological studies by von Nieding and Krekeler [17] revealed that the effects of NO_2 on airway resistance and arterial oxygen pressure were prevented by administration of an antihistamine, but not by drugs that block a parasympathetic reflex mechanism. These results suggest that the inhalation of low levels of NO_2 triggers the release of histamine, which appears to cause bronchoconstriction and ventilation/perfusion alterations at the alveolar regions.

Rokaw et al. [18], studied ten individuals—six healthy and four with chronic pulmonary disease—in a human exposure chamber. Airway resistance was measured in a resting state and while exercising on a bicycle. Recovery rates were measured 45 min after removal of NO_2 from the chamber. Nitrogen dioxide concentrations ranged from 0–3.0 ppm and increased sequentially by increments of 0.5 ppm. Airway resistance increased in most subjects during rest at 3.0 ppm and while exercising at 1.5 ppm. Subjects with chronic respiratory disease had increased airway resistance at rest beginning at 2.0 ppm and during exercise at 1.5 ppm.

Although these responses are dramatic in altering pulmonary function in healthy persons and those with chronic lung disease (both groups without demonstrated prior exposure to NO_2), they are reversible. Furthermore, none of these studies have included inured workers or those who have had a prolonged occupational exposure to NO_2. Thus, it is difficult to equate the acute physiological functional changes reported above to health hazards in the working environment.

ACUTE EFFECTS IN ANIMALS

Depending on the duration of exposure, nitrogen dioxide levels of 100 ppm are lethal to most animal species [19]. The LC_{50} (4-hr) for rats is 88 (79–99) ppm. Gray [20] also reported that a rise in ambient temperature of about 20°F increased the toxicity (or lowered the LC_{50}) by approximately 25%. One important observation from the LC_{50} data by Gray was that the product of concentration X time (CT) did not equal a constant, i.e., the LC_{50s} were 2890 and 21,120 ppm–minutes, respectively, for a high-dose, short-time vs lower dose, longer time. This difference in CT approximated eightfold and indicated for NO_2 that concentration was considerably more important than time in determining the mortality rate. The major acute toxic pathological response usually observed in animals is pneumonitis, with varying degrees of edema [21].

Effects other than mortality have been reported for short-term exposures. Inhalation of 1 ppm for 1 hr, or 0.5 ppm for 4 hr, led to degranulation of lung mast cells [22]. These effects, however, were reversible within 24 hours. Ehrlich [23] has reported increased susceptibility of mice to bacterial pneumonia following exposures for 2 hours at 1.5–3.5 ppm NO_2. In this same review, Ehrlich reported a reduction in clearance of bacteria from the lungs of mice exposed to *Klebsiella pneumonia* and increased mortality at 5.0 ppm NO_2 for 2 hr; however, mortality was not increased in hamsters under the same exposure conditions.

The effect of nitrogen dioxide as an etiological exacerbating agent in respiratory infections is one that has been studied extensively in animals. This response is clearly dose-related and has been demonstrated in short-term, continuous and intermittent exposures. Mice exposed continuously for 90 days to NO_2 concentrations as low as 0.5 ppm and then immediately given a laboratory-induced infection had a significant increase in mortality rate [23].

This increased risk to respiratory infection, however, usually requires higher concentrations in other animal species, e.g., the hamster and squirrel monkey. It remains to be unequivocally demonstrated that exposure to concentrations at or below 5 ppm will increase the worker's susceptibility to respiratory infections.

Murphy et al. [24] exposed guinea pigs at 5.2 ppm for 4 hr and reported increased respiratory rates and decreased tidal volumes. The pulmonary indices reverted to normal when the animals were returned to clean air. Similar results were obtained by Henry [25] in squirrel monkeys exposed at 15 ppm for 2 hr.

CHRONIC EFFECTS IN MAN

The data most applicable to establishing permissible occupational exposure are those that arise from human studies, particularly those related to the working environment. However, one of the major difficulties in assessing these data is reconstructing human industrial exposures, both quantitatively and qualitatively. Vigdortschik et al. [26] reported emphysema, chronic bronchitis and inflamed mucosa of the gums after 3–5 years of daily NO_2 below 2.8 ppm. These conclusions, however, are questionable in that the studies were conducted in sulfuric acid plants and etching operations and, along with the clinical signs and symptoms, cast doubt that the reported differences could be attributed merely to NO_2. Kennedy [27], using pulmonary residual volumes of greater than 150% of that predicted, classified 84 of 100 British coal miners as suffering from emphysema. The author attributed the emphysema to exposure to oxides of nitrogen, whose concentrations ranged up to 88 ppm following shot-firing and up to 167 ppm under conditions of misfires. The generalized or focal emphysema also could have arisen from other causes, such as pneumoconiosis resulting from exposure to coal mining dusts. Kosmider et al. [28] studied 70 chemical workers exposed 6–8 hours daily for 4–6 years at NO_x concentrations between 0.4 and 2.7 ppm. Spirometric measurements, blood gases and chest roentgenograms were comparable in the exposed and control populations. The exposed group showed a statistically significant increase in the excretion of hydroxyproline and acid mucopolysaccharides in their urines. Since these biochemical indices may be indicative of connective tissue destruction, the authors concluded that exposure to NO_x probably caused emphysema in humans. These findings were reinforced by clinical evidence of chronic bronchitis, dyspnea and pulmonary rales in an unstated number of the workers.

CHRONIC EFFECTS IN ANIMALS

Many studies to determine the effects of exposure to NO_2 at low levels over a long term have employed continuous or almost continuous exposure regimens, rather than intermittent exposures that mimic the occupational exposure situation. Chronic effects resulting from long-term exposures must be distinguished as to whether they are produced by continuous or intermittent exposures before the data can be used as criteria for recommending an occupational air standard. There are at least two reasons for stronger reliance on intermittent exposure data. First, the recommended occupational exposure standards are set on the basis of an 8–10 hr work day, with the assumption that a subsequent 16-hour period of nonexposure, coupled with 2 days per week of nonexposure, will aid in distributing and eliminating the toxic agent from the body and, more appropriately for NO_2, that the homeostatic, adaptive and repair mechanisms of the body will lessen the observed toxicity on an intermittent basis more so than under continuous insult. Furthermore, it is evident from experimental data on nitrogen dioxide that a concentration producing a given effect by continuous exposure is significantly lower than the factor applied to an intermittent exposure to equate dose on an equivalent CT basis.

In general, continuous exposure at a given concentration of NO_2 is more severe than from a corresponding intermittent exposure. Continuous exposure to NO_2 for 90 days at 5 ppm resulted in the death of 18% of the exposed rats and 13% of mice in a study reported by Back [29], and the death of 13% of the guinea pigs and 66% of the rabbits in a second study reviewed by Siegel [30]. This is to be compared to the intermittent exposure study by Wagner et al. [31] in which no increased mortality was observed in any of the species during the 18-month duration at levels of 5 and 25 ppm. A second example of lower toxicity in intermittent exposures vs continuous exposures lies in demonstrable pulmonary pathology reported by Freeman et al. [32] following continuous exposure, as compared to similar studies conducted by Hine et al. [33] and Boren [34], whereby several animal species exposed daily at 25 ppm for 1–4.5 months showed no diffuse fibrosis, obstructive respiratory disease or focal emphysema. Additional supportive data for this matter are found in the studies by Coffin et al. [35] utilizing the mouse infectivity model, which showed that mortality as a function of total time of intermittent exposure was lower than that observed for continuous exposure. Accordingly, reference will be made here primarily to recent studies and to those that approximate the industrial situation, i.e., intermittent exposure patterns.

In 1965; Wagner et al. [31] exposed male rabbits, guinea pigs, rats and hamsters at 1, 5 and 25 ppm of ppm NO_2, 6 hr/day, 5 days/wk for periods ranging from 15–18 months. A second animal population of three strains of mice and the same strain of rats were exposed at 5 ppm of NO_2 on an identical exposure schedule for durations of 10–16 months to assess specific toxicological responses, i.e., tolerance development studies and tumor-accelerating capacity of NO_2. Six dogs were exposed at 1 and 5 ppm of NO_2 intermittently for 18 months. Adequate numbers of control animals were included in the study. There was no increased mortality in any of the species, nor did changes in body weight, hematological values or biochemical indices differ significantly among the controls and animals exposed at the various levels. Studies on a spontaneous pulmonary tumor-susceptible strain of mice (CAF_1/Jax) suggested to the authors a possible tumorigenic accelerating capacity for NO_2 at an airborne concentration of 5 ppm. After 1 year's exposure, the tumor incidence was 7 of 10 in exposed vs 4 of 10 in controls; at 14 months, the incidence was 7 of 15 in exposed vs 8 of 15 in controls; at 16 months, the incidence was 15 of 24 in both exposed and control mice. This conclusion is quite tenuous since the maximum number of control and exposed animals examined at any one of the three durations of exposure . (12, 14 and 16 months) was 24. Statistical comparisons between the two groups at the three time intervals were not significantly different, and the final incidence of tumors was identical in the exposed and control populations.

Perhaps the most interesting aspect of this study, but one that received little attention, was the induction of tolerance to acute lethal effects of NO_2 in a large percentage of rodents exposed at 5 and 25 ppm. For example, rats exposed intermittently for 13 months at 5 ppm NO_2, followed by 6 weeks at 25 ppm and challenged subsequently to an LC_{50} dose, had 0% mortality compared with 67% mortality in the controls. Other evidence supportive of a role for tolerance in the toxicological manifestations of NO_2 is the marked decrease in dose–response relationships with increasing increments of time, whether the exposure regimen be repetitive or continuous. This biological response may explain why inured workers fail to exhibit clinical signs or symptoms when exposed to industrial levels of NO_2 over extended working periods.

The primary target of nitrogen dioxide is the respiratory system. Although emphysema has been reported in experimental animals, the majority of these studies have used NO_2 concentrations in the range of 5–50 ppm and/or continuous exposures [32,36–38]. At concentrations below 5 ppm, but again with continuous exposures, the results are more subtle than emphysema and include the reduction or loss of cilia [39], hypertrophy and focal hyperplasia in the epithelium of terminal bronchioles [40], and a replacement of type I with type II pneumonocytes in the alveoli [41]. Furthermore, evidence of irreversible pulmonary function tests has not been demonstrated in animals intermittently exposed to NO_2 on a 6- to 8-hr daily, 5 day/wk exposure regimen [42].

In concluding an evaluation of biological criteria for risk assessment to humans, one must always address the important matters of carcinogenicity, teratogenicity and mutagenicity. The inconclusivity of the data by Wagner et al. in 1965, suggesting that NO_2 might be a tumor-promoting agent, has been discussed earlier. Henschler and Ross [43,44] conducted two long-term studies (16–18 months)—one in mice and the other in hamsters—using 40 ppm of NO_2 and reported cellular proliferations in the terminal bronchioles but no carcinogenic action of NO_2 in the lung or other organs. Another concern of the potential carcinogenic hazard of nitrogen dioxide is by formation of nitrosamines with tissue amines; however, the limited data on this *in vivo* formation following inhalation of NO_2 are inconclusive.

Kuschner and Laskin [45,46] conducted two studies—one to assess the primary carcinogenesis of NO_2 and the second to assess the cocarcinogenicity of NO_2 with benzpyrene. In the carcinogenic study employing 100 rats and 96 hamsters exposed at 25 ppm, 6 hr/day, 5 days/wk for up to 646 days, an adenocarcinoma was found in one exposed rat; otherwise, the carcinogenic findings were similar between exposed and control animals. In the cocarcinogenicity study, a squamous cell carcinoma was seen in the lungs of one rat of 30 exposed intermittently to 25 ppm NO_2 with daily 1-hour exposures to 10 mg/m^3 of benzpyrene and 10 ppm of NO_2. These two tumors, particularly since the results of the cocarcinogenicity study are preliminary, must be considered as inconclusive evidence of a role of NO_2 in the etiology of cancer.

To these points it must be stated that there is no positive evidence that nitrogen dioxide is a carcinogen, cocarcinogen or a carcinogenic promoter in animal studies designed to assess such matters. There is no evidence of the mutagenicity of nitrogen dioxide per se, but nitrous acid, a product formed when NO_2 reacts with water, is a potent mutant for tobacco mosaic virus and *Esherichia coli*. Likewise, there is no evidence for nitrogen dioxide to be considered a teratogen.

BASES FOR RECOMMENDED STANDARDS—PAST AND PRESENT

The historical beginning of recommended standards for nitrogen dioxide began with the work of Lehmann and Hasegawa [47], published in 1913. The value of 39 ppm as a maximal allowable concentration (MAC) was based on acute effects to a single exposure, which could be withstood without undue discomfort by many men for several hours. This MAC for oxides of nitrogen was utilized until the 1940s.

In 1937, the state of Massachusetts suggested a MAC of 10 ppm NO_2 by analogy to the 10 ppm MAC for sulfur dioxide and hydrochloric acid [48]. In 1944, the American Standards Association approved a MAC of 25 ppm [48]. This value was adopted by the American Conference of Governmental Industrial Hygienists (ACGIH) as the MAC for "nitrogen oxides (other than N_2O)" [49]. In 1948, the ACGIH substituted the term Threshold Limit Value (TLV) for MAC without changing the 25-ppm value. It must be remembered that these recommended standards covered a mixture of oxides of nitrogen and inorganic nitrogen-containing acids and adequate analytical procedures for precise determination of the various components of the mixtures were not available.

In 1954, the TLV was lowered to 5 ppm, and the designation became specific for nitrogen dioxide [50], based largely on the studies by Gray et al. in 1952 [51] and 1954 [52], which reported no toxic responses in rats exposed at 4 ppm of red fuming nitric acid vapor, 4 hr/day, 5 days/wk for 6 months. In 1964, the ACGIH placed a ceiling value on the 5 ppm, which previously had been measured as an 8-hour time-weighted average (TWA) [53]. The prime reason for this change was the report by Wagner et al. [31] that NO_2 accelerated lung tumor development in a strain of mice that had spontaneous lung tumors. In 1971, the American National Standards Institute (ANSI) recommended a ceiling value of 5 ppm for industrial intermittent exposures and added a 5-min peak value of 15 ppm [54]. The peak exposure was permissible on an infrequent (not daily) basis. Until December 8, 1978, the federal standard for NO_2 was 5 ppm as an 8-hr TWA. When the 1968 TLVs were adopted as federal standards under the Occupational Safety and Health Act of 1970, the ceiling designation was erroneously omitted. This error was recently corrected by OSHA (29 CFR 1910.1000, published in the *Federal Register* 43:57603, December 8, 1978).

In 1976, NIOSH, in its criteria document on oxides of nitrogen, recommended a ceiling value of 1 ppm of NO_2 as a workplace environmental standard. The basis for the NIOSH recommendation is as follows: "evidence obtained since the time of this (TLV) documentation suggests that humans with normal respiratory function may be acutely affected by exposure at or below this level (5 ppm). Furthermore, the conditions of workers with chronic respiratory diseases, such as chronic bronchitis, may be aggravated by exposure to nitrogen dioxide at a concentration of approximately one-third of the current federal standard. Although much of the animal data are inconsistent, and thereby inconclusive at this time, some studies have indicated chronic effects on respiration, cellular morphology of the pulmonary system, reproduction, immune responses and weight gain in animals exposed to nitrogen dioxide at or below the current federal standard. In view of these results, it is concluded that the federal standard of 5 ppm should be reduced."

The reduction to a value of 1 ppm was predicated in large part on the acute studies by Abe [11] and von Nieding and associates [15,16], since animal data on pulmonary effects appeared "to be inconsistent or inconclusive below 2 ppm." The acute mechanical and gaseous exchange alterations reported by Abe in lungs of normal adult males exposed to nitrogen dioxide for 10–15 min at 4–5 ppm was viewed as having a threshold "obviously below 4 ppm." The changes in arterial oxygen partial pressure, alveolo-arterial pressure gradients and airway resistance observed by von Nieding et al. in healthy subjects exposed at 4–5 ppm and in persons with chronic bronchitis above 1.5 ppm were the bases for the recommendation of a ceiling value of 1 ppm. The latter value would "prevent acute irritant effects in the lungs of workers exposed to nitrogen dioxide. . . In addition, the prevention of repeated acute episodes of irritancy should lessen the risk of developing chronic obstructive lung disease." In April 1977, using essentially the same evaluation criteria, the Bureau of Chemical Hazards, Environmental Health Directorate of Canada, recommended the identical standard of 1 ppm for NO_2 with a ceiling notation [55].

Following this downward revision of the recommended workplace standard for NO_2, the ACGIH reevaluated its recommendation of 5 ppm for NO_2 with a ceiling notation. All of the literature cited in this chapter was reviewed during the reevaluation. The synopsis of this review reads as follows: "The majority of the Committee feel that the evidence presented is not unequivocal, or substantial enough to warrant reduction from the current 5 ppm. It is noteworthy that none of the other countries for which information is available, including the USSR, has seen fit to set a limit as low as 1 ppm for NO_2 (see below).

Other recommendations: Smyth (1956), Elkins (1959), ANSI (1961). East and West Germany (1973 and 1974), Sweden (1975) and Czechoslovakia (1969) 5 ppm or its equivalent in mg/cu m; USSR (1972) 2.6 ppm.

In recommending a workplace standard for NO_2, certain relevant factors are germane. The airborne concentration of NO_2 is much more significant in determining risk than the duration, at least through an 8- to 10-hour work day. Both animal and human data on health effects are inconsistent or inconclusive at exposure concentrations of less than 2 ppm. Intermittent exposure appears to be less hazardous than continuous exposure. A number of examples supporting this concept were presented. The phenomenon of tolerance to NO_2 demonstrated in several animal studies may be operative in workers with repetitive exposures and may be responsible for the lack of evidence of pulmonary disease in industrial workers or populations exposed at or below the federal industrial air standard. Several animal studies have demonstrated a healing of anatomical and biochemical lesions in the lung after cessation of exposure to nitrogen dioxide. This does not mean, however, that if the exposure is continued for a sufficient duration and airborne concentration, emphysema or other pulmonary diseases will not occur.

A second aspect that should be addressed is that when a ceiling value is placed on a federal industrial air standard it cannot be exceeded. In industrial hygiene practice, a ceiling value vs a time-weighted average value lowers the actual exposure of a worker by approximately 30–50%. In the industrial situation, airborne concentrations vary over the day depending on the operations performed and the respective duration of performance. Thus, a ceiling value places a maximum excursion limit, whereas a TWA permits excursions above the TLV within the limits of a weighted average.

One may encounter marked discrepancies between industrial and ambient air limits, even though the basis, i.e., the protection of human health, is the same in both cases. In the case of NO_2, the current differential between industrial and ambient standards is 100-fold, 5 ppm vs 0.05 ppm, respectively. Part of this differential is due to the population to be protected. The working population includes, by and large, mature, healthy individuals, while ambient air standards are designed to protect the health of the most sensitive element of the general population. This factor, coupled with job-selective processes, e.g., the exclusion of intrinsic and extrinsic asthmatics from occupations involving exposure to respiratory irritants because of their increased susceptibility, provides the worker with a greater capacity for exposure to airborne contaminants than the population at large. The magnitude of permissible occupational and ambient exposures varies according to the extent of the disease and age of the diseased.

The current federal standard (a ceiling of 5 ppm, 8 hr/day, 40 hr/wk) was based on prolonged industrial experience within the U.S. and abroad, and as previously described, was made more stringent as new criteria became available.

CONCLUSIONS

Differing recommended limits and/or standards exist for occupational exposure to nitrogen dioxide both nationally and internationally. These values range from a ceiling of 1 ppm to a ceiling of 5 ppm and are based on a value judgment of essentially the same animal and human data.

REFERENCES

1. Storlazzi, I. D. "Hygiene of Welding in U.S. Naval Shipyards," *Arch. Ind. Health* 19:307–311 (1959).
2. National Institute for Occupational Safety and Health. "Criteria for a Recommended Standard . . . Occupational Exposure to Oxides of Nitrogen (Nitrogen Dioxide and Nitric Oxide)," HEW Publ. No. (NIOSH) 76-149 (1976).
3. Wade, H. A., H. B. Elkins and B. P. W. Ruotolo. "Composition of Nitrous Fumes from Industrial Processes," *Arch. Ind. Hyg. Occup. Med.* 1:81–89 (1950).
4. Henschler, D., A. Stier, H. Beck and W. Neumann. "Olfactory Threshold of Some Important Irritant Gases and Effects in Man at Low Concentrations," *Arch. Gewerbepathol. Gewerbehyg.* 17:547–570 (1960) (in German).
5. Rumsey, D. W., and R. P. Cesta. "Oder Threshold Levels for UMDH and NO_2," *Am. Ind. Hyg. Assoc. J.* 31:339–342 (1970).
6. Bondareva, E. N. "Hygienic Evaluation of Low Concentrations of Nitrogen Oxides Present in Atmospheric Air," in *U.S.S.R. Literature on Air Pollution and Related Occupational Diseases, A Survey*, Vol. 8, B. S. Levine, Ed. (Washington, D.C.: U.S. Public Health Service, 1963), pp. 98–101.
7. Shalamberidze, O. P. "Reflex Effects of Mixtures of Sulfur and Nitrogen Dioxides," *Hyg. Sanit.* 32(7–9):10–15 (1967).
8. Meyers, F. H., and C. H. Hine. "Some Experiences of NO_2 in Animals and Man," paper presented at 5th Air Pollution Medical Research Conference, Los Angeles, CA, 1961.
9. Patty, F. A., Ed. *Industrial Hygiene and Toxicology*, Vol. 2, revised edition (New York: Wiley Interscience, 1962), p. 922.
10. Vigliani, E. C., and N. Zurlo. "Experiences of the del Lavoro Clinic with Several Maximal Workplace Concentrations (MAK) of Industrial Poisons," *Arch. Gewerbepathol. Gewerbehyg.* 13:528–534 (1955) (in German).
11. Abe, M. "Effects of Mixed NO_2-SO_2 Gas on Human Pulmonary Function," *Bull. Tokyo Med. Dent. Univ.* 14:415–433 (1967).
12. Nakamura, K. "Response of Pulmonary Airway Resistance by Interaction of Aerosols and Gases in Different Physical and Chemical Nature," *Japan J. Hyg.* 19:322–333 (1964) (in Japanese).
13. Suzuki, T., and K. Ishikawa. "A Study on the Effects of Smog on Man. Special Studies on Air Pollution Control," Report No. 2 (Science and Technology Agency, Japan) (1965) (in Japanese), pp. 199–221.
14. Yokoyama, E. "Effects of Acute Controlled Exposure to NO_2 on Mechanics of Breathing in Healthy Subjects," *Bull. Inst. Public Health* 17: 337–346 (1968) (in Japanese).

15. von Nieding, G., H. M. Wagner, H. Krekeler, U. Smidt and K. Muysers. "Absorption of NO_2 in Low Concentrations in the Respiratory Tract and its Acute Effects on Lung Function and Circulation," paper presented at the 2nd International Clean Air Congress of the International Union of Air Pollution Prevention Association, Washington, D.C., December 6–11, 1970.

16. von Nieding, G., H. Krekeler, R. Fuchs, M. Wagner and K. Koppenhagenn. "Studies of the Acute Effects of NO_2 on Lung Function: Influence on Diffusion, Perfusion and Ventilation in the Lungs," *Int. Arch. Arbeitsmed.* 31:61–72 (1973).

17. von Nieding, G., and H. Krekeler. "Pharmacologic Control of Acute NO_2-Exposure on Lung Function of Healthy Subjects and Patients with Chronic Bronchitis," *Int. Arch. Arbeitsmed.* 29:55–63 (1971).

18. Rokaw, S. N., H. E. Swann, Jr., R. L. Keenan and J. R. Phillips. "Human Exposure to Single Pollutants: NO_2: in a Controlled Environmental Facility," paper presented at the 9th Air Pollution Medical Research Conference, Denver, CO, 1968.

19. Latowsky, L. W., E. L. MacQuiddy and J. P. Tollman. "Toxicology of Oxides of Nitrogen. I. Toxic Concentrations," *J. Ind. Hyg. Toxicol.* 23: 129–133 (1941).

20. Gray, E. LeB., F. M. Patton, S. B. Goldberg and E. Kaplan. "Toxicology of the Oxides of Nitrogen. II. Acute Inhalation Toxicity of Nitrogen Dioxide, Red Fuming Nitric Acid, and White Fuming Nitric Acid," *Arch. Ind. Hyg. Occup. Med.* 10:418–422 (1954).

21. Carson, R. R., M. S. Rosenholtz, F. T. Wilenski and M. H. Weeks. "The Responses of Animals Inhaling Nitrogen Dioxide for Single, Short-term Exposures," *Am. Ind. Hyg. Assoc. J.* 23:457–462 (1962).

22. Thomas, H. V., P. K. Mueller and R. L. Lyman. "Lipoperoxidation of Lung Lipids in Rats Exposed to Nitrogen Dioxide," *Science* 159:532–534 (1968).

23. Ehrlich, R. "Effect of Nitrogen Dioxide on Resistance to Pulmonary Infection," *Bact. Rev.* 30:604–614 (1966).

24. Murphy, S. D., C. E. Ulrich, S. Frankowitz and C. Xintaras. "Altered Function in Animals Inhaling Low Concentrations of Ozone and Nitrogen Dioxide," *Am. Ind. Hyg. Assoc. J.* 25:246–253 (1964).

25. Henry, M. C., R. Ehrlich and W. H. Blair. "Effect of Nitrogen Dioxide on Resistance of Squirrel Monkeys to Klebsiella pneumoniae Infection," *Arch. Environ. Health* 18:580–587 (1969).

26. Vigdortschik, N. A., E. C. Andreeva, I. Z. Matussevitsch, M. M. Nikulina, L. M. Frumina and V. A. Striter. "The Symptomatology of Chronic Poisoning with Oxides of Nitrogen," *J. Ind. Hyg. Toxicol.* 19:463–473 (1937).

27. Kennedy, M. C. S. "Nitrous Fumes and Coal-Miners with Emphysema," *Ann. Occup. Hyg.* 15:285–301 (1972).

28. Kosmider, S., K. Ludyga, A. Misiewicz, M. Drozdz and J. Sagan. "Experimental and Clinical Investigations of Emphysematous Effects of Nitrogen Oxides," *Z. Arbeitsmed.* 22:363–368 (1972) (in German).

29. Back, K. C. "Review of Air Force Data from Long-Term Continuous Exposures at Ambient Pressure," (Wright-Patterson Air Force Base, Ohio) *Proc. 1st Ann. Conf. Atmosph. Contam. in Confined Spaces* (1965). pp. 124–133.

30. Siegel, J. "Review of Ambient Pressure Animal Exposure Data from Selected Navy Compounds," (Wright-Patterson Air Force Base, Ohio) *Proc. 1st Ann. Conf. Atmosph. Contam. in Confined Spaces* (1965), pp. 134–147.

31. Wagner, W. D., B. R. Duncan, P. G. Wright and H. E. Stokinger. "Experimental Study of Threshold Limit of NO_2," *Arch. Environ. Health* 10: 455–466 (1965).

32. Freeman, G., S. C. Crane, N. J. Furiosi, R. J. Stephens, M. J. Evans and W. D. Moore. "Covert Reduction in Ventilatory Surface in Rats During Prolonged Exposure to Subacute Nitrogen Dioxide," *Am. Rev. Resp. Dis.* 106:563–579 (1972).

33. Hine, C. H., R. D. Cavalli and R. R. Wright. Unpublished results cited by H. E. Stokinger and D. L. Coffin, in *Air Pollution*, 2nd ed., A. C. Stern, Ed. (New York: Academic Press, 1968), pp. 446–546.

34. Boren, H. G. "Carbon as a Carrier Mechanism for Irritant Gases," *Arch. Environ. Health* 8:119–124 (1964).

35. Coffin, D. L., D. E. Gardner and E. J. Blommer. "Time/Dose-Response for Nitrogen Dioxide Exposure in an Infectivity Model System," *Environ. Health Persp.* 13:11–15 (1976).

36. Haydon, G. B., J. T. Davidson, G. A. Lillington and K. Wasserman. "Nitrogen Dioxide-Induced Emphysema in Rabbits," *Am. Rev. Resp. Dis.* 95:797–805 (1967).

37. Kleinerman, J., and G. W. Wright. "The Reparative Capacity of Animal Lungs after Exposure to Various Single and Multiple Doses of Nitrate," *Am. Rev. Resp. Dis.* 83:423–424 (1961).

38. Riddick, H. H., Jr., K. I. Campbell and D. L. Coffin. "Histopathologic Changes Secondary to Nitrogen Dioxide Exposure in Dog Lungs," *Am. J. Clin. Pathol.* 49:239 (1968).

39. Freeman, G., S. C. Crane, R. J. Stephens and N. J. Furiosi. "Pathogenesis of the Nitrogen Dioxide-Induced Lesion in the Rat Lung—A Review and Presentation of New Observations," *Am. Rev. Resp. Dis.* 98:429–443 (1968).

40. Stephens, R. J., G. Freeman and M. J. Evans. "Early Responses of Lungs to Low Levels of Nitrogen Dioxide—Light and Electron Microscopy," *Arch. Environ. Health* 24:160–179 (1972).

41. Sherwin, R. P., J. Dibble and J. Weiner. "Alveolar Wall Cells of the Guinea Pig: Increase in Response to 2 ppm NO_2," *Arch. Environ. Health* 24:43–47 (1972).

42. Balchum, O. J., R. D. Buckley, R. Sherwin and M. Gardner. "Nitrogen Dioxide Inhalation and Lung Antibodies," *Arch. Environ. Health* 10: 274–277 (1965).

43. Henschler, D., and W. Ross. "Lung Cancer from Nitrous Gas," *Arch. Exp. Pathol. Pharmakol.* 250:256–257 (1965) (in German).

44. Ross, W., and D. Henschler. "Absence of any Carcinogenic Effect of Nitrous Gas in the Golden Hamster," *Experientia* 24:55 (1968) (in German).

45. Kuschner, M., and S. Laskin. "Inhalation Studies with Nitrogen Dioxide," in *Studies in Pulmonary Carcinogenesis—Summary Progress Report to the National Cancer Institute*, Contract No. NOI CP 33260, New York University Medical Center, Institute of Environmental Medicine (1973), pp. 37–67.

46. Kuschner, M., and S. Laskin. In *Studies in Pulmonary Carcinogenesis— Summary Progress Report to the National Cancer Institute*, Contract No. NOI CP 33260, New York University Medical Center, Institute of Environmental Medicine (1974), p. 8.

47. Lehmann, K. B., and Hasegawa. "Studies on the Effect of Technically and Hygenically Important Gases and Vapors on Man (31)—The Nitrous Gases—Nitric Oxide, Nitrogen Dioxide, Nitrous Acid, Nitric Acid," *Arch. Hyg.* 77:323–368 (1913) (in German).

48. Shrenk, H. H. "Controversial Toxicity of Oxides of Nitrogens Continues to Receive Attention—Hydrogen Peroxide Toxicity Reported," *Ind. Eng. Chem.* 47(2):93A–94A (1955).

49. American Conference of Governmental Industrial Hygienists. *Proc. Eighth Annual Meeting*, Chicago, April 7–13 (1946), pp. 54–55.

50. American Conference of Governmental Industrial Hygienists. Threshold Limit Values for 1954, Adopted at the Sixteenth Annual Meeting of ACGIH, Chicago, April 24–27, 1954. *Arch. Ind. Hyg. Occup. Med.* 9:530–534 (1954).

51. Gray, E. L., J. K. MacNamee and S. B. Goldberg. "Toxicity of NO_2 Vapor at Very Low Levels," *Arch. Ind. Hyg.* 6:20–21 (1952).

52. Gray, E. L., S. B. Goldberg and F. M. Patton. "Toxicity of the Oxides of Nitrogen. III. Effect of Chronic Exposure to Low Concentrations of Vapors from Red Fuming Nitric Acid," *Arch. Ind. Hyg. Occup. Med.* 10:423–425 (1954).

53. American Conference of Governmental Industrial Hygienists: Threshold Limit Values for 1964, Adopted at the 26th Annual Meeting of ACGIH, Philadelphia, April 25–28, 1964. *Arch. Environ. Health* 9:545–554 (1964).

54. American National Standards Acceptable Concentrations of Nitrogen Dioxide, Z37.13, American National Standards Institute, Inc., New York (1971).

55. Environmental Health Directorate, Health Protection Branch. "Nitrogen Dioxide and Nitric Oxide Recommendations for Occupational Exposure," Minister of National Health and Welfare, Canada (1977).

nitrous oxide N_2O

obtained by heating ammonium nitrate.